STUDENT SOLUTIONS MANUAL

for use with

NELSON

Physics

for Scientists and Engineers
An Interactive Approach

HAWKES   IQBAL   MANSOUR   MILNER-BOLOTIN   WILLIAMS

Prepared by

Naeem Syed Ahmed, Laurentian University • Mihai Gherase _____ t
Fresno • Santo D'Agostino, Brock University • Greg Stortz, U_____ •
Robert Hawkes, Mount Allison University • Javed Iqbal, Uni_____ ____mbia •
Firas Mansour, University of Waterloo • Marina Milner-Bol___ ___versity of British
Columbia • Peter Williams, Acadia University

Student Solutions Manual for use with
Physics for Scientists and Engineers: An Interactive Approach
Hawkes/Iqbal/Mansour/Milner-Bolotin/Williams
SSM Prepared By Naeem Syed Ahmed, *Laurentian University*, **Mihai Gherase, Santo D'Agostino,** *Brock University,* **Robert Hawkes,** *Mount Allison University,* **Javed Iqbal,** *University of British Columbia,* **Firas Mansour,** *University of Waterloo,* **Marina Milner-Bolotin,** *University of British Columbia,* **and Peter Williams,** *Acadia University*

Technical Editors: Greg Stortz, *University of British Columbia,* **and Santo D'Agostino,** *Brock University*

ISBN 0176691960

Table of Contents

Preface

The *Student Solutions Manual* accompanying the textbook *Physics for Scientists and Engineers: An Interactive Approach* was prepared to assist the student in mastering the skills required for an understanding of physics. The selected questions have been chosen by the authors of your text to allow you to discover the range and depth of your understanding.

To avoid loss of precision due to rounding in intermediate steps, many calculations in this manual use more significant figures than justified by the given data. Rounding the final answers to the correct number of significant figures has been left as an exercise for the student.

This *Student Solutions Manual* contains the worked-out solutions to all odd-numbered exercises at the end of each textbook chapter. It has been independently checked for accuracy.

We would like to thank Greg Stortz of the University of British Columbia for his detailed technical review of many of the solutions that appear in this solutions manual.

Naeem Syed Ahmed, *Laurentian University*
Mihai Gherase, *California State University at Fresno*
Santo D'Agostino, *Brock University*
Greg Stortz, *University of British Columbia*
Robert Hawkes, *Mount Allison University*
Javed Iqbal, *University of British Columbia*
Firas Mansour, *University of Waterloo*
Marina Milner-Bolotin, *University of British Columbia*
Peter Williams, *Acadia University*

Chapter 1—INTRODUCTION TO PHYSICS

1. c

3. $\text{mean} = \dfrac{5.8 + 2.2}{2} = 4.0 \text{ m}$

   $\sigma = \dfrac{5.8 - 4.0}{2} = 0.9 \text{ m}$

5. a

7. c

9. The graph does not state the difference between the measurement points represented by dots and circles. The y-axis should have the unit label (km/s) not KM. Since both measurements (distance and velocity) have associated uncertainties, the corresponding x and y error bars should be included. There is no indication on the graph as to what the two fitted lines (solid and dashed) represent and which data points (if any) were excluded for those fits.

11. The duration of 9 192 631 770 periods of the radiation corresponding to the transition between the two hyperfine levels of the ground state of a caesium 133 atom at rest at 0 K.

13. While the ampere is operationally equal to a current of 1 coulomb passing a point in 1 second, it is derived formally in the following manner (http://www.physics.nist.gov): "The ampere is that constant current which, if maintained in two straight parallel conductors of infinite length, of negligible circular cross section, and placed 1 meter apart in vacuum, would produce between these conductors a force equal to 2×10^{-7} newton per meter of length."

15. d

17. d

19. $N = \dfrac{\sigma^2}{\text{SDOM}^2} = \dfrac{2.4^2}{1.2^2} = 4$

21. (a) 4 (b) 2 (c) 4 (d) 2 (e) 3 (f) 4

23. (a) 2.452×10^3

    (b) 5.92×10^{-1}

    (c) 1.2×10^4

    (d) 4.5×10^{-5}

25. 25.8

27. (a)

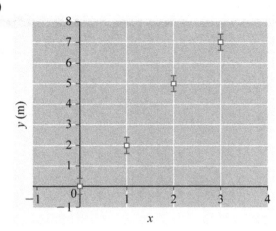

(b)

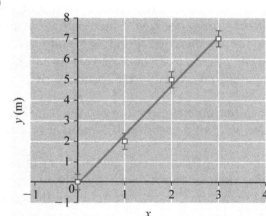

(c) ± 3.6 m (or 7.2 m total length)

29. We know that $[U] = \mathrm{kg} \cdot \mathrm{m}^2 \cdot \mathrm{s}^{-2}$, $[m] = \mathrm{kg}$, $[g] = \mathrm{m} \cdot \mathrm{s}^{-2}$ and $[h] = \mathrm{m}$
$\Rightarrow U \sim mgh$

31. $V \sim IR$
$\Rightarrow \mathrm{m}^2 \cdot \mathrm{kg} \cdot \mathrm{s}^{-3} \cdot \mathrm{A}^{-1} = (\mathrm{A})\,\mathrm{m}^2 \cdot \mathrm{kg} \cdot \mathrm{s}^{-3} \cdot \mathrm{A}^{-2}$

33. $[b] = \dfrac{[F]}{[v]} = \dfrac{\mathrm{kg} \cdot \mathrm{m} \cdot \mathrm{s}^{-2}}{\mathrm{m} \cdot \mathrm{s}^{-1}} = \mathrm{kg} \cdot \mathrm{s}^{-1}$

35. $V_{cell} = \frac{4}{3}\pi r^3 = 3.35 \times 10^{-17}$ m^3

$V_{total} = 25$ cm$^3 = 25 \times 10^{-6}$ m^3

Let us assume that the space between the cells is negligible.

$\Rightarrow N = \dfrac{V_{total}}{V_{cell}} = 7.5 \times 10^{11}$ cells

37. The population of Canada is about 35 million. If 2 percent of Canadians (of all ages) play hockey and if each team has 20 players of which 2 are goalies, that would imply there are approximately 70 000 goalies in Canada.

39. $V_{molecule} = \frac{4}{3}\pi r^3 = 1.15 \times 10^{-21}$ m^3

$\Rightarrow m = \rho V_{molecule} = 1000 \times 1.15 \times 10^{-21} = 1.15 \times 10^{-18}$ kg $= 1.15 \times 10^{-15}$ g $= 1.15 \times 10^{-3}$ pg

This would be written as 0.0015 pg to two significant figures.

41. Years 1990–1999: mean win percentage is 0.513 with a SDOM of 0.020

Years 2000–2009: mean win percentage is 0.497 with a SDOM of 0.012

Although the teams in years 1990–1999 had a somewhat better win–loss record, the difference is within the range of the SDOMs, so these data do not prove any significant difference between the two sets of teams.

43. (a) Since the surface area of Earth is 5.1×10^{14} m^2, we receive about

$\dfrac{20\,000 \times 1000}{5.1 \times 10^{14}} = 3.9 \times 10^{-8}$ kg per year per m^2.

(b) Twenty-five percent of the amount calculated in part (a) is 9.8×10^{-9} kg per year per m^2.

Using $\rho_m = 3400$ kg/m^3 we get

$m = \rho_m V = 3400 \times 8.18 \times 10^{-15} = 2.78 \times 10^{-11}$ kg

$\Rightarrow N = \dfrac{9.8 \times 10^{-9}}{2.78 \times 10^{-11}} = 450$ micrometeorites per year per m^2

45. (a) $t = \sqrt{\dfrac{\hbar G}{c^5}}$

(b) $t = 5.39 \times 10^{-44}$ s

Chapter 2—SCALARS AND VECTORS

1. (a) $F = 30$ N, $\theta = 315°$

   (b) $F_x = 21$ N, $F_y = -21$ N

   (c) $\vec{F} = \left(21\hat{i} - 21\hat{j}\right)$ N

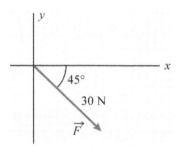

3. A parametric equation for the line is
$$\begin{cases} x = t \\ y = -3t + 4 \end{cases}$$
So a vector parallel to the given force vector will have the following components $(1, -3)$. And its magnitude is given by
$$\sqrt{1^2 + (-3)^2} = 3.162$$
A unit vector parallel to the force is therefore given by
$$\hat{n} = \frac{1}{3.162}\hat{i} - \frac{3}{3.162}\hat{j} = 0.316\hat{i} - 0.949\hat{j}$$
There are two possible force vectors, because the force can point in two possible directions along the line, either in the direction of $\hat{n}$ or in the opposite direction. That is,
$$\vec{F} = \pm 8\hat{n} \text{ N}$$
$$\vec{F} = \pm 8\left(0.316\hat{i} - 0.949\hat{j}\right) \text{ N}$$

5. $\left|\vec{D}\right| = \sqrt{(3-[-2])^2 + (0-2)^2 + (-2-0)^2} = 5.74$ m

7. (a) $(4, -2, 11)$

   (b) $(-4, 2, -11)$

   (c) $(-24, 11, 22)$

   (d) $(-20, 5, 10)$

9. (c) Vector magnitudes are invariant with respect to translations or rotations of coordinate systems.

11. Student 2: Consider a vector in each plane directed such that they both form a 45° angle with the line of intersection of the two orthogonal planes. These two vectors are not perpendicular to each other. You can also consider two vectors that are both parallel to the line of intersection of the two orthogonal planes. These two vectors are parallel (or anti-parallel) to each other and are not perpendicular to each other.

13. (a) $\vec{A_x} = (2,0,0)$, $\vec{A_y} = (0,2,0)$, $\vec{A_z} = (0,0,-2)$, $\vec{B_x} = (-1,0,0)$, $\vec{B_y} = (0,3,0)$, $\vec{B_z} = (0,0,-2)$

(b) $\vec{A} \bullet \vec{B} = 2 \cdot (-1) + 2 \cdot 3 + (-2) \cdot (-2) = 8$

(c) $\vec{A} \times \vec{B} = \hat{i}(-4 - (-6)) + \hat{j}(2 - (-4)) + \hat{k}(6 - (-2)) = 2\hat{i} + 6\hat{j} + 8\hat{k}$

15. Disagree; if more than one component equals 1, the magnitude will be greater than 1.

17. $\vec{F_1} = 1.00\hat{i} - 3.00\hat{j} + 1.00\hat{k}$

$\vec{F_2} = -2.00\hat{i} - 4.00\hat{j} + 3.00\hat{k}$

$\vec{F_3} = 1.00\hat{i} + 2.00\hat{j} + 0\hat{k}$

$|\vec{F_1}| = \sqrt{11.00}$ N, $|\vec{F_2}| = \sqrt{29.00}$ N, $|\vec{F_3}| = \sqrt{5.00}$ N

$\vec{F_R} = -5.00\hat{j} + 4.00\hat{k}$, $|\vec{F_R}| = \sqrt{41.00}$ N

19. Disagree. The magnitude of the resultant is not equal to the sum of the radius vectors of individual vectors. Similarly, the resultant angle is not equal to the sum of the individual vector angles.

21. Mathematically, the head-to-tail rule is equivalent to addition of vectors in Cartesian form. The associative property of vector addition, that is

$$\vec{V_1} + \vec{V_2} + \vec{V_3} = \left(\vec{V_1} + \vec{V_2}\right) + \vec{V_3} = \vec{V_1} + \left(\vec{V_2} + \vec{V_3}\right)$$

proves the head-to-tail rule for more than two vectors.

23. The scalar product of two vectors is defined as the product of the magnitudes of the two vectors and the cosine of the angle between them. Because the definition involves the product of three real numbers, and the product of real numbers is commutative, the scalar product of two vectors is commutative.

The vector product of two vectors is not commutative because the direction of the cross product is defined using the right-hand rule, and the direction is opposite if the two vectors are multiplied in opposite order.

25. $F_x = 5\cos(30°) = 4.33$ N

$F_y = 5\cos(-40°) = 3.83$ N

$F_z = 5\cos(-60°) = 2.50$ N

$\Rightarrow \vec{F} = \left(4.33\hat{i} + 3.83\hat{j} + 2.50\hat{k}\right)$ N

27. $\vec{F_A} = (2,0,-1);\ \vec{F_B} = (0,3,0);\ \vec{F_C} = (0,2,2);\ \vec{F_D} = (0,-3,3)$ or

$\vec{F_A} = 2\hat{i} - \hat{k};\ \vec{F_B} = 3\hat{j};\ \vec{F_C} = 2\hat{j} + 2\hat{k};\ \vec{F_D} = -3\hat{j} + 3\hat{k}$

Therefore:

$\vec{F_A} + \vec{F_B} + \vec{F_C} + \vec{F_D} = \left(2\hat{i} - \hat{k}\right) + 3\hat{j} + \left(2\hat{j} + 2\hat{k}\right) + \left(-3\hat{j} + 3\hat{k}\right) = 2\hat{i} + 2\hat{j} + 4\hat{k}$

It is impractical to add these vectors by hand, as it is difficult to accurately draw the vectors head-to-tail in three dimensions.

29. $\begin{cases} x = -\left(10\sin(20°) + 7 + 20\sin(30°)\right) = -20.4 \text{ m} \\ y = 10\cos(20°) - 20\cos(30°) = -7.9 \text{ m} \end{cases}$

$\Rightarrow D = \sqrt{20.4^2 + 7.9^2} = 21.9$ m

31. (a) $A_x = 20.0\cos(360° - 15°) = 19.3;\quad A_y = 20.0\sin(360° - 15°) = -5.18$

$B_x = 15.0\cos(35°) = 12.3;\qquad B_y = 15.0\sin(35°) = 8.60$

$C_x = 25.0\cos(125°) = -14.3;\qquad C_y = 25.0\sin(125°) = 20.5$

(b) $\vec{A} = \left(19.3\hat{i} - 5.18\hat{j}\right);\ \vec{B} = \left(12.3\hat{i} + 8.60\hat{j}\right);\ \vec{C} = \left(-14.3\hat{i} + 20.5\hat{j}\right)$

$\vec{U} = \vec{A} + \vec{B} + \vec{C} = \left(17.3\hat{i} + 23.9\hat{j}\right)$

$\left|\vec{U}\right| = \sqrt{17.3^2 + 23.9^2} = 29.5$

$\theta = \tan^{-1}\left(\frac{23.9}{17.3}\right) = 54.1°$

$\Rightarrow \vec{U} = (29.5,\ 54.1°)$

(c) $\vec{A} = \left(19.3\hat{i} - 5.18\hat{j}\right);\ \vec{B} = \left(12.3\hat{i} + 8.60\hat{j}\right);\ \vec{C} = \left(-14.3\hat{i} + 20.5\hat{j}\right)$

$\vec{D} = 2\vec{A} - 3\vec{B} + \vec{C} = \left(-12.6\hat{i} - 15.7\hat{j}\right)$

$\left|\vec{D}\right| = \sqrt{12.6^2 + 15.7^2} = 20.1$

$\theta = \tan^{-1}\left(\frac{15.7}{12.6}\right) = 51.3°$

33. (a) $\left|\vec{F}\right| = \sqrt{3^2 + 5^2 + 6^2} = 8.37$ N

$$\Rightarrow \hat{n} = \frac{3}{8.37}\hat{i} - \frac{5}{8.37}\hat{j} + \frac{6}{8.37}\hat{k} = 0.359\hat{i} - 0.598\hat{j} + 0.717\hat{k}$$

(b) Any unit vector $\vec{u}$ that satisfies the condition $\vec{u} \bullet \left(3\hat{i} - 5\hat{j} + 6\hat{k}\right) = 0$ is perpendicular to the given vector. This question thus has an infinite number of possible answers, such as $\dfrac{\left(3\hat{i} + 3\hat{j} + \hat{k}\right)}{\sqrt{19}}$

(c) There are an infinite number of unit vactors parallel to the plane. Any unit vector $\hat{u} = a\hat{i} + b\hat{j} + c\hat{k}$ whose coefficients satisfy the relation $3a + 2b - 4c = 0$ is a unit vector parallel to the given plane; an example is $\dfrac{1}{3}\left(2\hat{i} + \hat{j} + 2\hat{k}\right)$

(d) The vector $(3, 2, -4)$ is perpendicular to the plane; thus the following two unit vectors are perpendicular to the given plane:

$$\pm \frac{1}{\sqrt{29}}\left(3\hat{i} + 2\hat{j} - 4\hat{k}\right)$$

35. For this we compare the x, y, and z coordinates.

(a) $F_x\hat{i} + 3\hat{j} + \sqrt{2}\hat{i} - F_y\hat{j} + F_z\hat{k} - 5\hat{k} = 0$

$$\begin{cases} F_x\hat{i} + \sqrt{2}\hat{i} - = 0 \\ 3\hat{j} - F_y\hat{j} = 0 \\ F_z\hat{k} - 5\hat{k} = 0 \end{cases} \Rightarrow \begin{cases} F_x = -\sqrt{2} \\ F_y = 3 \\ F_z = 5 \end{cases}$$

The same reasoning applied to solving the second equation:

(b) $3\hat{i} - 5\hat{j} + F_x\hat{i} - 2F_y\hat{j} + F_z\hat{k} - 3\hat{k} = 0$

$$F_x = -3, \ F_y = -\frac{5}{2}, \ F_z = 3$$

37. Let the vector be $\hat{n} = x\hat{i} + y\hat{j}$.

Since this unit vector is located in the xy-plane, it must have a form of
$\hat{n} = x\hat{i} + y\hat{j} + 0\hat{k} = x\hat{i} + y\hat{j}$
Since it is a unit vector, it must satisfy the relation
$x^2 + y^2 = 1$ \hspace{2cm} (1)

Since it is perpendicular to the given vector $\vec{A} = 3\hat{i} - 2\hat{j} + 5\hat{k}$,

the scalar produce of the unit vector and vector $\vec{A}$ must be zero. Therefore,

$3x - 2y = 0$ (2)

Solving (1) and (2) simultaneously gives

$$x^2 + \frac{9}{4}x^2 = 1 \Rightarrow 13x^2 = 4 \Rightarrow x = \frac{2\sqrt{13}}{13}; y = \frac{3\sqrt{13}}{13}$$

the required unit vectors as

$$\frac{2\sqrt{13}}{13}\hat{i} + \frac{3\sqrt{13}}{13}\hat{j} \text{ and } -\frac{2\sqrt{13}}{13}\hat{i} - \frac{3\sqrt{13}}{13}\hat{j}$$

39. $\vec{A} \bullet \vec{B} = (10)(5)\cos(20°) = 47$

41. The line in parametric form is

$x = t$

$y = -2t + 5$

So a vector parallel to the line is $(1, -2)$ with magnitude given by

$$\sqrt{1^2 + (-2)^2} = \sqrt{5}$$

A unit vector parallel to the line is therefore given by

$$\vec{n} = \frac{1}{\sqrt{5}}\hat{i} - \frac{2}{\sqrt{5}}\hat{j}$$

The projection of the given vector $\vec{F}(3.00, 4.00)$ on the line is therefore

$$(3)\left(\frac{1}{\sqrt{5}}\right) + (4)\left(-\frac{2}{\sqrt{5}}\right) = -\sqrt{5}$$

43. $\vec{A} \times \vec{B} = \begin{pmatrix} \hat{i} & \hat{j} & \hat{k} \\ 3 & 4 & -5 \\ 2 & -2 & 4 \end{pmatrix}$

$= \hat{i}(4 \cdot 4 - (-5)(-2)) - \hat{j}(3 \cdot 4 - (-5)2) + \hat{k}(3 \cdot (-2) - 4 \cdot 2)$

$= 6\hat{i} - 22\hat{j} - 14\hat{k}$

45. $\left|\vec{A} \times \vec{B}\right| = (10)(5)\sin(20°) = 17$ pointing in the $+y$ or $-y$ direction depending on the order of vector product.

47. (a) The cross-product gives $-14\hat{i} + \dfrac{F_x}{2}\hat{j}$.

Hence, the given equation becomes

$$-14\hat{i} + \dfrac{F_x}{2}\hat{j} + F_y\hat{i} - \sqrt{2}\hat{j} = 0$$
$$\Rightarrow F_x = 2\sqrt{2},\ F_y = 14$$

(b) The cross-product gives $20\hat{i} - 4F_x\hat{k}$.

Hence, the given equation becomes

$$20\hat{i} - 4F_x\hat{k} + F_z\hat{k} + \sqrt{3}F_x\hat{i} = 0$$
$$\Rightarrow F_x = -\dfrac{20}{\sqrt{3}},\ F_z = -\dfrac{80}{\sqrt{3}}$$

49. (a) Translation matrix of a coordinate system in the xy-plane from the origin O to point O'(4,5) can be expressed using the following translation matrix:

$$T = \begin{pmatrix} 1 & 0 & 4 \\ 0 & 1 & 5 \\ 0 & 0 & 1 \end{pmatrix}$$

Rotation of a coordinate system around the z-axis counter-clockwise by $60°$ can be expressed using the following rotation matrix:

$$R = \begin{pmatrix} \cos(60°) & -\sin(60°) & 0 \\ \sin(60°) & \cos(60°) & 0 \\ 0 & 0 & 1 \end{pmatrix}$$

The transformation matrix for translation and rotation is therefore

$$\begin{pmatrix} \cos(60°) & -\sin(60°) & 4 \\ \sin(60°) & \cos(60°) & 5 \\ 0 & 0 & 1 \end{pmatrix}$$

The position of any point in the new coordinate system is then given by

$$\begin{pmatrix} x \\ y \\ 1 \end{pmatrix} = \begin{pmatrix} \cos(60°) & -\sin(60°) & 4 \\ \sin(60°) & \cos(60°) & 5 \\ 0 & 0 & 1 \end{pmatrix} \begin{pmatrix} x' \\ y' \\ 1 \end{pmatrix}$$

(b) From the above matrix we can see how the components of a vector will transform.

$$x = x'\cos(60°) - y'\sin(60°) + 4$$
$$y = x'\sin(60°) + y'\cos(60°) + 5$$

51. 1. This fact can be proved by definition of the vector product—see equation (2-26) and Figure 2-18.

2. This fact can also be proved using the algebraic definition of cross-product and the fact that the scalar product of two orthogonal vectors is zero. The proof below is for vector $\vec{A}$; the same can be done for vector $\vec{B}$:

$$\vec{A} \bullet \left(\vec{A} \times \vec{B}\right) = \left(A_x \hat{i} + A_y \hat{j} + A_z \hat{k}\right) \bullet \left(\left(A_y B_z - A_z B_y\right)\hat{i} - \left(A_x B_z - A_z B_x\right)\hat{j} + \left(A_x B_y - A_y B_x\right)\hat{k}\right) =$$

$$A_x A_y B_z - A_x A_z B_y - A_y A_x B_z + A_y A_z B_x + A_z A_x B_y - A_z A_y B_x = 0$$

Therefore: $\vec{A} \perp \left(\vec{A} \times \vec{B}\right)$

53. For simplicity we will assume that the z-components of both vectors are zero, that is $A_z = B_z = 0$. (There is no loss of generality, because the coordinate system can always be rotated so that two vectors can be situated in the xy-plane.)

Using the determinant formula, the magnitude of the cross product is then given by

$$|A \times B| = \sqrt{A_x^2 B_y^2 + A_y^2 B_x^2 - 2A_x A_y B_x B_y}$$

The scalar product between the two vectors is given by

$$\vec{A} \bullet \vec{B} = AB \cos\theta$$

$$\Rightarrow A_x B_x + A_y B_y = AB \cos\theta$$

Squaring both sides gives

$$\left(A_x B_x + A_y B_y\right)^2 = A^2 B^2 \cos^2\theta$$

$$\Rightarrow A^2 B^2 - \left(A_x B_x + A_y B_y\right)^2 = A^2 B^2 \left(1 - \cos^2\theta\right) = A^2 B^2 \sin^2\theta$$

With $A^2 = A_x^2 + A_y^2$ and $B^2 = B_x^2 + B_y^2$, the above expression becomes

$$A_x^2 B_y^2 + A_y^2 B_x^2 - 2A_x A_y B_x B_y = A^2 B^2 \sin^2\theta$$

But we saw earlier that the left-hand side is equal to $|A \times B|^2$.

$$\Rightarrow |A \times B| = AB \sin\theta$$

The same is true in general, even for vectors not in the xy-plane.

55. $\vec{A} \times \vec{B} = \begin{vmatrix} \hat{i} & \hat{j} & \hat{k} \\ 3 & 2 & -5 \\ 3 & 2 & -5 \end{vmatrix} = \hat{i}\left(-10 + 10\right) + \hat{j}\left(-15 + 15\right) + \hat{k}\left(6 - 6\right) = 0$

These two vectors are identical and their cross-product is zero. This should have been expected.

57. Let $\vec{r}$ represent a vector along the horizontal side of the triangle, with length equal to the radius of the circle. Let $\vec{R}$ represent a vector from the centre of the circle to the vertex of a triangle that is not located on the diameter of the circle.

The two vectors $\vec{A}$ and $\vec{B}$ shown in the figure represent the sides of the triangle. They are given by

$$\vec{A} = \vec{R} + \vec{r}$$
$$\vec{B} = \vec{R} - \vec{r}$$

Note that

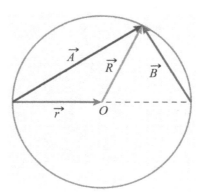

$$\vec{A} \cdot \vec{B} = \left(\vec{R} + \vec{r} \right) \cdot \left(\vec{R} - \vec{r} \right)$$
$$\vec{A} \cdot \vec{B} = \vec{R} \cdot \vec{R} + \vec{r} \cdot \vec{R} - \vec{R} \cdot \vec{r} - \vec{r} \cdot \vec{r} = R^2 - r^2$$
$$\vec{A} \cdot \vec{B} = R^2 - r^2$$
$$\vec{A} \cdot \vec{B} = 0$$

Since the magnitudes of vectors $\vec{r}$ and $\vec{R}$ are both equal to the radius of the circle, $\vec{A} \cdot \vec{B} = 0$. Thus, the vectors $\vec{A}$ and $\vec{B}$ are perpendicular to each other, and so the triangle containing these vectors as its sides must be a right triangle.

59. Any triangle can be oriented with respect to a coordinate system as shown in the figure. The sides of the triangles can be represented as vectors. The law of vector addition then gives

$$\vec{C} = \vec{B} - \vec{A}$$
$$\Rightarrow \vec{C} \cdot \vec{C} = \left(\vec{B} - \vec{A} \right) \cdot \left(\vec{B} - \vec{A} \right)$$

But $\vec{C} \cdot \vec{C} = C^2$ (and similarly for $\vec{A}$ and $\vec{B}$)

$$\Rightarrow C^2 = A^2 + B^2 - 2\vec{A} \cdot \vec{B}$$
$$\Rightarrow C^2 = A^2 + B^2 - 2AB \cos \theta$$

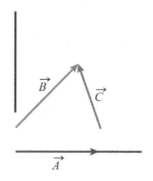

61. (a) Given: $\vec{v} = v_{0x}\hat{i} + \left(v_{0y} - gt \right) \hat{j}$

$$\Rightarrow v_x = v_{0x} \text{ and } v_y = v_{0y} - gt$$
$$v = \sqrt{v_x^2 + v_y^2} = \sqrt{v_x^2 + \left(v_{0y} - gt \right)^2}$$

(b) Assuming that v_{0x} is positive, a graph of the magnitude of the velocity will have a V-shape similar to the example shown below:

For $v_{0x} = 10$ m/s and $v_{0y} = 30$ m/s

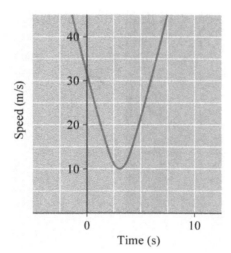

(c) The speed is smallest when the y-component of velocity vanishes; that is

$$v_{0y} - gt = 0$$

$$\Rightarrow t = \frac{v_{0y}}{g}$$

Under the conditions, mentioned in part (b), this would have happened when time $t = 3.06$ s. Notice, the speed at this point is 10 m/s (it equals the value of the horizontal component of the velocity).

(d) This model represents the motion of a projectile, assuming that there is no air resistance.

63. (a) $\vec{F}_{net} = \vec{F}_1 + \vec{F}_2 + \vec{F}_3 = 11\hat{j} - 14\hat{k}$

$$\Rightarrow \vec{a} = \frac{11\hat{j} - 14\hat{k}}{m} = \frac{11\hat{j} - 14\hat{k}}{0.1}$$

$$\Rightarrow \vec{a} = \left(110\hat{j} - 140\hat{k}\right)$$

$$\Rightarrow a = \sqrt{110^2 + 140^2} = 178 \text{ N/kg}$$

(b) $\alpha = \cos^{-1}\left(\frac{0}{178.04}\right) = 90°$

$\beta = \cos^{-1}\left(\frac{110}{178.04}\right) = 51.8°$

$\gamma = \cos^{-1}\left(\frac{-140}{178.04}\right) = 141.8°$

(c) As done in part (a)

$$\Rightarrow \vec{a} = \left(110\hat{j} - 140\hat{k}\right) \text{ N/kg}$$

(d) $a_{xy} = \vec{a} - \left(-140\hat{k}\right) = 110\hat{j}$ N/kg

$\qquad a_{xz} = \vec{a} - 110\hat{j} = -140\hat{k}$ N/kg

$\qquad a_{yz} = \vec{a} - 0\hat{i} = \left(110\hat{j} - 140\hat{k}\right)$ N/kg

65. The length of $\vec{B}$ in the figure below can be obtained from the coordinates of its two end points.

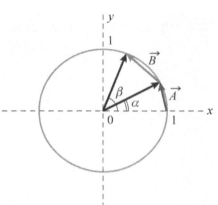

$$B^2 = \left(\cos\beta - \cos\alpha\right)^2 + \left(\sin\beta - \sin\alpha\right)^2$$
$$\Rightarrow B^2 = 2 - 2\cos\alpha\cos\beta - 2\sin\alpha\sin\beta \qquad (1)$$

The cosine law gives

$$B^2 = 1 + 1 - 2\cos\left(\beta - \alpha\right)$$

Substituting here value of B^2 from (1) we get

$$1 + 1 - 2\cos\left(\beta - \alpha\right) = 2 - 2\cos\alpha\cos\beta - 2\sin\alpha\sin\beta$$

Simplifying we get

$$\cos\left(\beta - \alpha\right) = \cos\alpha\cos\beta + \sin\alpha\sin\beta$$

67. Solution using Pythagorean Theorem:

Let us look at the triangle formed by one Cl-C-Cl combination. We can cut this triangle in half by dropping a perpendicular to the base from the vertex. Let us consider one of these triangles and find angle α it subtends at the vertex. Then we will find the required angle as $\theta = 2\alpha$.

If we assume each side to be of unit length, then the base of this triangle will be 0.5 units and height 0.707 units (by using the Pythagorean theorem).

$$\Rightarrow \alpha = \tan^{-1}\frac{0.707}{0.5} = 54.7°$$
$$\Rightarrow \theta = 2\alpha = 109.5°$$

Solution using vectors:

Let us imagine that a cube in Figure 2-25 has an edge length equal to 1. Let us imagine a coordinate system with the origin in the left bottom vertex of the cube. Then let us choose two vectors that coincide with the side of the triangle Cl-C-Cl (bottom left and right Cl atoms in the figure).

$$C(\frac{1}{2},\frac{1}{2},\frac{1}{2}); Cl_{bottom-left} = (0,0,0); Cl_{bottom-right} = (1,1,0)$$

$$\overrightarrow{CCl}_{left} = (-\frac{1}{2},-\frac{1}{2},-\frac{1}{2}); \overrightarrow{CCl}_{right} = (\frac{1}{2},\frac{1}{2},-\frac{1}{2})$$

$$\left|\overrightarrow{CCl}_{left}\right| = \sqrt{\frac{3}{4}} = \frac{\sqrt{3}}{2} = \left|\overrightarrow{CCl}_{left}\right|$$

$$\overrightarrow{CCl}_{left} \bullet \overrightarrow{CCl}_{right} = -\frac{1}{4}-\frac{1}{4}+\frac{1}{4} = -\frac{1}{4}$$

$$\overrightarrow{CCl}_{left} \bullet \overrightarrow{CCl}_{right} = \left|\overrightarrow{CCl}_{left}\right|\left|\overrightarrow{CCl}_{left}\right|\cos(\theta)$$

$$\cos(\theta) = \frac{-\frac{1}{4}}{\frac{3}{4}} = -\frac{1}{3} \Rightarrow \theta = \cos^{-1}(-\frac{1}{3}) = 109.5°$$

Chapter 3—MOTION IN ONE DIMENSION

1. (a) The two objects have the same average velocity.

   (b) B at $t = 30$ s; A at $t = 90$ s

   (c) Yes, when the slopes of their position-time graphs are equal, at about $t = 55$ s.

3. (a) no

   (b) yes, at $t = 0$

   (c) A. Maximum accelerations: A, about 100 m/s^2; B, 30 m/s^2

5. (a) B

   (b) B

   (c) A. Boat A has a higher (absolute) acceleration because the absolute value of the slope is higher.

   (d) A

   (e) Somewhere between $t = 50$ s and 100 s the objects have the same displacement, and they have the same acceleration at around $t = 130$ s.

7. (a) Yes, it is zero at the maxima and minima of the curve.

   (b) It is the highest at $t = 200$ s, and the lowest at $t = 30$ s and $t = 140$ s.

9. b

11. c. For the stone thrown straight up, the net displacement is zero when it lands.

$$v_f^2 = v_i^2 + 2\vec{a}\cdot\Delta\vec{x}$$
$$v_f^2 = v_i^2 + 2a\cdot(0)$$
$$|v_f| = |v_i|$$

The stone thrown at an angle has a net displacement in the horizontal direction when it lands. Because the displacement and acceleration vectors are perpendicular, their dot product is zero.

$$v_f^2 = v_i^2 + 2\vec{a}\cdot\Delta\vec{x}$$
$$v_f^2 = v_i^2 + 0$$
$$|v_f| = |v_i|$$

So both stones lands with a speed of 24 m/s.

13. b

15. b

$$v_1 = 10 \text{ m/s} + gt \text{ and } v_2 = 6 \text{ m/s} + gt$$
$$\Rightarrow v_{rel} = v_1 - v_2 = 4 \text{ m/s}$$

17. d

19. Yes, for example, an object moving at a constant acceleration has the same average and instantaneous acceleration.

21. At a point where the tangent to the position-time graph is parallel to the line segment joining the endpoints of the position-time graph on the time interval of interest, the average velocity for this time interval is equal to the instantaneous velocity.

23. a

25. a. The velocity to the right is a negative number. This gets smaller in magnitude in time; hence, less negative; hence, the velocity vector increases.

27. Only momentarily. If you are moving in one direction but have acceleration in the opposite direction, you will have zero speed for just the instant when you come to a stop before starting to move in the direction of the acceleration.

29. d

31. d

33. a

35. c (since the displacement is zero)

37. (a) $d = 1100 \text{ m} + 1300 \text{ m} = 2400 \text{ m}$

    (b) $\Delta \vec{x} = 1100 \text{ m} - 1300 \text{ m} = -200 \text{ m [down]}$

39. $d = 75 \text{cm} + 75 \text{ cm} + 80 \text{ cm} = 230 \text{ cm}$

    $\Delta \vec{x} = 0 - 80 \text{cm} = 80 \text{cm [down]}$

41. $v_{avg} = \dfrac{21 \text{m} + 20 \text{m}}{10 \text{s}} = 4.1 \text{m/s}$

    $\Delta \vec{x} = 20 \text{m} - 21 \text{m} = 1 \text{m [down]}$

    $\vec{v}_{avg} = \dfrac{1 \text{m}}{10 \text{s}} = 0.1 \text{m/s [down]}$

43. $v_{avg} = \dfrac{d}{\Delta t} = \dfrac{1.2\,\text{m}}{0.9\,\text{s}} = 1.3\,\text{m/s}$

$\vec{v}_{avg} = \dfrac{\Delta \vec{x}}{\Delta t} = \dfrac{1.2\,\text{m}}{0.9\,\text{s}} = 1.3\,\text{m/s}\,[\text{down}]$

45. Distance between Paris and new York is $d = 5800$ km

$t = \dfrac{d}{v} = \dfrac{5800\,\text{km}}{945\,\text{km/h}} = 6.14\,\text{h}$

47. Given: $r = 149\ 597\ 871$ km

$v = \dfrac{2\pi r}{T} = \dfrac{(2\pi)(149\,597\,871\,\text{km})}{365.25 \times 24 \times 60 \times 60\,\text{s}} = 29.8\,\text{km/s}$

49. Assuming the light arrives almost instantly:

$d = vT = (345\,\text{m/s})(3.00\,\text{s}) = 1035\,\text{m} = 1.04\,\text{km}$

51. (a) $\Delta \vec{x} = 10\,\text{kg} - 5\,\text{km} = 5\,\text{km}\,[\text{forward}]$

(b) $d = 10\,\text{kg} + 5\,\text{km} = 15\,\text{km}$

(c) $\vec{v}_{avg} = \dfrac{\Delta \vec{x}}{\Delta t} = \dfrac{5\,\text{km}}{7.5\,\text{min}} = 0.67\,\text{km/min} = 40\,\text{km/h}\,[\text{forward}]$

$v_{avg} = \dfrac{\Delta x}{\Delta t} = \dfrac{15\,\text{km}}{7.5\,\text{min}} = 2\,\text{km/min} = 120\,\text{km/h}$

53. Displacement is the area under the graph, and acceleration is the slope. By inspection:

(a) 105 m, 70 m, 40 m

(b) $-10\,\text{m/s}^2$

55. (a) $d = 149.6 \times 10^9\,\text{m}$

$t = \dfrac{d}{c} = \dfrac{149.6 \times 10^9\,\text{m}}{3.0 \times 10^8\,\text{m/s}} = 500\,\text{s} = 8.3\,\text{min}$

(b) $d = 8.14 \times 10^{16}\,\text{m}$

$t = \dfrac{d}{c} = \dfrac{8.14 \times 10^{16}\,\text{m}}{3.0 \times 10^8\,\text{m/s}} = 2.7 \times 10^8\,\text{s} = 8.6\,\text{years}$

(c) $d = 9.46 \times 10^{20}\,\text{m}$

$t = \dfrac{d}{c} = \dfrac{9.46 \times 10^{20}\,\text{m}}{3.0 \times 10^8\,\text{m/s}} = 3.15 \times 10^{12}\,\text{s} = 9.99 \times 10^4\,\text{years}$

57. By inspection: $v = 800 \, \text{m/s}, v = 800 \, \text{m/s} \, [\text{backward}]$

59. The distance covered before the stone hits the surface of the molasses can be calculated from
$$v_f^2 - v_i^2 = 2gd_1$$
$$\Rightarrow d_1 = \frac{v_f^2 - v_i^2}{2g} = \frac{(21 \, \text{m/s})^2 - (4 \, \text{m/s})^2}{2 \times 9.81 \, \text{m/s}^2} = 21.66 \, \text{m}$$

The time taken to cover this distance can be calculated from
$$v_f - v_i = gt_1$$
$$\Rightarrow t_1 = \frac{v_f - v_i}{g} = \frac{21 \, \text{m/s} - 4 \, \text{m/s}}{9.81 \, \text{m/s}^2} = 1.73 \, \text{s}$$

The deceleration in the molasses can be calculated from
$$v_f - v_i = at_2$$
$$\Rightarrow a = \frac{v_f - v_i}{t_2} = \frac{4 \, \text{m/s} - 21 \, \text{m/s}}{2 \, \text{s}} = -8.5 \, \text{m/s}^2$$

The distance covered in molasses can now be calculated from
$$v_f^2 - v_i^2 = 2ad_2$$
$$\Rightarrow d_2 = \frac{v_f^2 - v_i^2}{2g} = \frac{(4 \, \text{m/s})^2 - (21 \, \text{m/s})^2}{2 \times -8.5 \, \text{m/s}^2} = 25.0 \, \text{m}$$

The total distance covered by the stone is
$$d_{total} = d_1 + d_2 = 21.66 \, \text{m} + 25 \, \text{m} = 46.66 \, \text{m}$$

The total time is $t_t = t_1 + t_2 = 1.73 \, \text{s} + 2 \, \text{s} = 3.73 \, \text{s}$
$$\Rightarrow v_{avg} = \frac{d_{total}}{t_t} = \frac{46.66 \, \text{m}}{3.73 \, \text{s}} = 12.5 \, \text{m/s}$$

61. In simple harmonic motion, the position of a particle varies as a sinusoid. Acceleration, which is the second derivative of position, is then also a sinusoid. We will assume the acceleration is a cosine function with amplitude of 35 m/s² and period of 4 seconds. This can be integrated to yield the velocity function.

$$\vec{a}(t) = 35 \, \text{m/s}^2 \, \cos\left(\frac{\pi}{2\text{s}}t\right)$$
$$\vec{v}(t) = \int \vec{a}(t)dt = \frac{70}{\pi} \text{m/s} \, \sin\left(\frac{\pi}{2\text{s}}t\right)$$

(a) The average value of $\vec{v}(t)$ between $t = 2$ and $t = 3$ s is given as

$$\vec{v}_{avg} = \frac{\int_{2\text{s}}^{3\text{s}} \frac{70}{\pi} \text{m/s} \, \sin\left(\frac{\pi}{2\text{s}}t\right) dt}{3 \, \text{s} - 2 \, \text{s}} = -\frac{70}{\pi} \text{m/s} \times \frac{2\text{s}}{\pi} \times \frac{1}{1 \, \text{s}} \cos\left(\frac{\pi}{2 \, \text{s}}t\right)\Big|_{2\text{s}}^{3\text{s}} = 14.2 \, \text{m/s}$$

(b) $\quad \vec{v}(3 \text{ s}) = \dfrac{70}{\pi} \text{m/s} \ \sin\left(\dfrac{\pi}{2 \text{ s}} 3 \text{ s}\right) = -22.3 \text{ m/s}$

$\quad\quad v(3 \text{ s}) = |\vec{v}(3 \text{ s})| = 22.3 \text{ m/s}$

63. (a) $\quad a = \dfrac{v_f - v_i}{t} = \dfrac{21377 \text{ ft/s} - 7882.9 \text{ ft/s}}{384.22 \text{ s}} = 35.12 \text{ ft/s}^2 = 10.705 \text{ m/s}^2$

(b) $\quad d = \dfrac{v_f^2 - v_i^2}{2a} = \dfrac{(21377 \text{ ft/s})^2 - (7882.9 \text{ ft/s})^2}{2 \times 35.12 \text{ ft/s}^2} = 5.62 \times 10^6 \text{ ft} = 1713.3 \text{ km}$

65. We have (for simplicity we will leave out arrows to represent vectors)

$v_{12} = v_{012} + a_{12}t \quad\quad\quad (1)$

$\Delta x_{12} = v_{012}t + \dfrac{1}{2}a_{12}t^2 \quad\quad (2)$

$(1) \Rightarrow t = \dfrac{v_{12} - v_{012}}{a_{12}}$

Substituting the expression for t into (2) gives

$\Delta x_{12} = v_{012}\left(\dfrac{v_{12} - v_{012}}{a_{12}}\right) + \dfrac{1}{2}a_{12}\left(\dfrac{v_{12} - v_{012}}{a_{12}}\right)^2$

Simplifying this we get

$v_{12}^2 - v_{012}^2 = 2a_{12}\Delta x_{12}$

67. The average acceleration is equal to the acceleration due to gravity; that is, 9.81 m/s^2.

69. Given: $a = 1.45g$, $v_i = 0$, $v_f = 100 \text{ km/h} = 27.78 \text{ m/s}$

$t = \dfrac{v_f - v_i}{a} = \dfrac{27.78 \text{ m/s} - 0}{1.45 \times 9.81 \text{ m/s}^2} = 1.95 \text{ s}$

71. The ball's average acceleration is equal to the acceleration due to gravity; that is, 9.81 m/s^2. The ball's time of flight is

$t = \dfrac{v_f - v_i}{g} = \dfrac{-32.0 \text{ m/s} - 32.0 \text{ m/s}}{-9.81 \text{ m/s}^2} = 6.523 \text{ s}$

The distance travelled by the ball during the upward motion is

$d = \dfrac{v_f^2 - v_i^2}{2g} = \dfrac{0 - (32 \text{ m/s})^2}{-2 \times 9.81 \text{ m/s}^2} = 52.192 \text{ m}$

And the total distance is

$$d_t = 2 \times 52.192\,\text{m} = 104.38\,\text{m}$$

Hence, the average speed is

$$v_{avg} = \frac{d_t}{t_t} = \frac{104.38\,\text{m}}{6.523\,\text{s}} = 16.0\,\text{m/s}$$

Since the displacement is zero, the average velocity is also zero. $\vec{v}_{avg} = 0$

73. Given: $v_0 = 0$, $v_1 = 100$ km/h = 27.78 m/s, $v_2 = 240$ km/h = 66.67 m/s

   (a) $a_1 = \dfrac{v_1 - v_0}{t} = \dfrac{27.78\ \text{m/s} - 0}{2.46\ \text{s}} = 11.3\ \text{m/s}^2$

   (b) $d_1 = \dfrac{v_1^2 - v_0^2}{2a} = \dfrac{(27.78\ \text{m/s})^2 - 0}{2 \times 11.3\ \text{m/s}^2} = 34.1\ \text{m}$

   Acceleration to reach 240 km/h is

   $$a_2 = \frac{v_2 - v_1}{t} = \frac{66.67 - 27.78\ \text{m/s}}{7.34\ \text{s}} = 5.30\ \text{m/s}^2$$

   $$d_2 = \frac{v_2^2 - v_1^2}{2a} = \frac{(66.67\ \text{m/s})^2 - (27.78)^2}{2 \times 5.30\ \text{m/s}^2} = 346.5\ \text{m}$$

   The total distance travelled is

   $$d_{total} = d_1 + d_2 = 34.1\ \text{m} + 346.5\ \text{m} = 381\ \text{m}$$

   (c) Given: $v_3 = 415$ km/h = 115.28 m/s

   $$t_3 = \frac{v_3 - v_2}{a} = \frac{115.28\ \text{m/s} - 66.67}{5.30\ \text{m/s}^2} = 10.2\ \text{s}$$

   $$t_{total} = t_1 + t_2 + t_3 = 2.46\ \text{s} + 7.34\ \text{s} + 10.2\ \text{s} = 20.0\ \text{s}$$

75. (a) The speed of the ball should be 3 m/s on its way up as well. First, we calculate the initial velocity of the ball.

   $$v_f^2 - v_i^2 = 2gy$$

   $$\Rightarrow v_i = \sqrt{v_f^2 - 2gy} = \sqrt{3^2 - (2)(-9.81\ \text{m/s}^2)(14\ \text{m})} = 16.84\ \text{m/s}$$

   Now we can determine the flight time of the ball when it first reaches your friend.

   $$\Rightarrow t = \frac{v_f - v_i}{g} = \frac{3.0\ \text{m/s} - 16.83\ \text{m/s}}{-9.81\ \text{m/s}^2} = 1.4\ \text{s}$$

(b) Time to reach the apex, t, is

$$t = \frac{v_3 - v_2}{a} = \frac{0 - 3 \text{ m/s}}{-9.81 \text{ m/s}^2} = 0.31 \text{ s}$$

$$t_{total} = 1.4 \text{ s} + 2 \times 0.31 \text{ s} = 2.0 \text{ s}$$

Distance from friend to apex, d, is

$$d = \frac{v_3^2 - v_2^2}{2a} = \frac{(0)^2 - (3 \text{ m/s})^2}{2 \times -9.81 \text{ m/s}^2} = 0.46 \text{ m}$$

$$d_{total} = 14 \text{ m} + 2 \times 0.46 \text{ m} = 14.92 \text{ m}$$

$$v_{avg} = \frac{d_{total}}{t} = \frac{14.92 \text{ m}}{2.02 \text{ s}} = 7.4 \text{ m/s}$$

(c) $d_{max} = 14 \text{ m} + 0.46 \text{ m} = 14.5 \text{ m}$

$$t_{max} = 1.41 \text{ s} + 0.31 \text{ s} = 1.72 \text{ s}$$

$$v_{avg} = \frac{d_{max}}{t_{max}} = \frac{14.5 \text{ m}}{1.72 \text{ s}} = 8.4 \text{ m/s}$$

77. Given: $a = 46.2g$

$$v_i = \sqrt{v_f^2 - 2gd} = \sqrt{0 - (2)(-46.2 \times 9.81 \text{ m/s}^2)(1.6 \text{ m})} = 38 \text{ m/s}$$

$$t = \frac{v_f - v_i}{a} = \frac{0 - 38.08 \text{ m/s}}{-46.2 \times 9.81 \text{ m/s}^2} = 0.084 \text{ s} = 84 \text{ ms}$$

79. $v_f = v_i + at = 0 + (14 \times 9.81 \text{ m/s}^2)(1.3 \text{ s}) = 180 \text{ m/s}$

81. (a) Speed after first stage is

$$v_1 = a_1 t_1 = (5 \times 9.81 \text{ m/s}^2)(40 \text{ s}) = 1962 \text{ m/s}$$

Speed after second stage is

$$v_2 = v_1 + a_2 t_2 = 1962 \text{ m/s} + (3.5 \times 9.81 \text{ m/s}^2)(25 \text{ s}) = 2820 \text{ m/s}$$

(b) $\Delta x_3 = v_2 t_3 + \frac{1}{2} a_3 t_3^2 = (2820 \text{ m/s})(90 \text{ s}) + \frac{1}{2}(3.2 \times 9.81 \text{ m/s}^2)(90 \text{ s})^2 = 3.81 \times 10^5 \text{ m}$

(c) $\Delta x_1 = \frac{1}{2} a_1 t_1^2 = \frac{1}{2}(5.0 \times 9.81 \text{ m/s}^2)(40 \text{ s})^2 = 3.924 \times 10^4 \text{ m}$

$$\Delta x_2 = v_1 t_2 + \frac{1}{2} a_2 t_2^2 = (1962 \text{ m/s})(25 \text{ s}) + \frac{1}{2}(3.5 \times 9.81 \text{ m/s}^2)(25 \text{ s})^2 = 5.978 \times 10^4 \text{ m}$$

$$\Rightarrow \Delta x_{total} = 3.81 \times 10^5 \text{ m} + 3.924 \times 10^4 \text{ m} + 5.978 \times 10^4 \text{ m} = 4.80 \times 10^5 \text{ m}$$

83. $\Delta x = v_i t + \dfrac{1}{2} a t^2$

$\Rightarrow 300 \times 10^3 \text{ m} = 0 + \dfrac{1}{2}\left(4.5 \times 9.81 \text{ m/s}^2\right)t^2$

$\Rightarrow t = 117 \text{ s} = 1.94 \text{ min}$

85. Time to reach the highest point is $t = \dfrac{7 \text{ s}}{2} = 3.5 \text{ s}$.

This is also the time to return to the surface from the highest point.
The distance covered from the highest point to the suface is given by

$y = v_i t + \dfrac{1}{2} a t^2 = 0 + \dfrac{1}{2}\left(9.81 \text{ m/s}\right)\left(3.5 \text{ s}\right)^2 = 60.086 \text{ m}$

This is also the distance from surface to the highest point.
Hence, $y_{total} = 2 \times 60.086 \text{ m} = 120.17 \text{ m}$
Since the ball reaches back to the original position, the displacement is zero.

$v_{avg} = \dfrac{y_{total}}{t_{total}} = \dfrac{120.17 \text{ m}}{7 \text{ s}} = 17.2 \text{ m/s}$

The average acceleration is equal to the acceleration due to gravity; that is, $a = 9.81 \text{ m/s}^2$.

87. (a) The distance travelled by the ball to the highest point can be calculated from

$y_{max} = \dfrac{v_f^2 - v_i^2}{2g} = \dfrac{0 - v_i^2}{-2\left(9.81 \text{ m/s}^2\right)} = \dfrac{v_i^2}{19.62 \text{ m/s}^2}$

Also $v_f = v_i + gt$ gives $v_i = 9.81t$

$\Rightarrow y_{max} = \dfrac{\left(9.81 \text{ m/s}^2 \times t\right)^2}{19.62 \text{ m/s}^2} = 4.905 \text{ m/s}^2 \times t^2$

If the distance travelled by the marble is y_{marble}, then

$y_{max} + y_{marble} + 1 \text{ m} = 16 \text{ m}$

$\Rightarrow y_{max} = 15 \text{ m} - y_{marble}$

$\Rightarrow 15 \text{ m} - y_{marble} = 4.905 \text{ m/s}^2 \times t^2$ \qquad (1)

During time t the marble will travel a distance of

$y_{marble} = \dfrac{1}{2} g t^2 = 4.905 \text{ m/s}^2 \times t^2$

Substituting the expression in the previous equation into equation (1), we get

$4.905 \text{ m/s}^2 \times t^2 + 4.905 \text{ m/s}^2 \times t^2 = 15 \text{ m}$

$\Rightarrow t = 1.24 \text{ s}$

$\Rightarrow v_i = 9.81 \times 1.24 = 12.13 \text{ m/s}$

(b) For the marble, we have

$$v_f^2 - v_i^2 = 2gy$$

$$\Rightarrow v_{m,f}^2 - 0 = 2 \times 9.81 \text{ m/s}^2 \times 16 \text{ m}$$

$$\Rightarrow v_{m,f} = 17.71 \text{ m/s}$$

$$t_m = \frac{v_{m,f} - v_{m,i}}{g} = \frac{17.71 \text{ m/s} - 0}{9.81 \text{ m/s}^2} = 1.81 \text{ s}$$

The ball reaches the highest point in 1.24 s, after which it drops down. We need the velocity of the ball after $t = 1.805 - 1.236 = 0.569$ s.

$$v_{b,f} = 9.81 \times 0.569 = 5.58 \text{ m/s}$$

$$\Rightarrow v_{rel} = 5.58 \text{ m/s} - 17.71 \text{ m/s} = -12.13 \text{ m/s}$$

89. $\vec{v}_{rel} = 90 \text{ km/h} - 135 \text{ km/h} = -45 \text{ km/h}$ [backward]

$v_{rel} = 90 \text{ km/h} - 135 \text{ km/h} = -45 \text{ km/h}$

91. Since the initial velocity of the dropped ball is zero, its velocity after dropping a distance h is given by

$$v_1^2 = 2gh$$

For the ball thrown upwards, the height is $56 \text{ m} - h$. Its velocity is given by

$$v_2^2 = v_i^2 - 2g(56 \text{ m} - h)$$

Since $v_1^2 = v_2^2$

$$\Rightarrow v_i^2 - 2g(56 - h) = 2gh$$

$$\Rightarrow v_i = \sqrt{2gh + 2g(56 \text{ m} - h)} = \sqrt{2 \times 9.81 \text{ m/s}^2 \times 56 \text{ m}} = 33.1 \text{ m/s}$$

93. (a) Since the elevator is moving down with constant velocity, the acceleration in the elevator frame of reference is still g. Hence, the time for the apple to reach the maximum height is given by

$$t = \frac{v_i}{g} = \frac{6 \text{ m/s}}{9.81 \text{ m/s}^2} = 0.61 \text{ s}$$

Since the apple takes the same amount of time to reach back the thrower, the total time is

$$t_{total} = 2 \times 0.61 \text{ s} = 1.22 \text{ s}$$

(b) In the moving frame of reference:

$$v_2^2 = v_1^2 + 2ah$$

$$0 = (6 \text{ m/s})^2 + 2 \times -9.81 \text{ m/s}^2 \times h$$

$$h = 1.83 \text{ m}$$

The answer would stay the same if the elevator was moving up too. As long as the speed is constant, the acceleration in this frame of reference is g.

(c) The kinematics equations are the same in the elevator and ground frames of reference, where acceleration is 9.81 m/s², so like part (a), the time it takes is 1.22 s.

95. Given: $v_i = 360$ km/h $= 100$ m/s

$$x = v_i t + \frac{1}{2} a t^2 = 100 \text{ m/s} \times 1.2 \text{ s} - \frac{1}{2} \left(6 \times 9.81 \text{ m/s}^2 \right) (1.2 \text{ s})^2 = 77.6 \text{ m}$$

97. (a) $x(t) = x_0 \cos(\omega t + \phi)$

$x(0) = 0 \Rightarrow \cos(0 + \phi) = 0$

$\Rightarrow \cos \phi = 0$

$\Rightarrow \phi = \dfrac{n\pi}{2}; \quad n = 1, 3, 5, \ldots$

(b) (i)

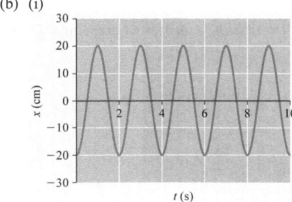

(ii)

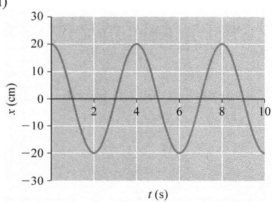

(iii)

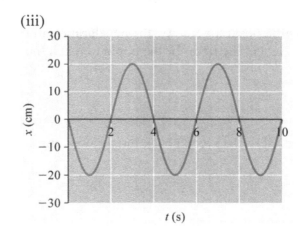

(c) Differentiating $x(t) = x_0 \cos(\omega t + \phi)$ with respect to time, we get

$\dot{x} = v(t) = -x_0 \omega \sin(\omega t + \phi)$

(d) Differentiating $\dot{x} = -x_0 \omega \sin(\omega t + \varphi)$ with respect to time, we get

$$\ddot{x} = a(t) = -x_0 \omega^2 \cos(\omega t + \varphi)$$

(e) $a(t) = -x_0 \omega^2 \cos(\omega t + \varphi)$

But $x_0 \cos(\omega t + \varphi) = x(t)$

$\Rightarrow a(t) = -\omega^2 x(t)$

99. (a) The speed of the speedboat is

$$v_{sb} = \frac{10 \text{ km}}{0.2 \text{ h}} = 50 \text{ km/h}$$

When the speedboat reaches the dock, the ocean liner has moved a distance of

$$x_{ol} = (20 \text{ km/h})(0.2 \text{ h}) = 4 \text{ km}$$

Therefore, the distance between the ocean liner and the speedboat is 6 km.

On the way back, the relative speed between the liner and the speedboat is

$$v_{rel} = v_{ol} + v_{sp} = 20 \text{ km/h} + 50 \text{ km/h} = 70 \text{ km/h}$$

The close the 6 km gap, the time taken is

$$t = \frac{6 \text{ km}}{70 \text{ km/h}} = 0.0857 \text{ h} = 5.1 \text{ min}$$

(b) The distance covered by the speedboat on its return to the liner is

$$d_2 = t_2 v_{sp} = 5.1 \text{ min} \times \frac{50 \text{ km}}{60 \text{ min}} = 4.3 \text{ km}$$

The total distance covered is then

$$d_{total} = d_1 + d_2 = 10 \text{ km} + 4.29 \text{ km} = 14.3 \text{ km}$$

The total displacement is

$$\Delta \vec{x} = 10 \text{ km} - 4.3 \text{ km} = 5.7 \text{ km} \text{ [toward the dock]}$$

101. When the apple is dropped, it is moving up at 9 m/s. The maximum height reached by the apple is given by

$$h_{max} = \frac{v^2}{2g} = \frac{(9 \text{ m/s})^2}{2 \times 9.81 \text{ m/s}^2} = 4.13 \text{ m}$$

Hence, the maximum height of the apple above ground is

$$h = 47 \text{ m} + 4.13 \text{ m} = 51.1 \text{ m}$$

The speed of the apple when it hits the ground is

$$v = \sqrt{2gh} = \sqrt{2 \times 9.81 \text{ m/s}^2 \times 51.13 \text{ m}} = 31.7 \text{ m/s}$$

103. The speed of the ball thrown up is given by

$$v_{ui} = \sqrt{v_{uf}^2 - 2gh}$$

Since the ball just reaches the balcony, $v_{uf} = 0$.

$$\Rightarrow v_{ui} = \sqrt{0 - 2(-9.81 \text{ m/s}^2)(30 \text{ m})} = 24.26 \text{ m/s}$$

Therefore, the final speed of the ball thrown down is

$$v_{df} = 2v_{ui} = 48.5 \text{ m/s}$$

$$\Rightarrow v_{di} = \sqrt{v_{df}^2 - 2gh} = \sqrt{(48.5 \text{ m/s})^2 - 2(9.81 \text{ m/s}^2)(30 \text{ m})} = 42.0 \text{ m/s}$$

The initial relative velocity seen by the ball moving down is

$$\vec{v}_{id} = \vec{v}_{ui} - \vec{v}_{di} = 24.26 \text{ m/s} - (-42.0 \text{ m/s}) = 66.3 \text{ m/s [up]}$$

Using equation (3-22), the relative acceleration between the two balls is

$$\vec{a}_{id} = g - g = 0$$

Because the relative acceleration is zero, the relative velocity is always 66.3 m/s.

105. Time taken by volleyball to reach the highest point is given by

$$t_v = \frac{v_{vi}}{g} = \frac{12 \text{ m/s}}{9.81 \text{ m/s}^2} = 1.22 \text{ s}$$

The vertical distance covered by the volleyball is

$$h = \frac{v_{vi}^2}{2g} = \frac{(12 \text{ m/s})^2}{2 \times 9.81 \text{ m/s}^2} = 7.34 \text{ m}$$

The time taken by the tennis ball to reach this height can be calculated from

$$s = v_i t + \frac{1}{2} gt^2$$

$$\Rightarrow 7.34 \text{ m} = 25 \text{ m/s} \times t - 4.905 \text{ m/s}^2 \times t^2$$

$$\Rightarrow t = 0.31 \text{ s (the earlier solution is the first time the ball is at this height)}$$

Hence, the tennis ball must be thrown after $1.22 \text{ m} - 0.31 \text{ m} = 0.91 \text{ s}$.

107. Initially, the relative speed between the bread and the lemon is

$$v_{rel} = v_{lemon,ground} - v_{bread,ground} = 30 \text{ m/s} - (-7.0 \text{ m/s}) = 37 \text{ m/s}$$

The relative acceleration between the bread and the lemon is

$$a_{rel} = a_{lemon,ground} - a_{bread,ground} = g - g = 0$$

Because the relative acceleration is zero, the relative speed stays constant at 37 m/s.

109. (a) The rate of change is less than 0.1 km/s per kilometre of altitude at 120 km and about 1.0 km/s per kilometre of altitude at 92 km.

(b) Since the meteor fragment is slowing, the force of atmospheric friction must be somewhat greater than the force of Earth's gravity at an altitude of 120 km, and much greater at an altitude of 90 km.

(c) The density of the atmosphere increases substantially as you descend in altitude from 120 to 85 km.

111. (a) about 30 m/s^2, 20 m/s^2, and 25 m/s^2

(b) about 13 m/s^2

(c) about 15 m/s^2, 12.5 m/s^2, and 12.3 m/s^2

113. Approximate shape shown below. Since this acceleration-time graph is based on values estimated from a small velocity graph, the resulting graph does not exactly match Figure 3-38 and does not show the details of the sudden changes in acceleration between the stages of the launch.

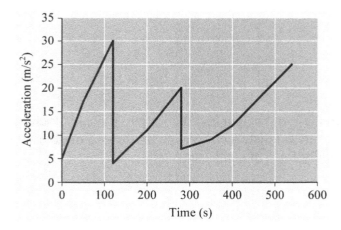

1. b

3. c

5. b

7. b. When the boat is pointed straight across the river:

$$\theta = \tan^{-1}\left(\frac{12}{60}\right) = 11.31°$$

Since $\tan\theta = \dfrac{v_{river}}{v_{boat}}$

$\Rightarrow v_{boat} = 5v_{river}$

In order for the boat to have a total velocity directly toward the target, its launch angle relative to the target should be

$$\theta' = \sin^{-1}\left(\frac{v_{river}}{v_{boat}}\right) = 11.54°$$

$\theta' > \theta$

9. d

11. d (Since both marbles are dropped from the same height and the initial vertical velocity of the rolled marble is zero. Therefore, the vertical distance travelled by the two marbles during the time the first marble moves horizontally for 1 m is the same.)

13. d. $d = 2.25 \times 2\pi \times \dfrac{4}{\pi}$ m $\neq 4$ m

$$A = \sqrt{\left(\frac{4}{\pi}\text{ m}\right)^2 + \left(\frac{4}{\pi}\text{ m}\right)^2} \neq \frac{2\sqrt{2}}{\pi}\text{ m}$$

15. d. In the apple's frame of reference, the relative acceleration of the bullet is zero. Thus, if the bullet was initially heading toward the apple, it always will be heading toward the apple.

17. a. The vertical distance covered by the projectile fired upward is

$$h = v\sin(30°)t - \frac{1}{2}gt^2$$

The vertical distance by which the second projectile will drop is

$$h' = -\frac{1}{2}gt^2$$

Hence, the vertical distance between the two projectiles is

$h - h' = v\sin(30°)t$

The horizontal distance covered by the first projectile is

$x = v\cos(30°)t$

The distance covered by the second projectile is

$x' = vt$

Hence, the horizontal distance between the two projectiles is

$x' - x = vt\left[1 - \cos(30°)\right]$

$\Rightarrow$ The horizontal distance between the two will increase with time.

19. a (since the ball will be travelling with the same horizontal speed)

21. Both will hit the ground at the same time. Both the apple and the arrow have the same (zero) initial vertical component of velocity, so they take the same amount of time to travel the same vertical distance.

23. tangential acceleration: tangent to the bowl and pointing in the direction of motion
radial acceleration: pointing towards the centre of the bowl

25. (a) $\left|\vec{v}_{avg}\right| = \dfrac{\sqrt{(3\text{ m})^2 + (4\text{ m})^2}}{2\text{ s} + 1.7\text{ s}} = 1.4\text{ m/s}$

(b) $v_{avg} = \dfrac{3\text{ m} + 4\text{ m}}{3.7\text{ s}} = 1.9\text{ m/s}$

(c) $v = \dfrac{3\text{ m}}{1.7\text{ s}} = 1.8\text{ m/s}$

(d) $v_{avg} = \dfrac{4\text{ m}}{2\text{ s}} = 2.0\text{ m/s}$

(e) No, these average speeds apply for only part of the time elapsed. The magnitude of the average velocity depends on the displacement.

27. (a) Given: $x = r\cos(8t)$, $y = r\sin(8t)$

We differentiate these to get x and y components of velocity.

$\Rightarrow \dot{x} = v_x = -8r\sin(8t)$

$\Rightarrow v_x(0.2) = -8r\sin(8\times0.2) = -7.99r$

Similarly

$\dot{y} = v_y = 8r\cos(8t)$

$\Rightarrow v_y(0.2) = 8r\cos(8\times0.2) = -0.23r$

(b) $|\vec{v}| = \sqrt{v_x^2 + v_y^2} = \sqrt{(-7.99r)^2 + (-0.23r)^2} = 8.0r$

$\theta = \tan^{-1}(0.23 / 7.99) = 1.6°$

$\vec{v} = 8.0r$ [1.6° below the negative x-axis]

(c) The particle is moving in a circle at a constant rate of rotation.

The acceleration can be found from differentiating velocity

$a_x = -64r \cos(8t)$

$a_y = -64r \sin(8t)$

In Cartesian notation:

$\vec{x}(t) = r \cos(8t)\hat{i} + r \sin(8t)\hat{j}$

$\vec{v}(t) = -8r \sin(8t)\hat{i} + 8r \cos(8t)\hat{j}$

$\vec{a}(t) = -64r \cos(8t)\hat{i} - 64r \sin(8t)\hat{j}$

29. Trajectory of the particle is given below.

Given: $x = 8t, \; y = \dfrac{1}{16t}$

$\Rightarrow v_x = 8, \; v_y = -\dfrac{1}{16t^2}$

$v(t) = \sqrt{8^2 + \left(-\dfrac{1}{16t^2}\right)^2} = \sqrt{64 + \dfrac{1}{256t^4}}$

$\Rightarrow v(3) = 8.0$ m/s

To calculate acceleration, we differentiate the velocity to get

$a_x = 0, \; a_y = \dfrac{1}{8t^3}$

$a(t) = \sqrt{0 + \left(\dfrac{1}{8t^3}\right)^2} = \dfrac{1}{8t^3}$

$\Rightarrow a(3) = 0.0046$ m/s^2

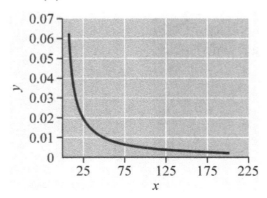

31. (a) Assume the ramp is in the x-z plane, and after turning, the car is moving in the y-direction. To calculate the distance travelled on the ramp, we first calculate the acceleration.

$$a_r = \frac{v_1 - v_0}{t} = \frac{23 \text{ m/s} - 0}{5 \text{ s}} = 4.6 \text{ m/s}^2$$

$$d_r = \frac{v_1^2 - v_0^2}{2a_r} = \frac{(23 \text{ m/s})^2 - 0}{2 \times 4.6 \text{ m/s}^2} = 57.5 \text{ m}$$

The displacement vector is

$$\Delta \vec{x}_r = 57.5 \text{ m} \times (\cos(37.0°)\hat{i} + \sin(37.0°)\hat{k})$$

$$\Delta \vec{x}_r = 45.9 \text{ m}\hat{i} + 34.6 \text{ m}\hat{k}$$

The additional displacement after moving through the curve is

$$\Delta \vec{x}_c = 121 \text{ m}\hat{i} + 121 \text{ m}\hat{j}$$

The total displacement is

$$\Delta \vec{x}_{total} = \Delta \vec{x}_r + \Delta \vec{x}_c$$

$$\Delta \vec{x}_{total} = 166.9 \text{ m}\hat{i} + 121 \text{ m}\hat{j} + 34.6 \text{ m}\hat{k}$$

(b) Next, we need to determine the time taken on the curve. First, we calculate acceleration. The distance travelled on the curve is 190.1 m, one quarter of the circle's perimeter.

$$a_c = \frac{v_2^2 - v_1^2}{2x_c} = \frac{(7 \text{ m/s})^2 - (23 \text{ m/s})^2}{2 \times 190.1 \text{ m}} = -1.26 \text{ m/s}^2$$

$$\Rightarrow t_c = \frac{v_2 - v_1}{a_c} = \frac{7 \text{ m/s} - 23 \text{ m/s}}{-1.26 \text{ m/s}^2} = 12.7 \text{ s}$$

$$\Rightarrow t_{total} = 5 \text{ s} + 12.7 \text{ s} = 17.7 \text{ s}$$

$$\vec{v}_{avg} = \frac{\Delta \vec{x}_{total}}{t_{total}} = \frac{166.9 \text{ m}\hat{i} + 121 \text{ m}\hat{j} + 34.6 \text{ m}\hat{k}}{17.7 \text{ s}}$$

$$\vec{v}_{avg} = 9.43 \text{ m/s}\hat{i} + 6.84 \text{ m/s}\hat{j} + 1.95 \text{ m}\hat{k}$$

(c) $$\vec{a}_{avg} = \frac{v_f - v_i}{t_{total}} = \frac{7 \text{ m/s}\hat{j} - 0}{17.7 \text{ s}} = 0.40 \text{ m/s}^2 \hat{j}$$

33. $$v_f^2 - v_i^2 = 2\vec{a} \cdot \Delta \vec{x} = 2gh$$

$$\Rightarrow v_i = \sqrt{(7 \text{ m/s})^2 - 2 \times 9.81 \text{ m/s}^2 \times 0.7 \text{ m}} = 5.9 \text{ m/s}$$

35. $$v_f^2 - v_i^2 = 2\vec{a} \cdot \Delta \vec{x} = 2gh$$

$$v_f = \sqrt{v_i^2 + 2gh} = \sqrt{(12 \text{ m/s})^2 + 2 \times 9.81 \text{ m/s}^2 \times 7 \text{ m}} = 16.7 \text{ m/s}$$

37. (a) Each time zone is 15° longitude apart (1/24th of 360°).

Given: $d = 2570$ km

$$\Rightarrow \theta = \frac{d}{R_{earth}} = \frac{2570 \text{ km}}{6371 \text{ km}} = 0.4034 \text{ rad} = 23.1°$$

The time zone difference between the two cities is either 1 hour or 2 hours.

(b) Assuming that the time zone difference between the two cities is 2 hours, the plane must reach Libreville within 2 hours. If we stop our watch when we board the plane, we do not need to adjust it when we land.

$$\Rightarrow v = \frac{2570 \text{ km}}{2 \text{ h}} = 1285 \text{ km/h}$$

39. Given: $v = 830$ km/h $= 230.6$ m/s

$$a_r = \frac{v^2}{r}$$

$$\Rightarrow r_{min} = \frac{v^2}{a_r} = \frac{v^2}{13g}$$

$$\Rightarrow r_{min} = \frac{(230.6 \text{ m/s})^2}{13 \times 9.81 \text{ m/s}^2} = 417 \text{ m}$$

41. Given: $\theta = 30° = 0.524$ rad

The distance travelled along the circumfrence is $d = \theta r = 0.524 \text{ rad} \times 0.32 \text{ m} = 0.17 \text{ m}$

Assuming constant tangential acceleration:

$$v_f^2 = v_i^2 + 2a_t d$$

$$a_t = \frac{v_f^2 - v_i^2}{2d} = \frac{(0.92 \text{ m/s})^2 - 0}{2 \times 0.17 \text{ m}} = 2.49 \text{ m/s}^2$$

$$a_r = \frac{v^2}{r} = \frac{(0.92 \text{ m/s})^2}{0.32 \text{ m}} = 2.65 \text{ m/s}^2$$

$$a = \sqrt{a_t^2 + a_r^2} = \sqrt{(2.49 \text{ m/s}^2)^2 + (2.65 \text{ m/s}^2)^2} = 3.63 \text{ m/s}^2$$

$$\theta = \tan^{-1}\left(\frac{2.49}{2.65}\right) = 43.2° \text{ with respect to radial direction}$$

43. $$a_r = \frac{v^2}{r}$$

$$\Rightarrow v = \sqrt{a_r r} = \sqrt{2 \times 10^6 \times 9.81 \text{ m/s}^2 \times 0.045 \text{ m}} = 940 \text{ m/s}$$

The speed divided by the circumfrence gives the rate of rotation

$$\text{rotation rate} = \frac{v}{2\pi r} = \frac{940 \text{ m/s}}{0.045 \text{ m}} = 3300 \text{ rotations/s}$$

45. Given: $\theta = 180° - 30° = 2.618$ rad

The distance covered along the perimeter is $d = \theta r = 2.62 \text{ rad} \times 4.0 \text{ m} = 10.5 \text{ m}$

$$a_t = \frac{v_f^2 - v_i^2}{2d} = \frac{(8.0 \text{ m/s})^2 - (12 \text{ m/s})^2}{2 \times 10.5 \text{ m}} = -3.82 \text{ m/s}^2$$

$$a_r = \frac{v^2}{r} = \frac{(8.0 \text{ m/s})^2}{4.0 \text{ m}} = 16.0 \text{ m/s}^2$$

$$a = \sqrt{a_t^2 + a_r^2} = \sqrt{(3.82 \text{ m/s}^2)^2 + (16.0 \text{ m/s}^2)^2} = 16.4 \text{ m/s}^2$$

$$\theta = \tan^{-1}\left(\frac{3.82}{16.0}\right) = 13.43° \text{ with respect to radial direction}$$

47. Tangential acceleration is given by

$$a_t = \frac{v_f^2 - v_i^2}{2S} = \frac{(4 \text{ m/s})^2 - 0}{2(3 \times 2\pi \times 0.72 \text{ m})} = 0.59 \text{ m/s}^2$$

Required condition is $a_r = a_t$

$$\Rightarrow \frac{v_2^2}{r} = 0.59 \text{ m/s}^2$$

$$\Rightarrow v_2 = 0.65 \text{ m/s}$$

Time to achieve this velocity is

$$t = \frac{v_2}{a_t} = \frac{0.65 \text{ m/s}}{0.59 \text{ m/s}^2} = 1.1 \text{ s}$$

$$x = \frac{v_2^2 - v_i^2}{2a_t} = \frac{(0.65 \text{ m/s})^2 - 0}{2 \times 0.59 \text{ m/s}^2} = 0.36 \text{ m}$$

This is equivalent to $\dfrac{0.36 \text{ m}}{2\pi(0.72 \text{ m})} = 0.08$ turns.

49. Since the horizontal speed of the acrobat is equal to the speed of the car, she will land in the parade car.

Given: $v_x = 10 \text{ km/h} = 2.78 \text{ m/s}$

Time to land back in the parade car can be found from

$$\Delta y = v_{y,0}t - \frac{1}{2}gt^2 = 0$$

$$t = 2 \times \frac{14 \text{ m/s}}{9.81^2 \text{ m/s}} = 2.9 \text{ s}$$

$$\Rightarrow x = v_x t = 2.78 \text{ m/s} \times 2.9 \text{ s} = 7.9 \text{ m}$$

51. The speed of the swimmer relative to ground during each interval is

$$v_{s,g,1} = v_{c,g} + v_{s,c}$$

$$v_{s,g,2} = v_{c,g} - v_{s,c}$$

The displacement of the swimmer relative to ground after 10 minutes is

$$x_{s,g} = v_{s,g,1} \times 300 \text{ s} + v_{s,g,2} \times 300 \text{ s} = 500 \text{ m}$$

$$(v_{c,g} + v_{s,c}) \times 300 \text{ s} + (v_{c,g} - v_{s,c}) \times 300 \text{ s} = 500 \text{ m}$$

$$\Rightarrow v_{c,g} = \frac{500 \text{ m}}{2 \times 300 \text{ s}} = 0.83 \text{ m/s}$$

53. Describe the motion of the arrow relative to the watermelon. The initial velocity of the arrow relative to the melon was such that it would pass a point 5.0 m above the melon after travelling a horizontal distance of 90 m. Because the acceleration of the arrow relative to the melon is zero, the melon will see the arrow travel in a perfectly straight line and pass 5.0 m above it.

55. Initially, the velocity of the teddy bear with respect to the ground has components:

$$v_x = 6 \text{ m/s} \times \cos(37°) + 2.7 \text{ m/s} = 7.5 \text{ m/s}$$

$$v_y = 6 \text{ m/s} \times \sin(37°) = 3.6 \text{ m/s}$$

$$v = \sqrt{v_x^2 + v_y^2} = \sqrt{(7.5 \text{ m/s})^2 + (3.6 \text{ m/s})^2} = 8.3 \text{ m/s}$$

At the time the teddy bear hits the ground, its horizontal velocity is unchanged, and its vertical velocity is

$$v_{y,f} = \sqrt{v_{y,i}^2 + 2\vec{a} \cdot \Delta \vec{y}} = \sqrt{(3.6 \text{ m/s})^2 + 2 \times 9.81 \text{ m/s}^2 \times 3 \text{ m}} = 8.5 \text{ m/s}$$

So when the bear hits the ground, its speed is

$$v = \sqrt{v_x^2 + v_y^2} = \sqrt{(7.5 \text{ m/s})^2 + (8.5 \text{ m/s})^2} = 11.3 \text{ m/s}$$

57. Given: $v_e = 34 \text{ km/h} = 9.44 \text{ m/s}$

The distance travelled by the eagle in 1 s is 9.44 m. Switching to the frame of reference of the eagle, a person is throwing a stone and trying to hit a point 23 m up and 9.94 m (9.94 m + 50 cm) over. The launch velocity in the eagle's frame of reference is

$$v_{x,0} = v\cos(38°) - 9.44 \text{ m/s}$$

$$v_{y,0} = v\sin(38°)$$

In order for the stone to hit the point 50 cm in front of the eagle

$$v_{x,0}t = 9.94 \text{ m} = (v\cos(38°) - 9.44 \text{ m/s})t \qquad (1)$$

$$v_{y,0}t - \frac{1}{2}gt^2 = 23 \text{ m} = v\sin(38°)t - \frac{1}{2}gt^2 \qquad (2)$$

One can use a graphing calculator to find that there is no solution to this problem. This means that there is no speed that can give the necessary altitude and horizontal displacement at the same time for a given angle.

59. Given: $v_b = 39$ km/h $= 10.8$ m/s

At the moment the gun is shot, the bird has flown 10.8 m/s × 4 s = 43.3 m

So that we're shooting at a stationary target, switch to the bird's frame of reference.

The x and y components of the bullet's initial velocity relative to the bird are

$v_{x,0} = 378$ m/s $\times \cos(\theta) - 10.8$ m/s

$v_{y,0} = 378$ m/s $\times \sin(\theta)$

The time it takes for the bullet to move a horizontal distance of 43.3 m is

$$t = \frac{43.3 \text{ m}}{(378 \text{ m/s} \times \cos(\theta) - 10.8 \text{ m/s})} \qquad (1)$$

In this time, the bullet reaches a height of 25 m

$$y_b = 25 \text{ m} = 378 \text{ m/s} \times \sin(\theta)t - \frac{1}{2}9.81 \text{ m/s}^2 t^2 \quad (2)$$

Substituting equation (1) into (2) and using a graphing calculator to solve for θ, we find if the bullet is fired at 29.3°, it will hit the bird after 0.14 s.

61. The time taken by the papaya to hit the ground is

$$t = \sqrt{\frac{2 \times 14 \text{ m}}{9.81 \text{ m/s}^2}} = 1.7 \text{ s}$$

Therefore, the stone hits the papaya after 1.7 s − 0.9 s = 0.8 s

To hit the papaya, which dropped from rest, aim directly at it assuming zero acceleration. Because there is not relative acceleration between the stone and the papaya, they will collide. For the monkey 11 m from the tree:

$$v_{x,i} = \frac{11 \text{ m}}{0.8 \text{ s}} = 13.8 \text{ m/s}$$

$$v_{y,i} = \frac{14 \text{ m}}{0.8 \text{ s}} = 17.5 \text{ m/s}$$

When the stone hits the papaya, its vertical speed and total speed are

$v_y = 17.5$ m/s $- 9.81$ m/s$^2 \times 0.8$ s $= 9.7$ m/s

$v = \sqrt{(13.8 \text{ m/s})^2 + (9.7 \text{ m/s})^2} = 16.8$ m/s

Similarily, the monkey throwing from 16 m throws with with a speed of 22.2 m/s.

63. (a) Given: $y = 60t^2 - 7t^3 + 4t + 120$

For vertical velocity we differentiate this with respect to time.

$\Rightarrow v_y = 120t - 21t^2 + 4$

At maximum height, vertical velocity is zero; that is

$120t - 21t^2 + 4 = 0$

$\Rightarrow t = 5.75$ s

$\Rightarrow y_{max} = 60(5.75)^2 - 7(5.75)^3 + 4(5.75) + 120 = 796$ m

(b) The time taken by the missile to travel 120 m below the point of launch can be calculated from

$y = 60t^2 - 7t^3 + 4t + 120 = 0$

$\Rightarrow t = 8.85$ s

The horizontal speed is a constant 3 m/s, and the vertical speed is

$v_y(8.85$ s$) = 120(8.85) - 21(8.85)^2 + 4 = -579$ m/s

$v_{total} = \sqrt{(3 \text{ m/s})^2 + (-579 \text{ m/s})^2} = 579$ m/s

At $t = 8.85$ s, the horizontal displacement is

$x(8.85$ s$) = 1400 + 3(8.85) = 1430$ m

(c) We see that the acceleration in the x-direction is zero. Therefore, the total acceleration is the acceleration in the y-direction, which can be obtained by differentiating the expression for velocity in the y-direction.

$\Rightarrow a_y = 120 - 42t$

$a(0) = 120$ m/s^2

$a(8.85$ s$) = -252$ m/s^2

Since acceleration varies linearly with time, it will have its maximum (absolute) value right before it hits the ground.

65. The component of velocity in the xy-plane is

$v_{xy} = 140 \text{ m/s} \times \cos(90° - 29°) = 140 \text{ m/s} \times \cos(61°)$

The x-component is then given by

$v_x = 140 \text{ m/s} \times \cos(61°)\cos(40°) = 52.0$ m/s

Since there is no acceleration in the x-direction, the x-component is

$x = 52.0 \text{ m/s} \times t$

Similarly, for the y-direction we have

$v_y = 140 \text{ m/s} \times \cos(61°)\sin(40°) = 43.6$ m/s

$\Rightarrow y = 43.6 \text{ m/s} \times t$

The initial velocity in the z-direction is

$v_z = 140 \text{ m/s} \times \sin(61°) = 122.4$ m/s

Since there is acceleration in the z-direction, we have

$z = 122 \text{ m/s} \times t - \frac{1}{2} 9.81 \text{ m/s}^2 \times t^2$

67. (a) The maximum height reached above the tree house is

$$h_1 = \frac{v_y^2}{2g} = \frac{(35\sin(37°))^2}{2\times9.8} = 22.6 \text{ m}$$

Hence, the maximum height from ground is $h_{max} = 4.0 + 22.6 = 26.6$ m

(b) The time needed to reach the ground can be found using the quadratic formula:

$$y(t) = -\frac{1}{2}9.81 \text{ m/s}^2 \times t^2 + 35 \text{ m/s} \times \sin(37°)t + 4 = 0$$

$$t = 4.48 \text{ s}$$

The horizontal distance covered is $x(4.48 \text{ s}) = 35 \text{ m/s} \times \cos(37°)(4.48 \text{ s}) = 125$ m

(c) $v_f = \sqrt{v_i^2 + 2\vec{a}\cdot\Delta\vec{x}} = \sqrt{(35.0 \text{ m/s})^2 + 2\times9.81 \text{ m/s}^2 \times 4 \text{ m}} = 36.1$ m/s

69. (a) The distance between the two points at calculated from the haversine formula is 10 494 km.

The plane moves with a speed of

$$v = \frac{10\,494 \text{ m}}{11 \text{ h}} = 954 \text{ km/h}$$

(b) Beijing is 11 time zones away from Toronto (during standard time), so the passenger will have to adjust the watch by 11 hours (12 hours during daylight savings time).

71. The plane has passed $7 \times 15° = 105°$ of longitude.

The distance covered by the plane can be calculated using the haversine formula

$$s = 9403 \text{ km}$$

Since passengers had to adjust their watches by one hour, the plane must have taken 6 or 8 hours to get there (assuming the passsengers stopped their watches on the plane)

$$\Rightarrow v = \frac{9403 \text{ km}}{6 \text{ h}} = 1567 \text{ km/h or } v = \frac{9403 \text{ km}}{8 \text{ h}} = 1175 \text{ km/h}$$

73. (a) $\vec{r}(t) = (2 + 0.5\sin(0.4t))\hat{i} + (3 + 0.5\sin(0.4t))\hat{j} + 2t^2\hat{k}$

$$\Rightarrow \vec{r}(0) = 2\hat{i} + 3\hat{j}$$

(b) $\vec{v} = \frac{d\vec{r}}{dt}$

$$\Rightarrow \vec{v}(t) = 0.2\cos(0.4t)\hat{i} + 0.2\cos(0.4t)\hat{j} + 4t\hat{k}$$

$$\Rightarrow \vec{v}(0.2) = 0.2\cos(0.4\times0.2)\hat{i} + 0.2\cos(0.4\times0.2)\hat{j} + 4\times0.2\hat{k}$$

$$\Rightarrow \vec{v}(0.2) = 0.1994\hat{i} + 0.1994\hat{j} + 0.8\hat{k}$$

$$\Rightarrow |\vec{v}(0.2)| = \sqrt{0.1994^2 + 0.1994^2 + 0.8^2} = 0.85 \text{ m/s}$$

(c) $\vec{r}(0.2) = (2 + 0.5\sin(0.4 \times 0.2))\hat{i} + (3 + 0.5\sin(0.4 \times 0.2))\hat{j} + 2(0.2)^2\hat{k}$

$\Rightarrow \vec{r}(0.2) = 2.040\hat{i} + 3.040\hat{j} + 0.08\hat{k}$

$\vec{r}(0.4) = (2 + 0.5\sin(0.4 \times 0.4))\hat{i} + (3 + 0.5\sin(0.4 \times 0.4))\hat{j} + 2(0.4)^2\hat{k}$

$\Rightarrow \vec{r}(0.4) = 2.080\hat{i} + 3.080\hat{j} + 0.32\hat{k}$

$\Rightarrow \vec{v}_{avg} = \dfrac{\vec{r}(0.4) - \vec{r}(0.2)}{0.2} = \dfrac{(2.080 - 2.040)\hat{i} + (3.080 - 3.040)\hat{j} + (0.32 - 0.08)\hat{k}}{0.2}$

$\Rightarrow \vec{v}_{avg} = (0.2\hat{i} + 0.2\hat{j} + 1.2\hat{k})$ m/s

(d) $\vec{v}(0.4) = 0.2\cos(0.4 \times 0.4)\hat{i} + 0.2\cos(0.4 \times 0.4)\hat{j} + 4 \times 0.4\hat{k}$

$\Rightarrow \vec{v}(0.4) = 0.1974\hat{i} + 0.1974\hat{j} + 1.6\hat{k}$

$\Rightarrow \vec{a}_{avg} = \dfrac{\vec{v}(0.4) - \vec{v}(0.2)}{0.4 - 0.2} = \dfrac{(0.1974 - 0.1994)\hat{i} + (0.1974 - 0.1994)\hat{j} + (1.6 - 0.8)\hat{k}}{0.2}$

$\Rightarrow \vec{a}_{avg} = (-0.01\hat{i} + -0.01\hat{j} + 0.8\hat{k})$ m/s

(e) $r(0) = 2\hat{i} + 3\hat{j}$

$r(1.1) = (2 + 0.5\sin(0.4 \times 1.1))\hat{i} + (3 + 0.5\sin(0.4 \times 1.1))\hat{j} + 2(1.1)^2\hat{k}$

$\Rightarrow r(1.1) = 2.21\hat{i} + 3.21\hat{j} + 2.42\hat{k}$

$\Rightarrow \Delta r = r(1.1) - r(0) = (0.21\hat{i} + 0.21\hat{j} + 2.42\hat{k})$ m

(f) The particle oscillates with a magnitude of 0.5 m on a plane 45° between the x- and y-axes as it speeds up in the z-direction at a uniform acceleration.

75. Given: $\omega = 3$ rev/s $= 6\pi$ rad/s

Define points A and B such that at $t = 0$, $|A - B| = 2r$

$A(t) = r(\cos(6\pi t)\hat{i} + \sin(6\pi t)\hat{j})$

$B = 3r\hat{i}$

We need to find the angle $\theta = 6\pi t$ at which $|A - B| = 3r$

Picture a triangle connecting A, B, and the origin O.

$OB = 3r$, $OA = r$, and $AB = 3r$

This is an isosceles triangle, and $\cos(\theta) = \dfrac{1}{6}$ and $\sin(\theta) = \sqrt{\dfrac{5}{6}}$

When the point makes this angle, the relative position $A - B$ is

$$\vec{x}_{rel} = r((\cos(6\pi t) - 3)\hat{i} + \sin(6\pi t)\hat{j}) = -\frac{17r}{6}\hat{i} + \sqrt{\frac{5}{6}}r\hat{j}$$

$$\vec{v}_{rel} = 6\pi r(-\sin(6\pi t)\hat{i} + \cos(6\pi t)\hat{j}) = -\sqrt{\frac{5}{6}}6\pi r\hat{i} + \frac{6\pi r}{6}\hat{j}$$

$$\vec{a}_{rel} = 36\pi^2 r(-\cos(6\pi t)\hat{i} - \sin(6\pi t)\hat{j}) = -6\pi^2 r\hat{i} - 36\pi^2 r\sqrt{\frac{5}{6}}\hat{j}$$

77. Given:

$R(t) = 1426.5e^{0.029t}$

$h(t) = 658.8 - 0.154t^3 + 18.3t^2 - 345t$

$\Rightarrow R(60) = 1426.5e^{0.029 \times 60} = 8.13$ km

$h(60) = 658.8 - 0.154 \times 60^3 + 18.3 \times 60^2 - 345 \times 60 = 12.6$ km

For velocity, we differentiate the given position functions with respect to time.

$v_x(t) = 41.37e^{0.029t}$

$\Rightarrow v_x(60) = 41.37e^{0.029 \times 60} = 235.7$ m/s

$v_y(t) = -0.462t^2 + 36.6t - 345$

$\Rightarrow v_y(60) = -0.462 \times 60^2 + 36.6 \times 60 - 345 = 187.8$ m/s

$\Rightarrow v = \sqrt{235.7^2 + 187.8^2} = 301$ m/s

For acceleration, we differentiate the velocity functions with respect to time.

$a_x(t) = 1.2e^{0.029t}$

$\Rightarrow a_x(60) = 1.2e^{0.029 \times 60} = 6.8$ m/s^2

$a_y(t) = -0.924t + 36.6$

$\Rightarrow a_y(60) = -0.924t + 36.6 = -18.8$ m/s^2

$\Rightarrow v = \sqrt{6.8^2 + 18.8^2} = 20.0$ m/s^2

79. Given: $v_c = 43$ km/h $= 11.94$ m/s

Choose the coordinate system such that the car travels in the x-direction, and z is up. In the frame of reference to the car, the acorn's velocity has the following components:

$v_{a,z} = 4.0$ m/s $\times \sin(26°) = 1.75$ m/s

$v_{a,x} = 4.0$ m/s $\times \cos(26°)\cos(34°) = 2.98$ m/s

$v_{a,y} = 4.0$ m/s $\times \cos(26°)\sin(34°) = 2.01$ m/s

(a) To find the x-distance from the car, use the relative x speed:

$$x_{acorn,car} = v_{a,x}t = 2.98 \text{ m/s} \times 1 \text{ s} = 2.98 \text{ m}$$

(b) In the frame of reference relative to the ground, at the instant the acorn is thrown:

$$v_{a,z} = 1.75 \text{ m/s}$$
$$v_{a,x} = 2.98 \text{ m/s} + 11.94 \text{ m/s} = 14.92 \text{ m/s}$$
$$v_{a,y} = 2.01 \text{ m/s}$$
$$v_a = \sqrt{(1.75 \text{ m/s})^2 + (14.92 \text{ m/s})^2 + (2.01 \text{ m/s})} = 15.2 \text{ m/s}$$

(c) The speed in the x-y plane stays constant. Relative to the ground, this speed is

$$v_{xy} = \sqrt{(14.92 \text{ m/s})^2 + (2.01 \text{ m/s})} = 15.1 \text{ m/s}$$

The ground is 1.0 m below where the acorn was thrown from

$$v_z = \sqrt{(1.75 \text{ m/s})^2 + 2(9.81 \text{ m/s}^2)(1.0 \text{ m})} = 4.76 \text{ m/s}$$

$$\theta_{horizontal} = \tan^{-1}\left(\frac{v_z}{v_{xy}}\right) = \tan^{-1}\left(\frac{4.76}{15.1}\right) = 17.5°$$

$$\theta_{road} = \tan^{-1}\left(\frac{2.01}{14.92}\right) = 7.7°$$

The acorn lands at an angle of 72° with respect to the horizontal plane. The projection of the acorn's velocity makes an angle of 7.7° with the road.

81. The cannon ball must clear Sir Burnalot, who is 4 m in front of the cannon.

$$v_x = -7 \text{ m/s} \times \cos(\theta)$$
$$v_{y,i} = 7 \text{ m/s} \times \sin(\theta)$$

The time taken to for the ball to reach Sir Burnalot is

$$t = \frac{\Delta x}{v_x} = \frac{-4.0 \text{ m}}{-7 \text{ m/s} \times \cos(\theta)} \qquad (1)$$

In this time, the ball must be 0.5 m high to clear Sir Burnalot.

$$y = 7 \text{ m/s} \times \sin(\theta) \times t - \frac{1}{2} 9.81 \text{ m/s}^2 \times t^2 = 0 \qquad (2)$$

By substituting equation (1) into (2), we have an equation in θ.

A graphing calculator was used to find that there is no solution. Any angle between 36.9° and 60.2° will result in the cannon ball clearing Sir Burnalot.

If the angle were 36.9°, the time taken for the ball to hit the ground would be

$$y = 7\sin(36.9°)t - \frac{1}{2} 9.81 \text{ m/s}^2 t^2 = 0 \Rightarrow t = 0.86 \text{ s}$$

The horizontal distance covered would be

$\Delta x = -7\cos(36.9°)t = -4.8$ m

At this time, Sir Crunchalot would only have travelled 1.1 m/s $\times 0.86$ s $= 0.95$ m.

For the cannon to hit Sir Crunchalot, it should pause for long enough for Sir Crunchalot to travel the extra 9.2 m before firing.

$$t_{pause} = \frac{\Delta x}{v} = \frac{9.2 \text{ m}}{11 \text{ m/s}} = 0.84 \text{ s}$$

83. (a) The particle is at $x = 20$, $y = 0$ at $t = 0$.

By differentiating the formulas in question 82, we have

$$v_x(t) = -8 \text{ m/s} \times \sin(0.4 \times t)$$
$$v_y(t) = 8 \text{ m/s} \times \cos(0.4 \times t)$$
$$\vec{v}(12 \text{ s}) = (7.97\hat{i} + 0.67\hat{j}) \text{ m/s}$$

(b) The average velocity is given by

$$\vec{v}_{avg} = \frac{\vec{x}(20 \text{ s}) - \vec{x}(12 \text{ s})}{20 \text{ s} - 12 \text{ s}}$$

$$\vec{v}_{avg} = \frac{20(\cos(0.4 \times 20)\hat{i} + \sin(0.4 \times 20)\hat{j}) \text{ m} - 20(\cos(0.4 \times 12)\hat{i} + \sin(0.4 \times 12)\hat{j}) \text{ m}}{8 \text{ s}}$$

$$\vec{v}_{avg} = (-4.66\hat{i} - 0.13\hat{j}) \text{ m/s}$$

For the average acceleration:

$$\vec{v}(20 \text{ s}) = (-7.91\hat{i} - 1.16\hat{j}) \text{ m/s}$$

$$\vec{a}_{avg} = \frac{\vec{v}(20 \text{ s}) - \vec{v}(12 \text{ s})}{20 \text{ s} - 12 \text{ s}} = \frac{(-7.91\hat{i} - 1.16\hat{j}) \text{ m/s} - (-7.97\hat{i} + 0.67\hat{j}) \text{ m/s}}{8 \text{ s}}$$

$$\vec{a}_{avg} = (7.5 \times 10^{-3}\hat{i} + 1.83\hat{j}) \text{ m/s}^2$$

85. From the geometry of the problem, the angles pointing from the original direction of the plane to 1) the airfield and 2) the point 34 km North of the airfield are

$$\theta_1 = \sin^{-1}\left(\frac{210}{300}\right) = 34.99°$$

$$\theta_2 = \sin^{-1}\left(\frac{244}{300}\right) = 39.12°$$

The original velocity of the plane was toward the airfield, with a speed of 270 km/h.

$$\vec{v}_p = 270 \text{ km/h} \times (\cos(34.99°)\hat{i} + \sin(34.99°)\hat{j})$$

If the pilot chose a new direction, α, her velocity combined with the wind would direct her toward the airport in just 98 min (1.63 hours).

We have three unknowns: new bearing α, windspeed v_{wind}, and wind direction θ_{wind}

$\vec{v}_{total} = \vec{v}_p + v_{wind}(\sin(\theta_{wind})\hat{i} + \cos(\theta_{wind})\hat{j})$ has angle of 39.12° $\qquad$ (1)

At the new bearing, the new total velocity is

$\vec{v}'_{total} = 120 \text{ km/h} \times (\sin(\alpha)\hat{i} + \cos(\alpha)\hat{j}) + v_{wind}(\sin(\theta_{wind})\hat{i} + \cos(\theta_{wind})\hat{j})$

$\vec{v}'_{total}$ has an angle of 34.99° $\qquad$ (2)

$\left|\vec{v}'_{total}\right| = \dfrac{\sqrt{(210 \text{ km})^2 + (300 \text{ km})^2}}{1.63 \text{ h}} = 244.7 \text{ km/h}$ $\qquad$ (3)

Solving equations (1), (2), and (3) simultaneously, we find that the wind velocity is

$\vec{v}_{wind} = -30.9 \text{ km/h} \, \hat{j}$

and the bearing that the plane should take to the airport is 31.2° North of East.

87. So that the pheasant sees the bullet heading directly upward toward it, the x-component of the bullet's velocity should match the speed of the pheasant.

$332 \text{ m/s} \times \cos(\theta) = 28 \text{ m/s}$

$\Rightarrow \theta = 85.2°$

89. (a) By inspection: 0.36 km/s

(b) By inspection: 0.42 km/s, 0.14 km/s, 2.0 km/s, 2.5 km/s

(c) The small increase in the slope of the graph for the next 250 s indicates some radial acceleration; after that the curve is almost straight, so there is little or no radial acceleration after 1750 s.

(d) Since the graph shows altitude versus time, the slope does not reflect any tangential component of the velocity.

91. The horizontal distance the ball must cover to clear the tree is

$v \cos(47°)t = 40 \text{ m}$ $\qquad$ (1)

The vertical distance the ball must cover to clear the tree is

$v \sin(47°)t - 4.9 \text{ m/s}^2 \times t^2 = 12 \text{ m}$ $\qquad$ (2)

Solving equations (1) and (2) simultaneously gives

$v = 23.4 \text{ m/s}$

The ball will clear the log with any velocity higher than 23.4 m/s. The result is that the ball will clear the tree by greater and greater distances.

1. a

3. b

5. b (since acceleration is negative)

7. b (since the friend does not move)

9. $\Delta x_2 = -2\Delta x_1$

$\Rightarrow a_2 = -2a_1$ (from differentiating the above twice)

The intended solution was to show the above relations between accelerations.

For completeness:

$$\frac{2T - m_1 g}{m_1} = -2\frac{T - m_2 g}{m_2}$$

$$\Rightarrow T = 58.9 \text{ N}$$

$$\Rightarrow a_1 = \frac{2(58.9 \text{ N}) - (16 \text{ kg})(9.81 \text{ m/s}^2)}{16 \text{ kg}} = -2.45 \text{ m/s}^2$$

$$\Rightarrow a_2 = \frac{(58.9 \text{ N}) - (4 \text{ kg})(9.81 \text{ m/s}^2)}{4 \text{ kg}} = 4.92 \text{ m/s}^2$$

The 16 kg mass accelerates down while the 4 kg mass accelerates up.

11. d

13. b (since the person being pulled is moving with constant speed)

15. d (since the net force is downward)

17. c

19. c

21. c. Since the sled does not move, there is no net force acting on it. This means there must be another force, such as friction, acting on the sled.

23. b. $F_{net} = ma = 120 \text{ kg} \times 5 \text{ m/s}^2 = 600 \text{ N}$

25. (a) In the vertical direction:

$F_y = mg = 60 \text{ kg} \times 9.81 \text{ m/s}^2 = 589 \text{ N}$

The car acts on you to pull you through the curve. Newton's third law says the you act on the car in the opposite direction.

So in the radial direction:

$$F_{radial} = \frac{mv^2}{r} = \frac{(60 \text{ kg})(20.0 \text{ m/s})^2}{40.0 \text{ m}} = 600 \text{ N}$$

$$F_{net} = \sqrt{(589 \text{ N})^2 + (600 \text{ N})^2} = 841 \text{ N}$$

(b) $F = \dfrac{Mv^2}{r} = \dfrac{1300 \text{ kg}(20.0 \text{ m/s})^2}{40.0 \text{ m}} = 1300 \text{ N} = 13.0 \text{ kN}$

(c) $F = \dfrac{mv^2}{r} = \dfrac{60 \text{ kg}(20.0 \text{ m/s})^2}{40.0 \text{ m}} = 600 \text{ N}$

(d) The force that the car exerts on you *is* the centripetal force calculated in part (c), 600 N.

(e) In the vertical direction, the road pushes the car with a force equal to the weight of you and the car. The force that the road exerts on the car is the centripetal force that moves you and the car through the curve. Newton's third law says the car will push the road with equal and opposite forces:

$$F_y = (M + m)g = (1300 \text{ kg} + 60.0 \text{ kg})(9.81 \text{ m/s}^2) = 13.3 \text{ kN}$$

$$F_{radial} = \frac{(M + m)v^2}{r} = \frac{(1300 \text{ kg} + 60 \text{ kg})(20.0 \text{ m/s})^2}{40.0 \text{ m}} = 13600 \text{ N} = 13.6 \text{ kN}$$

$$F_{net} = \sqrt{(13.3 \text{ kN})^2 + (13.6 \text{ N})^2} = 19.0 \text{ kN}$$

27. a (since no net force is acting on the buckets)

29. c

31. a. T_1, T_2, and T_3 are the tensions in the strings closest to, intermediate from, and furthest from the hand. $T_1 > T_2$ because the centripetal force on the mass *m* is $T_1 - T_2$ and must be positive. Likewise, $T_2 > T_3$ so that the centripetal force on mass *2m* is positive.

33. a (since the mass on table moves half of the distance covered by the hanging mass because of the movable pulley)

35. b and c are both correct. In the stationary frame of reference, the bug will move in a straight line as no force acts on it. In the rotating frame of reference of the turntable, there is a fictitious force radially outward (*centrifugal force*).

37. (a) $6N: \vec{F}_1 = 6 \text{ N} \times (\cos(31°)\hat{i} + \sin(31°)\hat{j}) = (5.14\hat{i} + 3.09\hat{j}) \text{ N}$

$8N: \vec{F}_2 = 8 \text{ N} \times (-\cos(52°)\hat{i} + \sin(52°)\hat{j}) = (-4.93\hat{i} + 6.30\hat{j}) \text{ N}$

$5N: \vec{F}_3 = 5 \text{ N} \times (-\sin(22°)\hat{i} - \cos(22°)\hat{j}) = (-1.87\hat{i} - 4.64\hat{j}) \text{ N}$

(b) $\vec{F} = \vec{F}_1 + \vec{F}_2 + \vec{F}_3 = (-1.66\hat{i} + 4.76\hat{j})$ N

$\Rightarrow |\vec{F}| = \sqrt{(1.66 \text{ N})^2 + (4.76 \text{ N})^2} = 5.04$ N, $\theta = \tan^{-1}\left(\dfrac{4.76}{-1.66}\right) = 109°$

39. (a) $\vec{F} = (-4+0+3)\text{N}\hat{i} + (-2+3+2)\text{N}\hat{j} + (0.5-4+0)\text{N}\hat{k} = \left(-\hat{i} + 3\hat{j} - 3.5\hat{k}\right)$ N

(b) $\vec{F}_1 = 14\cos(57°)\text{N}\hat{i} + 14\sin(57°)\text{N}\hat{j} = 7.6\text{N}\hat{i} + 11.7\text{N}\hat{j}$

$\vec{F}_3 = 8\cos(236°)\text{N}\hat{i} + 8\sin(236°)\text{N}\hat{j} = -4.5\text{N}\hat{i} - 6.6\text{N}\hat{j}$

$\vec{F} = \vec{F}_1 + \vec{F}_2 + \vec{F}_3$

$\Rightarrow \vec{F} = 6.2\text{N}\hat{i} + 1.1\text{N}\hat{j}$

(c) $\vec{F} = (4-9)\text{N}\hat{i} + (2+14)\text{N}\hat{j} + (-5+11)\text{N}\hat{k} = -5\text{N}\hat{i} + 16\text{N}\hat{j} + 6\text{N}\hat{k}$

(d) $\vec{F}_1 = 21\sin(32°)\cos(27°)\text{N}\hat{i} + 21\sin(32°)\sin(27°)\text{N}\hat{j} + 21\cos(32°)\text{N}\hat{k}$

$= 9.9\text{N}\hat{i} + 5.1\text{N}\hat{j} + 17.8\text{N}\hat{k}$

$\vec{F}_2 = 17\sin(160°)\cos(295°)\text{N}\hat{i} + 17\sin(160°)\sin(295°)\text{N}\hat{j} + 17\cos(160°)\text{N}\hat{k}$

$= 2.5\text{N}\hat{i} - 5.3\text{N}\hat{j} - 16.0\text{N}\hat{k}$

$\vec{F} = \vec{F}_1 + \vec{F}_2 = 12.4\text{N}\hat{i} - 0.2\text{N}\hat{j} + 1.8\text{N}\hat{k}$

41. $a = \dfrac{F}{m} = \dfrac{130}{9} = 14.4$ m/s

$v_f^2 - v_i^2 = 2a\Delta x$

$\Rightarrow v_f^2 - 0 = 2 \times 14.4 \times 14.0$

$\Rightarrow v_f = 20.1$ m/s

$a = \dfrac{F}{m} = \dfrac{130}{9} = 14.4$ m/s^2

$v_f^2 - v_i^2 = 2a\Delta x$

$\Rightarrow v_f = \sqrt{2 \times 14.4 \text{ m/s}^2 \times 14.0 \text{ m}} = 20.1$ m/s

43. Given: $v_i = 70$ km/s $= 19.4$ m/s

$\Delta x = \dfrac{v_f^2 - v_i^2}{2a} = \dfrac{0 - (19.4 \text{ m/s})^2}{-2 \times 36 \times 9.81 \text{ m/s}^2} = 0.54$ m $= 54.0$ cm

45. (a) Given: $v = 110$ km/h $= 30.56$ m/s

$F = \dfrac{mv^2}{r} = \dfrac{(1300 \text{ kg} + 71 \text{ kg} + 79 \text{ kg})(30.56 \text{ m/s})^2}{63.0 \text{ m}}$

$= 21.5$ kN [toward centre of circle]

(b) $F = \dfrac{mv^2}{r} = \dfrac{(71\ \text{kg})(30.56\ \text{m/s})^2}{63.0\ \text{m}} = 1.05\ \text{kN}$ [away from centre of circle]

(c) $F = \dfrac{mv^2}{r} = \dfrac{(1300\ \text{kg})(30.56\ \text{m/s})^2}{63.0\ \text{m}} = 19.3\ \text{kN}$ [toward centre of circle]

47. (a) Define x-axis as pointing down the slide, and y-axis perpendicular.
    Acceleration is in the x-direction.

$$\Sigma F_y = N - mg\cos(\theta) = 0$$
$$\Sigma F_x = ma_x = mg\sin(\theta)$$
$$a_x = g\sin(\theta)$$

(b) Define x-axis as pointing down the slide, and y-axis perpendicular.
    Acceleration is in the x-direction.

$$\Sigma F_y = N - mg\cos(\theta) = 0$$
$$\Sigma F_x = ma_x = mg\sin(\theta) - \mu_k mg\cos(\theta)$$
$$a_x = g(\sin(\theta) - \mu_k\cos(\theta))$$

(c) The axis of the pendulum is in the z-direction, the angle θ is measured relative to the negative z-axis, and the length of the string is L.
    Acceleration is opposite to the radial direction:

$$\Sigma F_z = T\cos(\theta) - mg = 0$$
$$\Rightarrow T = \dfrac{mg}{\cos(\theta)}$$
$$\Sigma \vec{F}_r = m\vec{a}_r = T\sin(\theta) = mg\tan(\theta) \text{ [opposite to radial direction]}$$
$$\vec{a}_r = g\tan(\theta) = -\dfrac{g}{L\cos(\theta)}r$$

(d) Define the x-axis as pointing up the slide, and y-axis perpendicular.
    Acceleration is in the x-direction.

$$\Sigma F_y = N - mg\cos(\theta) - F\sin(\theta) = 0$$
$$\Sigma F_x = ma = F\cos(\theta) - mg\sin(\theta)$$
$$a_x = \dfrac{F}{m}\cos(\theta) - g\sin(\theta)$$

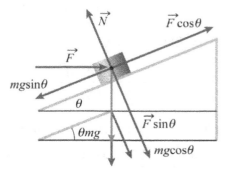

(e) Define the x-axis as parallel to the floor, and the y-axis perpendicular. Acceleration is in the x-direction.

$$\Sigma F_y = N + F\sin(\theta) - mg = 0$$
$$\Rightarrow N = mg - F\sin(\theta)$$
$$\Sigma F_x = ma = F\cos(\theta) - \mu_k(mg - F\sin(\theta))$$
$$a_x = \frac{F}{m}\cos(\theta) - \mu_k(g - \frac{F}{m}\sin(\theta))$$

(f) The motion takes place on the x-y plane. Positive x is at the top of the circle. Acceleration is opposite to the radial direction.
Because the speed is constant, there are no forces acting in the tangential direction. In the radial direction, the net force is

(i) $\Sigma F_y = ma_y = -T - mg$

$$a_y = -\frac{T}{m} - g$$

(ii) $\Sigma F_y = ma_y = T - mg$

$$a_y = \frac{T}{m} - g$$

(g) The motion takes place in the Cartesian plane on a circle on the x-y plane centred at the origin. Acceleration is opposite to the radial direction.

$$\Sigma F_z = N\cos(\theta) - mg = 0$$
$$\Rightarrow N = \frac{mg}{\cos(\theta)}$$
$$\Sigma \vec{F}_r = m\vec{a}_r = N\sin(\theta) = mg\tan(\theta) \text{ [opposite to radial direction]}$$
$$a_r = -g\tan(\theta)$$

(h) The motion takes place in the Cartesian plane on a circle on the x-y plane centred at the origin. Acceleration is opposite to the radial direction. The force of friction acts parallel to the ramp with the x-y component pointing to the centre of the circle.

$$\Sigma F_z = N\cos(\theta) - mg - \mu_s N\sin(\theta) = 0$$
$$\Rightarrow N = \frac{mg}{\cos(\theta) - \mu_s\sin(\theta)}$$
$$\Sigma \vec{F}_r = m\vec{a}_r = N\sin(\theta) + \mu_s N\cos(\theta) \text{ [opposite to radial direction]}$$
$$\Sigma \vec{F}_r = N(\sin(\theta) + \mu_s\cos(\theta)) = mg\frac{\sin(\theta) + \mu_s\cos(\theta)}{\cos(\theta) - \mu_s\sin(\theta)} \text{ [opposite to radial direction]}$$
$$a_r = -g\frac{\sin(\theta) + \mu_s\cos(\theta)}{\cos(\theta) - \mu_s\sin(\theta)}$$

49. (a) $a = \dfrac{v_f^2 - v_i^2}{2d} = \dfrac{0 - (9 \text{ m/s})^2}{2 \times (-4.0 \text{ m})} = 10.1 \text{ m/s}^2$

By Newton's second law, the net force acting on you is

$F = m_1 a = 81 \text{ kg} \times (10.1 \text{ m/s}^2) = 820 \text{ N}$

(b) $F = m_2 a = 2300 \text{ kg} \times (10.1 \text{ m/s}^2) = 23.2 \text{ kN}$

(c) In part (a), it was found the net force on you is 820 N. So the force of the elevator on you is

$F_{elevator \to you} = F_{net} + mg = 820 \text{ N} + 81 \text{ kg} \times 9.81 \text{ m/s}^2 = 1.61 \text{ kN}$

$F_{you \to elevator} = -F_{elevator \to you} = -1.61 \text{ kN}$

(d) $F = m_2 g = 2300 \text{ kg} \times 9.81 \text{ m/s}^2 = -22.6 \text{ kN}$

51. The net force acting on the cat is

$\Sigma F = F_{woman \to cat} - mg = ma$

$F_{woman \to cat} = ma + mg = m(a + g) = (2.7 \text{ kg})(3.2 \text{ m/s}^2 + 9.81 \text{ m/s}^2) = 35 \text{ N}$

$F_{cat \to woman} = -F_{woman \to cat} = -35 \text{ N}$

53. $a = \dfrac{v_f^2 - v_i^2}{2\Delta x} = \dfrac{(3.2 \text{ m/s})^2 - 0}{2 \times 15 \text{ m}} = 0.34 \text{ m/s}^2$

Acceleration in the direction of motion is caused by gravity:

$\Sigma F = mg \sin \theta = ma$

$\Rightarrow \theta = \sin^{-1}\left(\dfrac{a}{g}\right) = \sin^{-1}\left(\dfrac{0.34}{9.81}\right) = 2.0°$

55. $a = \dfrac{v_f^2 - v_i^2}{2\Delta x} = \dfrac{(4.7 \text{ m/s})^2 - 0}{2 \times 31 \text{ m}} = 0.36 \text{ m/s}^2$

Acceleration is caused by the net force parallel to the slide (gravity and friction).

$\Sigma F = mg \sin \theta - \mu mg \cos \theta = ma$

$\Rightarrow \mu = \dfrac{g \sin \theta - a}{g \cos \theta} = \dfrac{9.81 \text{ m/s}^2 \sin(37°) - 0.36 \text{ m/s}^2}{9.81 \text{ m/s}^2 \cos(37°)} = 0.71$

57. (a) Applying Newton's second law to both masses:

$\Sigma F_2 = M_2 g - T = M_2 a$

$\Sigma F_1 = T = M_1 a$

Solving these equations simultaneously gives

$$T = \frac{M_1 M_2}{M_1 + M_2} g$$

(b) The analysis would be no different, so the tension is the same.

(c) $T = \left(\dfrac{2 \text{ kg} \times 12 \text{ kg}}{2 \text{ kg} + 12 \text{ kg}} \right) 9.81 \text{ m/s}^2 = 16.8 \text{ N}$

59. (a) Applying Newton's second law to m_2 gives

$$m_2 g - T = m_2 a$$

$$\Rightarrow m_2 g - \frac{m_2 g}{2} = m_2 \textcircled{a} \nearrow^{-ve}$$

$$\Rightarrow a = \frac{g}{2}$$

(b) Setting $T = 0$ in part (a), we obtain $a = g$. Thus, the system falls freely, and so the tension in the upper rope is also zero.

61. The spring constant k can be calculated as

$$k = \left| \frac{F}{\Delta x} \right| = \frac{120 \text{ N}}{0.1 \text{ m}} = 1200 \text{ N/m}$$

If the force to stretch the spring were now the weight of a 10 kg mass:

$$\Delta x = \left| \frac{F}{k} \right| = \frac{mg}{k} = \frac{10 \text{ kg} \times 9.81 \text{ m/s}^2}{1200 \text{ N/m}} = 0.082 \text{ m} = 8.2 \text{ cm}$$

63. The spring pulls the mass up with a force kx. Applying Newton's second law:

$$\Sigma F = mg - kx = ma$$

$$\Rightarrow k = \frac{mg - ma}{x} = \frac{12 \text{ kg} \times 9.81 \text{ m/s}^2 - 12 \text{ kg} \times 3.0 \text{ m/s}^2}{0.3 \text{ m}} = 272 \text{ N/m}$$

65. The centripetal force is provided by the force of static friction:

$$\frac{mv^2}{r} = \mu_s mg$$

$$\Rightarrow v = \sqrt{\mu_s gr} = \sqrt{0.45 \times 9.81 \text{ m/s}^2 \times 3.0 \text{ m}} = 3.6 \text{ m/s}$$

67. (a) The normal force acts perpendicular to the road. The component in the vertical direction balances gravity. The component in the radial direction is the centripetal force:

$$\Sigma F_y = N\cos(\theta) - mg = 0 \quad (1)$$

$$\Sigma F_y = N\sin(\theta) = \frac{mv^2}{r} \quad (2)$$

Combining equations (1) and (2), we have

$$\Rightarrow v = \sqrt{gr\tan(\theta)} = \sqrt{9.81 \text{ m/s}^2 \times 72 \text{ m} \times \tan(23°)} = 17.3 \text{ m/s} = 62.3 \text{ km/h}$$

(b) The net force is the centripetal force:

$$\Sigma \vec{F} = \frac{mv^2}{r} = \frac{72.0 \text{ kg} \times (17.3 \text{ m/s})^2}{17.0 \text{ m}} = 300 \text{ N [toward the centre of the turn]}$$

69. In the train's frame of reference, there is a fictitious acting in the opposite direction of acceleration. Friction will act in the forward direction to counteract this force.

$$\vec{F}_{fictitious} = -m\vec{a}$$

$$\Sigma F = \mu_s mg - ma = 0$$

$$\mu_s = \frac{a}{g} = \frac{5.0 \text{ m/s}^2}{9.81 \text{ m/s}^2} = 0.51$$

In the train station's frame of reference, the force of static friction accelerates the passenger forward.

$$\Sigma F = ma = \mu_s mg$$

$$\mu_s = \frac{a}{g} = \frac{5.0 \text{ m/s}^2}{9.81 \text{ m/s}^2} = 0.51$$

71. Given: $v = 67 \text{ km/h} = 18.6 \text{ m/s}$

The frame of reference has the following acceleration:

$$a_{truck} = \frac{v^2}{r}$$

The forces ballance on the bag, so:

$$\left| F_{wall} \right| = \left| F_{fictitious} \right| = \left| -m_{bag} a_{truck} \right| = \frac{mv^2}{r} = \frac{30 \text{ kg} \times (18.6 \text{ m/s}^2)^2}{23 \text{ m}} = 450 \text{ N}$$

73. The net force in the radial direction is the centripetal force. In the vertical direction, the force of gravity balances with the vertical component of tension.

$$\Sigma F_z = T\cos\theta - mg = 0$$

$$\Sigma F_{radial} = T\sin\theta = \frac{mv^2}{r}$$

$$\Rightarrow v = \sqrt{gr\tan\theta} = \sqrt{(9.81 \text{ m/s}^2)(6.0 \text{ m})\left(\frac{6}{\sqrt{17^2 - 6^2}}\right)} = 4.7 \text{ m/s}$$

75. Given: $v = 30$ km/h $= 8.33$ m/s

The horizontal component of tension provides the centripetal acceleration.

The vertical component of tension balances gravity.

$$T\sin(\theta) = ma_{ship} = \frac{mv^2}{r} \quad (1)$$

$$T\cos(\theta) = mg \quad (2)$$

Dividing equation (1) by (2) gives

$$\tan(\theta) = \frac{v^2}{gr}$$

$$\Rightarrow \theta = \tan^{-1}\left(\frac{v^2}{gr}\right) = \tan^{-1}\left(\frac{(8.33 \text{ m/s})^2}{9.81 \text{ m/s}^2 \times 150 \text{ m}}\right) = 2.7°$$

77. The net force acting on the spider is the centripetal force, toward the centre of the circle.
Summing forces perpendicular to the wall:

$$\Sigma F_{perpendicular} = mg\sin(\theta) - N = \frac{mv^2}{r}\cos(\theta)$$

$$\Rightarrow N = m\left(g\sin(\theta) - \frac{v^2}{r}\cos(\theta)\right)$$

Summing forces parallel to the wall:

$$\Sigma F_{parellel} = \mu m\left(g\sin(\theta) - \frac{v^2}{r}\cos(\theta)\right) - mg\cos(\theta) = \frac{mv^2}{r}\sin(\theta)$$

$$\Rightarrow \frac{v^2}{r}(\sin(\theta) + \mu\cos(\theta)) = g(\mu\sin(\theta) - \cos(\theta))$$

$$\Rightarrow v = \sqrt{gr\frac{\mu\sin(\theta) - \cos(\theta)}{\sin(\theta) + \mu\cos(\theta)}} = \sqrt{(9.81 \text{ m/s}^2 \times 72 \text{ m})\frac{0.91\sin(60°) - \cos(60°)}{\sin(60°) + 0.91\cos(60°)}} = 12.4 \text{ m/s}$$

79. (a) If the positive direction is up for m_1 and down for m_2, both masses have the same speed and acceleration. By applying Newton's second law:

$$m_2 g - T = m_2 a \quad (1)$$

$$T - m_1 g = m_1 a \quad (2)$$

$$\Rightarrow a = \frac{m_2 - m_1}{m_1 + m_2}g = \frac{14.0 \text{ kg} - 12.0 \text{ kg}}{12.0 \text{ kg} + 14.0 \text{ kg}}9.81 \text{ m/s}^2 = 0.75 \text{ m/s}^2$$

The velocity of each mass is

$$v = \sqrt{2ad} = \sqrt{2 \times 0.75 \text{ m/s}^2 \times 4.00 \text{ m}} = 2.46 \text{ m/s}$$

(b) $T = m_1(a + g) = 12.0 \text{ kg} \times (0.75 \text{ m/s}^2 + 9.81 \text{ m/s}^2) = 127 \text{ N}$

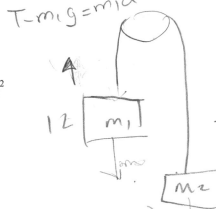

81. The angle swept by the marble is

$$\theta = \sin^{-1}\frac{R}{2R} = 30°$$

$$a_t = g\cos(30°) = \frac{\sqrt{3}}{2}9.81 \text{ m/s}^2$$

$$a_r = \frac{v^2}{R} = \frac{(2.3 \text{ m/s})^2}{R}$$

$$|a_{total}| = \sqrt{\frac{3(9.81 \text{ m/s}^2)^2}{4} + \frac{(2.3 \text{ m/s})^4}{R^2}}$$

83. The acceleration of the entire system is given as

$$a = \frac{F}{m+M}$$

The block has the same acceleration:

$$\Sigma F_y = N\cos(\theta) - mg = 0$$

$$\Sigma F_x = N\sin(\theta) = m\frac{F}{m+M}$$

$$\Rightarrow F = (m+M)g\tan(\theta)$$

85. Summing forces acting on the wedge:

$$\Sigma F_x = F - \mu N\cos(\theta) - N\sin(\theta) = MA \qquad (1)$$

Summing forces for the block:

$$\Sigma F_y = N\cos(\theta) - \mu N\sin(\theta) - mg = ma\sin(\theta) \qquad (2)$$

$$\Sigma F_x = N\sin(\theta) + \mu N\cos(\theta) = m(A - a\cos(\theta)) \qquad (3)$$

$$(2) \Rightarrow N = \frac{m(g + a\sin(\theta))}{\cos(\theta) - \mu\sin(\theta)}$$

$$(1)+(3) \Rightarrow A = \frac{F + ma\cos(\theta)}{M+m}$$

Inserting the above expressions for N and A into equation (3) gives

$$\frac{m(g + a\sin(\theta))}{\cos(\theta) - \mu\sin(\theta)}(\mu\cos(\theta) + \sin(\theta)) = m\left(\frac{F + ma\cos(\theta)}{M+m} - a\cos(\theta)\right)$$

$$\Rightarrow F = \frac{(M+m)(g + a\sin(\theta))}{\cos(\theta) - \mu\sin(\theta)}(\mu\cos(\theta) + \sin(\theta)) + Ma\cos(\theta)$$

87. Because the $3m$ mass does not move vertically, tension balances gravity:

$T = 3mg$

The acceleration of the $1m$ mass is caused by tension alone:

$a_{1m} = \dfrac{T}{m} = 3g$

The whole system must then have an acceleration of $3g$, so

$F = m_{Total}a = (M + m + 3m)(3g) = 3g(M + 4m)$

89. (a) We use Newton's second law to find the tension needed to accelerate it up at 3.00 m/s².

$\Sigma F = ma = T - mg$

$T = m(a + g) = 1.00 \text{ kg} \times (3.00 \text{ m/s}^2 + 9.81 \text{ m/s}^2) = 12.81 \text{ N}$

This same tension causes block A to accelerate forward:

$a_A = \dfrac{T}{m_A} = \dfrac{12.81 \text{ N}}{2.00 \text{ kg}} = 6.41 \text{ m/s}^2$

Relative to block B, block A accelerates backwards at 3.00 m/s², so

$a_B = a_A + 3.00 \text{ m/s}^2 = 9.41 \text{ m/s}^2$

In the horizontal direction, blocks B and C accelerate together at 9.41 m/s².

$\Sigma F_{B+C} = (m_B + m_C)a_{B+C}$

$F - 12.81 \text{ N} = (4.00 \text{ kg} + 1.00 \text{ kg})(9.41 \text{ m/s}^2)$

$F = 59.9 \text{ N}$

The reason we had to subtract 12.81 N to get the net force on the system B + C is because block A pulls the system in the reverse direction by the tension in the string.

(b) $F_A = \vec{T} = 12.8 \text{ N [right]}$

$\vec{F}_B = m_B\vec{a}_B = (4.00 \text{ kg})(9.41 \text{ m/s}^2) = 37.6 \text{ N [right]}$

$\vec{F}_C = m_C\vec{a}_C = 1.00 \text{ kg} \times (9.41\text{m/s}^2\hat{i} + 3.00\text{m/s}^2\hat{j}) = 9.41 \text{ N}\hat{i} + 3 \text{ N}\hat{j}$

The net force on block B is different than the force we use to push it because it is pushed back by the step that attaches the pulley.

91. (a) The normal force and the normal component of gravity cancel, so

$N = Mg \cos\theta$

(b) At this speed, the normal force grows as the car presses into the track. The vertical component of the normal force cancels the force of gravity.

$N = \dfrac{Mg}{\cos(\theta)}$

(c) We now have a friction force of magnitude $\mu_s N$ parallel to the road pointing toward the centre of the circle. The sum of forces in the vertical direction is zero.

$$N\cos(\theta) = mg + \mu N \sin(\theta)$$

$$\Rightarrow N = \frac{mg}{\cos(\theta) - \mu\sin(\theta)}$$

93. The ramp has centripetal acceleration of a toward the centre of the circle. The astronaut (mass m) will feel a fictitious force ma in the opposite direction (called centrifugal force). Four forces act on the astronaut in his frame of reference. Gravity acts down, centrifugal force acts backward, normal force is perpendicular to the ramp, and static friction acts parallel to the ramp (toward the centre of the circle). Balance all forces in the direction of the normal:

$$N = ma\sin(\theta) + mg\cos(\theta) \qquad (1)$$

Now balance all forces on the plane of the ramp:

$$ma\cos(\theta) = mg\sin(\theta) + \mu_s N \qquad (2)$$

Combining the two finds the maximum acceleration of the ramp before a slip:

$$a_{max} = g\frac{\sin(\theta) + \mu_s\cos(\theta)}{\cos(\theta) - \mu\sin(\theta)} = 9.81 \text{ m/s}^2 \times \frac{\sin(45°) + 0.25\cos(45°)}{\cos(45°) - 0.25\sin(45°)} = 16.35 \text{ m/s}^2$$

The tension at the end of the rope is causing the acceleration of the ramp. The combined mass of the astronaut and ramp is M, so

$$T = Ma_{max} = 1000 \text{ kg} \times 16.35 \text{ m/s}^2 = 16.35 \text{ kN}$$

Because the rope has mass, we must consider that the tension in the rope is higher at the centre of the circle. If each differential section of the rope with mass dm is accelerating, the increase in tension dT is in the direction of acceleration and is given by

$$dT = a(r)dm$$

$$a(r) = -\frac{v^2}{r} = -\left(\frac{v}{r}\right)^2 r$$

$\frac{v}{r}$ is a constant. Because we know that at the end of the rope $a(10.0 \text{ m}) = -13.5 \text{ m/s}^2$,

$$a(r) = -\frac{13.5 \text{ m/s}^2}{10.0 \text{ m}}r = -1.35\text{s}^{-2}r$$

We can compute the change in tension from the centre of the circle to 10.0 m away.

dm can be written in terms of the mass density of the rope $\lambda = 3.0$ kg/m.

$$\Delta T = \int a(r)\lambda dr = -1.35 \text{ s}^{-2} \times 3.0 \text{ kg/m} \int_0^{10\text{m}} r dr = -202.5 \text{ N}$$

Therefore, the tension at the centre of the circle is

$$T_{max} = 202.5 \text{ N} + 16350 \text{ N} = 16.6 \text{ kN}$$

95. Working in the frame of reference of the moving block which accelerates at a_M to the left, the mass m feels a fictitious force acting to the right, ma_M.

Adding all forces normal to the curve gives the centripetal acceleration:

$$N(\theta) + ma_M \sin\theta - mg\cos\theta = \frac{mv^2}{R} \qquad (1)$$

The sum of forces tangent to the curve causes the tangential acceleration:

$$mg\sin\theta + ma_M\cos\theta = ma_t \qquad (2)$$

The acceleration of the block is caused by the reaction force to N

$$a_M = \frac{N(\theta)}{M}\sin\theta \qquad (3)$$

Using the result of question 106 in Chapter 3, we can integrate the tangential acceleration to yield

$$2\int a_t ds = -2\int a_t R d\theta = v^2 \quad \text{(because of how } \theta \text{ is measured, } ds = -Rd\theta\text{)}$$

$$\Rightarrow \frac{mv^2}{R} = -2\int ma_t d\theta$$

This is equivalent to equation (1). Using equation (3) to eliminate $N(\theta)$ from equation (1), we have

$$-2\int ma_t d\theta = a_M\left(M\csc\theta + m\sin\theta\right) - mg\cos\theta$$

Taking a derivative with respect to θ gives

$$-2\left(mg\sin\theta + ma_M\cos\theta\right)$$

$$= \frac{da_M}{d\theta}\left(M\csc\theta + m\sin\theta\right) + a_M\left(-M\csc\theta\cot\theta + m\cos\theta\right) + mg\sin\theta$$

$$\frac{da_M}{d\theta} = -\frac{3g\sin\theta}{\sin\theta + \frac{M}{m}\csc\theta} + a_M\frac{\frac{M}{m}\csc\theta\cot\theta - 3\cos\theta}{\sin\theta + \frac{M}{m}\csc\theta}$$

The solution to this ordinary differential equation (found using software) is

$$a_M(\theta) = \frac{c\sin\theta + g\left(5\sin 2\theta + 6\frac{M}{m}\sin 2\theta - \sin 2\theta\cos 2\theta\right)}{\left(2\frac{M}{m} - \cos 2\theta + 1\right)^2}; \quad c \text{ is a constant}$$

When $\theta = 90°$, $a_M = 0$ (when the ball was dropped, the normal force was zero), so $c = 0$.

When the small mass is 0.2R from the bottom of the bowl, $\theta = 36.9°$

$$a_M = 9.81 \text{ m/s}^2 \times \frac{5.76\frac{M}{m} + 4.53}{\left(2\frac{M}{m} + 0.72\right)^2}$$

97. Summing forces on the small piece of ice in the vertical and horizontal directions gives

$$\Sigma F_y = N\cos(\theta) - mg = 0$$

$$\Sigma F_x = N\sin(\theta) = ma$$

Combining these gives

$$a = g\tan(\theta) = g\frac{\sqrt{R^2 - h^2}}{h}$$

The force needed to push the dome is then

$$F = (M+m)a = (M+m)g\frac{\sqrt{R^2 - h^2}}{h}$$

99. We sum the forces and write Newton's second law for each friend. We note that the pulley system constrains the friend in the seat to have half the acceleration as the hanging friend.

$$\Sigma F = m_1 g - T = ma_1 \qquad (1)$$

Similarly, for the movable pulley we have

$$\Sigma F = 2T = m_2\left(\frac{a_1}{2}\right) \qquad (2)$$

Solving equations (1) and (2) simultaneously gives

$$a_1 = \frac{m_1}{m_1 + \dfrac{m_2}{4}}g = \frac{56.0\text{ kg}}{56.0\text{ kg} + \dfrac{78.0\text{ kg}}{4}}\times 9.81\text{ m/s}^2 = 7.28\text{ m/s}^2$$

$$a_2 = \frac{1}{2}a_1 = 3.64\text{ m/s}^2$$

The speed after a 19.0 m drop is

$$v_f = \sqrt{2a_1 y} = \sqrt{2\times 7.90\text{ m/s}^2 \times 19.0\text{ m}} = 16.6\text{ m/s}$$

101. (a) The force of friction is responsible for circular motion.

$$\Rightarrow f = \frac{mv^2}{r} = \frac{69\text{ kg}\times(11\text{ m/s})^2}{54\text{ m}} = 154.6\text{ N}$$

(b) The force that you exert on the truck is a combination of friction in the horizontal and vertical due to your weight.

$$\left|\vec{F}_{total}\right| = \sqrt{f^2 + F_G^2} = \sqrt{(154.6\text{ N})^2 + (9.81\text{ m/s}^2 \times 69\text{ kg})^2} = 694.3\text{ N}$$

$$\theta = \tan^{-1}\left(\frac{154.6}{9.81\times 69}\right) = 12.9°\text{ above vertical}$$

(c) The net force on you is the centripetal force, 154.6 N.

103. Given: $v = 110$ km/h $= 30.56$ m/s

$$a = \frac{v_f^2 - v_i^2}{2x} = \frac{0 - (30.56 \text{ m/s})^2}{2 \times 120 \text{ m}} = -3.89 \text{ m/s}^2$$

Since the book is just starting to move: $-ma = \mu mg$

$$\Rightarrow \mu = -\frac{a}{g} = \frac{3.89 \text{ m/s}^2}{9.81 \text{ m/s}^2} = 0.40$$

1. Yes. The work–kinetic energy theorem was derived using equation (3-20). Question 106, Chapter 3, showed that equation (3-20) is valid for nonconstant forces.

3. True, since work is being done on the spring to stretch it.

5. Negative, since the spring applies the force in a direction opposite to the direction of motion.

7. Yes, if the displacement is in a direction opposite to the applied force.

9. It depends on the *net* work done on the car by yourself and other forces. For example, if you push a moving car on level ground, it will speed up. If you push a moving car up a hill, it may slow down. Although you did positive work, gravity did even more negative work.

11. The total work done by Earth is zero, since the satellite is moving along the line of constant potential. The Earth only interacts with the satellite through gravity, so the work done by gravity must also be zero.

13. Zero, since the work done to stop the puck is exactly equal to the work done to hit the puck back.

15. The weight began with zero speed, and ended with zero speed, so

    $$W = K_f - K_i = 0 - 0 = 0$$

17. d

19. b. The projectile still has some kinetic energy at the apex due to its horizontal velocity.

21. b

23. Since the child is moving at constant speed, we must have

    $$\mu mg \cos\theta = mg \sin\theta$$
    $$\Rightarrow \mu = \tan\theta$$

    Since the child is moving at constant speed, the work done by gravity must be equal and opposite to the work done by friction.

25. a and c

27. e

29. c and d

31. c (since the weight starts from rest and stops at the end)

33. e

35. (a) $W_g = \vec{F} \cdot \Delta \vec{y} = -mgh = -(0.400 \text{ kg})(9.81 \text{ m/s}^2)(0.45 \text{ m}) = -1.77 \text{ J}$

(b) $W_{hand} = \vec{F} \cdot \Delta \vec{y} = mgh = (0.400 \text{ kg})(9.81 \text{ m/s}^2)(0.45 \text{ m}) = 1.77 \text{ J}$

(c) $W_{total} = W_g + W_{hand} = 0$

37. $W = \vec{F} \cdot \Delta \vec{x} = -20 \text{ N} \times 0.90 \text{ m} = -18 \text{ J}$

39. (a) As the ball drops 50 cm, the total work done by you and gravity is enough to bring the kinetic energy to zero.

$$W_{Total} = \frac{1}{2}mv_f^2 - \frac{1}{2}mv_i^2 = 0 - \frac{1}{2}(8)(3)^2 = -36 \text{ J}$$
$$W_g = \vec{F}_g \cdot \Delta \vec{y} = -(8.0 \text{ kg} \times 9.81 \text{ m/s}^2)(-0.50 \text{ m}) = 39.2 \text{ J}$$
$$W_{you} = W_{total} - W_g = -36 \text{ J} - 39.2 \text{ J} = -75.2 \text{ J}$$

(b) As calculated in part (a), the work done by gravity is 39.2 J.

41. (a) Using conservation of mechanical energy:

$$v_1 = \sqrt{2gh} = \sqrt{(2)(9.81)(7-1.5)} = 10.4 \text{ m/s}$$
$$W = U_f + K_f = mgh + 0 = (20.0 \text{ kg})(9.81 \text{ m/s}^2)(7.00 \text{ m}) = 1.37 \text{ kJ}$$

(b) $W_g = \vec{F}_g \cdot \Delta \vec{y} = -mgh = -(20.0 \text{ kg})(9.81 \text{ m/s}^2)(1.5 \text{ m}) = -294 \text{ J}$

43. (a) By summing the forces acting on each mass, and noting they have the same acceleration, we can solve for tension. Note that when M moves down a distance x, m moves up the same distance, and the spring extends the same distance too.

$$\Sigma F_M = Mg - T = Ma \qquad (1)$$
$$\Sigma F_m = T - mg - kx = ma \qquad (2)$$

Solving (1) and (2) for tension, we find:

$$T(x) = \frac{2mMg + kMx}{m + M}$$

Substituting $T(x)$ into (2), we find the net force on m, which varies with x.

$$\Sigma F_m(x) = \frac{mg(M - m) - kmx}{m + M}$$

The work done on m to move to 45 cm above its starting point is

$$W_m = \int_0^h \Sigma F_m(x)\,dx = \frac{2mg(M-m)h - kmh^2}{2(m+M)} = \frac{1}{2}mv^2$$

$$\Rightarrow v = \sqrt{\frac{2(M-m)(gh) - kh^2}{(M+m)}} = \sqrt{\frac{2(16.8\text{ kg} - 4.8\text{ kg})(9.81\text{ m/s}^2 \times 0.45\text{ m}) - k(0.45\text{ m})^2}{(16.8\text{ kg} + 4.8\text{ kg})}}$$

$$v = \sqrt{4.91\text{ m}^2/\text{s}^2 - 9.38 \times 10^{-3}\text{m}^2/\text{kg} \times k}$$

(b) This can be obtained by maximizing v^2 with respect to h.

$$\frac{dv^2}{dx} = \frac{2(M-m)g - 2kh}{(M+m)} = 0$$

$$\Rightarrow h = (M-m)\frac{g}{k}$$

(c) $v = \sqrt{4.91\text{ m}^2/\text{s}^2 - 9.38 \times 10^{-3}\text{m}^2/\text{kg} \times 96\text{ N/m}} = 2.0\text{ m/s}$

45. (a) Force balance gives $F = f$, where $f = \mu mg$ is the force of friction.

$$\Rightarrow \mu mg = F$$

$$\Rightarrow \mu = \frac{F}{mg}$$

(b) $W = Fd$

(c) If the same force is applied over the same distance, the work is the same as found in part (b). The work creates gravitational potential energy instead of heat.

(d) Again, if the force and distance are the same, the work is the same. In this case, the work creates a mix of gravitational potential energy and kinetic energy.

(e) The work done is the same. The work creates a mix of gravitational potential energy, kinetic energy, and heat.

47. $W = \frac{1}{2}mv_1^2 - \frac{1}{2}mv_0^2 = 0 - \frac{1}{2}(22\text{ kg})(4.0\text{ m/s})^2 = -176\text{ J}$

$W = Fd\cos(\theta)$

$$\Rightarrow F = \frac{W}{d\cos\theta} = \frac{-176\text{ J}}{1.3\text{ m} \times \cos(180° - 42°)} = 180\text{ N}$$

49. The component of the force in the direction of motion is

$F_x = F\cos(27°)\cos(41°) = 60\ \text{N} \times \cos(27°)\cos(41°)$

The component of the normal force in the z-direction is

$N_z = mg - F\sin(41°)$

The component of the normal force in the y-direction is

$N_y = F\cos(41°)\sin(27°)$

$\Rightarrow N = \sqrt{N_y^2 + N_z^2} = \sqrt{\left(F\cos 41°\sin 27°\right)^2 + \left(mg - F\sin 41°\right)^2}$

Forces balance because speed is constant:

$\mu N = F_x$

$\Rightarrow \mu_k = \dfrac{60\ \text{N}\cos 41°\cos 27°}{\sqrt{\left(60\ \text{N}\cos 41°\sin 27°\right)^2 + \left(34\ \text{kg}\times 9.81\ \text{m/s}^2 - 60\ \text{N}\sin 41°\right)^2}}$

$\Rightarrow \mu_k = 0.14$

51. (a) The displacement vector is

$\vec{d} = (5-3)\text{m}\ \hat{i} + (-4-2)\text{m}\ \hat{j} + (3+1)\text{m}\ \hat{k} = 2\text{m}\ \hat{i} - 6\text{m}\ \hat{j} + 4\text{m}\ \hat{k}$

Components of the force vector are

$F_x = F\cos(32°)\cos(195°) = 170\ \text{N}\times\cos(32°)\cos(195°) = -139.26\ \text{N}$

$F_x = F\cos(32°)\sin(195°) = 170\ \text{N}\times\cos(32°)\sin(195°) = -37.31\ \text{N}$

$F_z = F\sin(32°) = 170\ \text{N}\times\sin(32°) = 90.09\ \text{N}$

Hence, the work done is

$W = \vec{F}\cdot\vec{d} = \left(-139.26\hat{i} - 37.31\hat{j} + 90.09\hat{k}\right)\cdot\left(2\text{m}\ \hat{i} - 6\text{m}\ \hat{j} + 4\text{m}\ \hat{k}\right) = 306\ \text{J}$

Since the particle starts from rest, the work done is

$W = \dfrac{1}{2}mv^2$

$\Rightarrow v = \sqrt{\dfrac{2W}{m}} = \sqrt{\dfrac{2\times 306\ \text{J}}{1.1\ \text{kg}}} = 24\ \text{m/s}$

(b) To compute the normal force, it is necessary to know the component of the applied force parallel to the normal. Another vector on the track is s = 6 mî + 2 mĵ, so

$\Rightarrow \tilde{N} = (2\hat{i} - 6\hat{j} + 4\hat{k})\times(6\hat{i} + 2\hat{j}) = -8\hat{i} + 24\hat{j} + 40\hat{k}$

The component of F parallel to this direction is

$F_N = \hat{N}\cdot\vec{F} = \left(\dfrac{-8\hat{i} + 24\hat{j} + 40\hat{k}}{\sqrt{8^2 + 24^2 + 40^2}}\right)\cdot\left(-139.26\hat{i} - 37.31\hat{j} + 90.09\hat{k}\right)\text{N} = 80.8\ \text{N}$

The normal force must balance the force of gravity and the applied force in the direction perpendicular to the surface. The component of the force of gravity into the track can be calculated from the direction cosine. Note that the normal force actually pulls the particle into the track (i.e., the particle is held into the track by a rail).

$$\cos \gamma = \frac{4}{\sqrt{2^2 + 6^2 + 4^2}} = 0.53$$
$$\Rightarrow \gamma = 57.7°$$
$$N = F_N - mg \sin(\gamma) = 80.8 \text{ N} - 1.1 \text{ kg} \times 9.81 \text{ m/s}^2 \times \sin(57.7°) = 71.7 \text{ N}$$

The work done by the force of friction is

$$W_f = -\mu_k N |\vec{d}| = -0.15 \times 71.7 \text{ N} \times \sqrt{(2 \text{ m})^2 + (6 \text{ m})^2 + (4 \text{ m})^2} = -80.5 \text{ J}$$

The work done by the force of gravity is

$$W_g = -mg \Delta z = -1.1 \text{ kg} \times 9.81 \text{ m/s}^2 \times 4 \text{ m} = -43.2 \text{ J}$$

The total work done is

$$W_T = 306 \text{ J} - 81 \text{ J} - 43 \text{ J} = 182 \text{ J} = \frac{1}{2} mv^2$$

$$\Rightarrow v = \sqrt{\frac{2W}{m}} = \sqrt{\frac{2 \times 182 \text{ J}}{1.1 \text{ kg}}} = 18 \text{ m/s}$$

53. $W = mgh$

55. The block will gain energy as it moves 3.00 m down the incline. Once it meets the spring, the speed will decrease once the work done by the net force of the spring and gravity take all of the energy from the block.

The work done by the force of gravity before meeting the spring is
$$W_1 = mgh = mgd \sin(\theta)$$

Once contacting the spring and moving an additional distance x, the work done is

$$W_2 = mgx \sin(\theta) - \frac{1}{2} kx^2$$

The block will stop when the total work done is zero, so

$$W_1 + W_2 = mgd \sin(\theta) + mgx \sin(\theta) - \frac{1}{2} kx^2 = 0$$

$$\Rightarrow -170 \text{ N/m } x^2 + 49.1 \text{ N } x + 147.2 \text{ Nm} = 0$$
$$\Rightarrow x = 1.09 \text{ m}$$

57. (a) $\Delta U_g = mgh = 27 \text{ kg} \times 9.81 \text{ m/s}^2 \times 15 \text{ m} \times \sin(19°) = 1293 \text{ J}$

(b) $W_{total} = \frac{1}{2}mv_2^2 - \frac{1}{2}mv_1^2 = \frac{1}{2}(27 \text{ kg})(1.7 \text{ m/s})^2 - 0 = 39 \text{ J}$

Assuming the crate moved in a straight line up the incline, the work done by friction is

$W_f = -fd = -\mu mg \cos(\theta)d = -0.15 \times 27 \text{ kg} \times 9.81 \text{ m/s}^2 \times \cos(19°) \times 15 \text{ m} = -563 \text{ J}$

$W_g = -\Delta U_g = -1293 \text{ J}$

$W_{Force} = W_{total} - W_f - W_g = 39 \text{ J} + 563 \text{ J} + 1293 \text{ J} = 1.90 \text{ kJ}$

59. The potential energy stored in the spring is the difference in the mass's gravitational potential energy and the work done by the friction force.

$U_{spring} = U_{gravity} - W_{friction}$

$\frac{1}{2}kx^2 = mgx\sin(\theta) - \mu mgx\cos(\theta)$

$\Rightarrow x = \frac{2mg(\sin(\theta) - \mu\cos(\theta))}{k} = \frac{2(0.900 \text{ kg})(9.81 \text{ m/s}^2)(\sin(63°) - 0.0800\cos(63°))}{110 \text{ N/m}}$

$x = 0.137 \text{ m} = 13.7 \text{ cm}$

61.

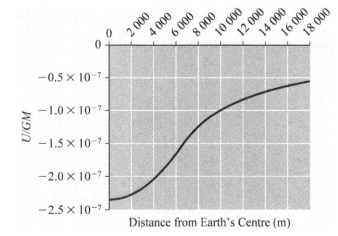

63. (a) $U_1 = \frac{KQq}{R}; \quad U_2 = \frac{kQq}{R + \Delta r}$

$\frac{1}{2}mv^2 = \Delta K = -\Delta U = KQq\left(\frac{1}{R} - \frac{1}{R + \Delta r}\right)$

$\Rightarrow v = \sqrt{\frac{2KQq}{m}\left(\frac{1}{R} - \frac{1}{R + \Delta r}\right)}$

(b) $v = \sqrt{\frac{2(8.99 \times 10^9 \text{ Nm/C}^2)(47 \times 10^{-6} \text{ C})(16 \times 10^{-6} \text{ C})}{0.051 \text{ kg}}\left(\frac{1}{0.011 \text{ m}} - \frac{1}{0.021 \text{ m}}\right)} = 107 \text{ m/s}$

65. 800 kN (since the thrust must at least balance the resistance for constant velocity)

67. Given: $P = 490 \text{ hp} = 365.1 \text{ kW}; \ v = 320 \text{ km/h} = 88.89 \text{ m/s}$

Air resistance balances the applied force, so

$$F = \frac{P}{v} = \frac{365.1 \text{ kW}}{88.89 \text{ m/s}} = 4.1 \text{ kN}$$

69. (a) $\theta = \tan^{-1}\left(\frac{l}{2\pi r}\right) = \tan^{-1}\left(\frac{1 \text{ cm}}{2\pi \times 17 \text{ cm}}\right) = 0.54°$

(b) Energy conservation gives

$$\frac{1}{2}mv^2 = mgh$$

$$\Rightarrow v = \sqrt{2gh} = \sqrt{2 \times 9.81 \text{ m/s}^2 \times 0.07 \text{ m}} = 1.17 \text{ m/s}$$

71. Conservation of energy gives

$$mg(h-r) = \frac{1}{2}mv^2$$

$$\Rightarrow \frac{mg(h-r)}{r} = \frac{1}{2}\frac{mv^2}{r} = \frac{1}{2}F_c$$

At half the circle, the normal force supplies all of the centripetal force:

$$\Rightarrow \frac{mg(h-r)}{r} = \frac{1}{2}N$$

Given condition: $N = 3mg$

$$\Rightarrow \frac{mg(h-r)}{r} = \frac{3}{2}mg$$

$$\Rightarrow h = \frac{5}{2}r$$

73. $\Delta K = -\Delta U$

$$\frac{1}{2}mv^2 = mg(r-h)$$

$$\Rightarrow \frac{mv^2}{r} = \frac{2mg(r-h)}{r} \qquad (1)$$

The centripetal force at height h from the bottom is

$$\frac{mv^2}{r} = N - mg\cos\theta$$

$$\frac{mv^2}{r} = N - mg\frac{(r-h)}{r} \qquad (2)$$

Equating equations (1) and (2) and using $N = mg$

$$h = \frac{2}{3}r$$

The kinetic energy at the bottom of the bowl is the potential energy at the top

$$\frac{1}{2}mv^2 = mgr$$

$$\Rightarrow \frac{mv^2}{r} = 2mg \qquad (3)$$

The centripetal force is a combination of the contact force and gravity:

$$\frac{mv^2}{r} = N - mg \qquad (4)$$

Solving equations (3) and (4) simultaneously gives

$$N = 3mg$$

75. The spring will compress a distance x. The work done by friction is

$$W_{nc} = -\mu mg \cos\theta \left(\frac{0.92 \text{ m}}{\sin\theta} + x \right)$$

The change in mechanical energy is

$$\Delta E_m = U_2 - U_1 = \frac{1}{2}kx^2 - mg(h + x\sin\theta)$$

$$W_{nc} = \Delta E_m$$

$$\frac{1}{2}kx^2 - (mg\sin\theta - \mu mg\cos\theta)x + mgh(\mu\tan\theta - 1) = 0$$

$$10 \text{ N/m } x^2 - 1.18 \text{ N} x - 1.49 \text{ N} = 0$$

$$\Rightarrow x = 0.45 \text{ m} = 45 \text{ cm}$$

77. (a) The sum of forces in the radial direction is the centripetal force:

$$F_c = \frac{mv^2}{r} = T - mg\cos(23°) \qquad (1)$$

The kinetic energy can be found from energy conservation:

$$\frac{1}{2}mv^2 = mgL\cos(23°)$$

$$\Rightarrow \frac{mv^2}{r} = \frac{2mgL}{r}\cos(23°) \qquad (2)$$

Equating (1) and (2) and solving for T:

$$T = mg\cos(23°)\left(\frac{2L}{r} + 1 \right)$$

(b) $W_g = mgh = mgL\cos(23°)$

No work is done by tension, so the total work is $mgL\cos(23°)$.

(c) Zero, since there is no movement in the direction of tension.

79. (a) At the fully stretched position, balancing forces gives

$$mg = k|x_1|$$

$$\Rightarrow x_1 = -\frac{mg}{k} \text{ (negative because it is below the unstretched position)}$$

(b) Define zero potential energy at the unstretched position:

$$E = 0 = mgx_2 + \frac{1}{2}kx^2$$

$$\Rightarrow x_2 = -\frac{2mg}{k}$$

Kinetic energy is maximum where force changes directions (equilibrium position).

$$E = 0 = mgx_1 + \frac{1}{2}kx_1^2 + \frac{1}{2}mv^2$$

$$\frac{1}{2}mv^2 = mg\left(\frac{mg}{k}\right) - \frac{1}{2}k\left(-\frac{mg}{k}\right)^2 = \frac{(mg)^2}{2k}$$

$$\Rightarrow v = g\sqrt{\frac{m}{k}}$$

81. If the system can move together without the top block slipping, the acceleration is

$$a = \frac{37\text{ N}}{3.2\text{ kg} + 16.4\text{ kg}} = 1.89\text{ m/s}^2$$

The maximum acceleration of the top block before is slips is

$$a_{max} = \frac{f_s^{max}}{m} = \mu_s g = 0.23 \times 9.81\text{ m/s}^2 = 2.26\text{ m/s}^2$$

Therefore the block on top will not slip, and no energy is lost to friction.

83. (a) The force of friction acting over distance Δx reduces the mechanical energy of the system. Both blocks move with the same velocity, and block m drops Δx.

$$W_{nc} = \Delta E_m$$

$$-\mu_k Mg\Delta x = \frac{1}{2}(m+M)v^2 - mg\Delta x$$

$$v = \sqrt{\frac{2g\Delta x(m - \mu_k M)}{m+M}} = \sqrt{\frac{2(9.81\text{ m/s}^2)(0.60\text{ m})(11\text{ kg} - 0.20 \times 37\text{ kg})}{11\text{ kg} + 37\text{ kg}}} = 0.94\text{ m/s}$$

(b) It can be shown that the force on mass m is constant. Consider the net work done on the mass by the forces of gravity and tension:

$$W = F_{net}d$$

$$\frac{1}{2}mv^2 = mgh - \mu_k Mgh$$

Differentiating this equation with respect to time gives

$$mva = mgv - \mu_k Mgv$$

$$\Rightarrow a = \frac{g(m - \mu_k M)}{m + M} = \frac{(9.81 \text{ m/s}^2)(11 \text{ kg} - 0.20 \times 37 \text{ kg})}{11 \text{ kg} + 37 \text{ kg}} = 0.74 \text{ m/s}^2$$

85. (a) $\Delta U = -\int_{r_1}^{r_2}\left(-\frac{A}{r^3}\right)dr = -\frac{A}{2r^2}\bigg|_{r_1}^{r_2} = \frac{A}{2}\left(\frac{1}{r_1^2} - \frac{1}{r_2^2}\right)$

(b) $W_{nc} = \Delta U + \Delta K = \frac{A}{2}\left(\frac{1}{r_1^2} - \frac{1}{r_2^2}\right)$

(c) $W_c = -\Delta U_c = -\frac{A}{2}\left(\frac{1}{r_1^2} - \frac{1}{r_2^2}\right)$

(d) The expression for the force is negative, so the force is attractive.

(e) Because the force is attractive, the particles final position will be 40 cm.

$$W_c = -\frac{A}{2}\left(\frac{1}{r_1^2} - \frac{1}{r_2^2}\right) = -\frac{A}{2}\left(\frac{1}{(1.10 \text{ m})^2} - \frac{1}{(0.40 \text{ m})^2}\right) = 0.800 \text{ m}^{-2}A$$

$$W_c = \frac{1}{2}mv^2$$

$$\Rightarrow v = \sqrt{\frac{2W_c}{m}} = \sqrt{\frac{2(0.800 \text{ m}^{-2}A)}{0.012 \text{ kg}}} = 8.14 \text{ m}^{-1}\sqrt{A}$$

87. $\vec{F} = -\frac{\partial U}{\partial r}\hat{r} = -\frac{\partial}{\partial r}\left(-\frac{7}{r^2 + 6}\right)\hat{r} = -\frac{14r}{(r^2 + 6)^2}\hat{r}$

$$\vec{a} = \frac{\vec{F}}{m} = -\frac{14r}{m(r^2 + 6)^2}\hat{r}$$

The solution to $\vec{r}(t)$ does not have an analytic form. If we assume the initial position and velocity are $\vec{r}(0) = 1$ m$\hat{i}$ and $\vec{v} = 0.5$ m/s$\hat{j}$, the velocity after 3.0 seconds can be found using numerical methods. Using Microsoft Excel, it was found that $\vec{v}(3 \text{ s}) = 0.68$ m/s$\hat{i} - 0.46$ m/s$\hat{j}$.

89. The potential created by this force is

$$U(x) = -\int F(x)dx = \frac{k}{4}x^4$$

$$\Delta U = -\Delta K$$

$$\frac{k}{4}x^4 = \frac{1}{2}mv^2$$

$$x = \sqrt[4]{\frac{2mv^2}{k}} = \sqrt[4]{\frac{2 \times 0.310 \text{ kg} \times (120 \text{ m/s})^2}{320 \text{ N/m}^3}} = 2.30 \text{ m}$$

91. The $x, y,$ and z component of force gives the rate of decent of U in the each direction.

$$\vec{F}_c = -\frac{\partial U}{\partial x}\hat{i} - \frac{\partial U}{\partial y}\hat{j} - \frac{\partial U}{\partial z}\hat{k}$$

The rate of decent in any arbitrary direction can be found by projecting onto a unit vector, $\hat{r}$:

$$-\frac{\partial U}{\partial r} = \vec{F}_c \cdot \hat{r} = \left|\vec{F}_c\right|\cos\theta$$

This is maximum when the unit vector is in the direction of the force ($\theta = 0$).

∴ Force points in the direction of steepest descent.

93. The centripetal force is the sum of radial forces:

$$mg\cos\theta - N = \frac{mv^2}{R}$$

$$\Rightarrow N = mg\cos\theta - \frac{mv^2}{R} = mg\cos\theta - 2\frac{K}{R} \quad (1)$$

After moving over an angle θ, the energy lost to friction is

$$\Delta E = -\int_0^s f(s)ds = -\int_0^\theta \mu N(\theta)Rd\theta$$

$$K = mgh - \Delta E = mgR(1 - \cos\theta) - \int_0^\theta \mu N(\theta)Rd\theta \quad (2)$$

Substituting equation (2) into (1):

$$N = mg\cos\theta - \frac{2}{R}\left(mgR(1 - \cos\theta) - \int_0^\theta \mu N(\theta)Rd\theta\right)$$

$$N(\theta) = 3mg\cos\theta - 2mg - 2\mu\int_0^\theta N(\theta)\,d\theta$$

Propose: $N(\theta) = A\cos\theta + B\sin\theta + Ce^{-D\theta}$

$$\int_0^\theta N(\theta)\,d\theta = \left(A\sin\theta - B\cos\theta - \frac{C}{D}e^{-D\theta}\right)\Bigg|_0^\theta = A\sin\theta - B\cos\theta + B - \frac{C}{D}e^{-D\theta} + \frac{C}{D}$$

$$\Rightarrow \cos\theta(A - 3mg - 2\mu B) + \sin\theta(B + 2\mu A) + e^{-D\theta}\left(C - 2\mu\frac{C}{D}\right) + 1\left(2mg + 2\mu B + 2\mu\frac{C}{D}\right) = 0$$

$A = 3mg + 2\mu B$ (3)

$B = -2\mu A$ (4)

By solving equations (3) and (4):

$$A = \frac{3mg}{1 + 4\mu^2} \quad\text{and}\quad B = -\frac{6\mu mg}{1 + 4\mu^2}$$

$$\Rightarrow C = 2\mu\frac{C}{D} \Rightarrow D = 2\mu$$

$$\Rightarrow 2\mu\frac{C}{D} = -2mg - 2\mu B \Rightarrow C = -2mg + 2\mu\frac{6\mu mg}{1 + 4\mu^2} = -2mg\left(\frac{1 - 2\mu^2}{1 + 4\mu^2}\right)$$

$$N(\theta) = mg\left(\frac{3}{1 + 4\mu^2}\cos\theta - \frac{6\mu}{1 + 4\mu^2}\sin\theta - 2\left(\frac{1 - 2\mu^2}{1 + 4\mu^2}\right)e^{-2\mu\theta}\right)$$

$$N(\theta) = mg\left(2.59\cos\theta - 1.03\sin\theta - 1.59e^{-0.4\theta}\right)$$

(a)   You slide off the cliff when $N = 0$

$$\frac{N}{mg} = 2.59\cos\theta - 1.03\sin\theta - 1.59e^{-0.4\theta} = 0$$

$$\Rightarrow \theta = 43.3°$$

(b) $\dfrac{N}{mg} = 2.59\cos\theta - 1.03\sin\theta - 1.59e^{-0.4\theta} = \dfrac{1}{3}$

$$\Rightarrow \theta = 33.5°$$

95. Given: $w = 62430$ lb $= 62430 \times 4.45 = 277800$ N, $v_{max} = 2000$ mph $= 894$ m/s

    $h = 58500$ ft $= 17800$ m

    $U = mgh = 277800 \times 17800 = 4.95 \times 10^9$ J

(a)   Given: $F = 2 \times 23450$ lb $= 208700$ N

      $P = Fv_{max} = 208700$ N $\times 894$ m/s $= 1.87 \times 10^8$ W

(b) Given: $v = 701$ mph $= 313.4$ m/s, $m = \dfrac{277\,800\text{ N}}{9.81\text{ m/s}^2} = 28\,350$ kg

$F_{drag} = \alpha v^2$

At maximum cruising speed we will have the drag force balances thrust:

$\alpha = \dfrac{F}{v^2} = \dfrac{208\,700\text{ N}}{\left(894\text{ m/s}\right)^2} = 0.261\text{ Ns}^2/\text{m}^2$

At cruising speed:

$F_{thrust} - F_{drag} = ma$

$\Rightarrow a = \dfrac{\alpha v^2}{m} = \dfrac{208\,700\text{ N} - 0.261\text{ Ns}^2/\text{m}^2 \times \left(313.4\text{ m/s}\right)^2}{28\,350\text{ kg}} = 6.46\text{ m/s}^2$

(c) Given a vertical climbing rate v_{climb} the time taken to climb is

$t_{climb} = \dfrac{h}{v_{climb}}$

$W = Pt = \dfrac{1.87 \times 10^8\text{ W} \times 17\,800\text{ m}}{v_{climb}} = 3.22 \times 10^{12}\text{ J}$

Chapter 7—LINEAR MOMENTUM, COLLISIONS, AND SYSTEMS OF PARTICLES

1. d. c is only true if kinetic energy is quadrupled by doubling the speed.

3. c

5. d. The final momentum is

$$\vec{p} = mv\sin(30°)\hat{i} + mv\cos(30°)\hat{j} = \frac{1}{2}mv\hat{i} + \frac{\sqrt{3}}{2}mv\hat{j}$$

7. $\Delta\vec{p} = 2mv\sin(30°) = mv$ [away from the wall]

$$F_{avg} = \frac{\Delta\vec{p}}{\Delta t} = \frac{0.100 \text{ kg} \times 10 \text{ m/s}}{1 \times 10^{-3} \text{ s}} = 1000 \text{ N [away from the wall]}$$

9. b. Conservation of linear momentum in the horizontal direction gives

$$p_1 \cos(60°) = p_2 \cos(30°)$$

$$\Rightarrow \frac{p_2}{p_1} = \frac{1}{\sqrt{3}}$$

11. The centre of mass moves down at an acceleration of 9.81 m/s².

13. Yes, see solution to question 12.

15. The speed of the ball just before impact is

$$v = \sqrt{2gh} = \sqrt{2 \times 9.81 \text{ m/s}^2 \times 11 \text{ m}} = 14.7 \text{ m/s}$$

$$\Delta p = 2mv = F\Delta t$$

$$\Rightarrow F = \frac{2mv}{\Delta t} = \frac{2 \times 0.030 \text{ kg} \times 14.7 \text{ m/s}}{20 \times 10^{-3} \text{ s}} = 44.1 \text{ N}$$

If the rebound speed were less than the impact speed, this is an inelastic collision, so energy is dissipated to the surroundings.

17. d. Because no external force acts on the binary star system, the centre of mass will not change (in a frame of reference where the net momentum was zero before the explosion). The force on gravity on the intact star does not change as long as the exploded star still has spherical symmetry, and the radius of explosion plume is smaller than the distance between the stars.

19. c

21. a

23. a. The force of friction at the pivot provides an impulse.

25. b

27. c

29. b

31. True

33. a

35. c. The forces that act on the rocket are thrust from ejected fuel, air resistance, and the gravity of Earth. If the fuel, atmosphere, and Earth are all included in the system, these forces are all internal so energy is conserved. There is little interaction with the Milky Way galaxy.

37. a

39. The speed of the stone before hitting the ground is

$$v = \sqrt{2gh} = \sqrt{2 \times 9.81 \text{ m/s}^2 \times 20 \text{ m}} = 19.8 \text{ m/s}$$
$$\Rightarrow p = 0.060 \text{ kg} \times 19.8 \text{ m/s} = 1.2 \text{ kg m/s}$$

41. (a) $\vec{p}_i = m\vec{v}_i = 74 \text{ kg} \times 9.0 \text{ m/s} = 666 \text{ kg m/s [forward]}$

When the althlete stops, she and the train move at the same speed.
$$\vec{p}_f = (m + M)\vec{v}_f = \vec{p}_i$$
$$\Rightarrow \vec{v}_f = \frac{\vec{p}_i}{m + M} = \frac{666 \text{ kg m/s}}{74 \text{ kg} + 2100 \text{ kg}} = 0.31 \text{ m/s [forward]}$$

(b) $\vec{F}_{avg} = \frac{\Delta \vec{p}_{train}}{\Delta t} = \frac{0.31 \text{ m/s} \times 2100 \text{ kg} - 0}{0.2 \text{ s}} = 3.2 \text{ kN [forward]}$

43. (a) Given: $R = 55 \text{ balls/min} = \frac{55}{60} \text{ balls/s}, \quad v = 117 \text{ km/h} = 32.5 \text{ m/s}$

The change in momentum of each ball is $2mv$, and $\Delta t = R^{-1}$
$$\vec{F}_{avg} = \frac{\Delta \vec{p}}{\Delta t} = 2mvR = 2(0.058 \text{ kg})(32.5 \text{ m/s})\left(\frac{55 \text{ balls}}{60 \text{ s}}\right) = 3.5 \text{ N [forward]}$$

(b) $\vec{F}_{avg} = \frac{\Delta \vec{p}}{\Delta t} = \frac{2mv}{\Delta t} = \frac{2(0.058 \text{ kg})(32.5 \text{ m/s})}{25 \times 10^{-3} \text{ s}} = 151 \text{ N [forward]}$

45. The speed of the ball just before it hit the ground is

$$v_1 = \sqrt{2gh} = \sqrt{2 \times 9.81 \text{ m/s}^2 \times 10 \text{ m}} = 14.0 \text{ m/s}$$

The impulse is

$$\Delta \vec{p} = \vec{p}_2 - \vec{p}_1 = 0.99 m v_1 - (-m v_1) = 1.99 \times 14.0 \text{ m/s} \times m = 27.9 \text{ m/s} \times m \text{ [up]}$$

$$\Delta t = \frac{\Delta \vec{p}}{\vec{F}_{avg}} = \frac{27.9 \text{ m/s} \times 0.025 \text{ kg}}{45 \text{ N}} = 0.016 \text{ s} = 16 \text{ ms}$$

47. The centre of mass is zero when all 16 masses are around the circle. Let's say we remove the 16th mass, which is located on the positive x-axis.

$$r_{cm,16} = \frac{\sum_{i=1}^{16} m \vec{R}_i}{\sum_{i=1}^{16} m} = \frac{\sum_{i=1}^{15} m \vec{R}_i + m R \hat{i}}{16m} = 0$$

$$\Rightarrow \sum_{i=1}^{15} m \vec{R}_i = -m R \hat{i}$$

The centre of mass of the remaining 15 objects is

$$r_{cm,15} = \frac{\sum_{i=1}^{15} m \vec{R}_i}{\sum_{i=1}^{15} m} = \frac{-m R \hat{i}}{15m} = -\frac{R}{15} \hat{i}$$

The centre of mass shifts 1/15th of the radius away from the missing mass.

49. We will fix the origin of our coordinate system at the centre of the largest mass.

$$r_m = (0, L), \quad r_{2m} = (L, L), \quad r_{3m} = (L, 0), \quad r_{4m} = (0, 0)$$

$$r_{cm} = \frac{m r_m + 2m r_{2m} + 3m r_{3m} + 4m r_{0m}}{m + 2m + 3m + 4m} = (0.5L, 0.3L)$$

51. The centre of mass cannot move, because no external force acts on the system, so

$$m_d \Delta x_d = m_c \Delta x_c \quad (1)$$

The distance that the dog moves in the ground frame of reference is

$$\Delta x_d = l_c - \Delta x_c \quad (2)$$

Solving equations (1) and (2) simultaneously gives

$$\Delta x_c = l_c \frac{m_d}{m_d + m_c} = 6.7 \text{ m} \times \frac{4.0 \text{ kg}}{4.0 \text{ kg} + 21 \text{ kg}} = 1.1 \text{ m}$$

53. The centre of mass moves with projectile motion. First find the time it takes for r_{cm} to reach the ground. The vertical velocity is initially zero.

$$\Delta y = \frac{1}{2} a t^2$$

$$\Rightarrow t = \sqrt{\frac{2\Delta y}{a}} = \sqrt{\frac{2(-0.90 \text{ m})}{-9.81 \text{ m/s}^2}} = 0.43 \text{ s}$$

The vertical distance r_{cm} moves in this time is

$\Delta x = v_x \Delta t = 2.3$ m/s $\times 0.43$ s $= 0.99$ m

The centre of mass of the water is 0.99 m in front of the child when he hits the ground.

55. Momentum conservation in the East direction gives

$m_1 v_1 = (m_1 + m_2) v \cos \theta$ (1)

Momentum conservation in the North direction gives

$m_2 v_2 = (m_1 + m_2) v \sin \theta$ (2)

Solving (1) and (2) simultaneously gives

$$v = \frac{\sqrt{(m_1 v_1)^2 + (m_2 v_2)^2}}{m_1 + m_2} = \frac{\sqrt{(27 \text{ kg} \times 6.5 \text{ m/s})^2 + (21 \text{ kg} \times 4.0 \text{ m/s})^2}}{27 \text{ kg} + 21 \text{ kg}} = 4.1 \text{ m/s}$$

57. The height reached by the pendulum is

$h = 0.92$ m$(1 - \cos(27°)) = 0.10$ m

The speed needed to reach this height is determined by conservation of energy.

$$\frac{1}{2} mv^2 = mgh$$

$$\Rightarrow v = \sqrt{2gh} = \sqrt{2 \times 9.81 \text{ m/s}^2 \times 0.10 \text{ m}} = 1.4 \text{ m/s}$$

Conservation of momentum gives

$$m_b v_b = (m_b + m_p) v$$

$$\Rightarrow v_b = \left(\frac{0.011 \text{ kg} + 68 \text{ kg}}{0.011 \text{ kg}} \right) (1.4 \text{ m/s}) = 8.7 \times 10^3 \text{ m/s} = 8.7 \text{ km/s}$$

59. The fractional loss in kinetic energy is given by

$$\varepsilon = \frac{\frac{1}{2} mv_i^2 - \frac{1}{2} mv_f^2}{\frac{1}{2} mv_i^2} = 1 - \frac{v_f^2}{v_i^2}$$

Applying momentum conservation in the x-direction we get

$mv_i \cos \theta = mv_f \cos \theta'$

$$\Rightarrow \frac{v_f}{v_i} = \frac{\cos \theta}{\cos \theta'}$$

Substituting this in the above expression for fractional energy loss, we get

$$\varepsilon = 1 - \left(\frac{\cos \theta}{\cos \theta'} \right)^2$$

61. $K_1' = \dfrac{3}{4}K_1$

$\Rightarrow v' = \dfrac{\sqrt{3}}{2}v$

The second ball makes an angle of $(43° - 21°)$ $22°$ to the initial velocity v.

Conservation of momentum perpendicular to the initial velocity gives

$mv'\sin(21°) = m_2 v_2' \sin(22°)$

$m\left(\dfrac{\sqrt{3}}{2}v\right)\sin(21°) = \left(\dfrac{5}{4}m\right)v_2'\sin(22°)$

$\Rightarrow v_2' = \dfrac{2\sqrt{3}}{5}\dfrac{\sin(21°)}{\sin(22°)}v = 0.66v$

63. Energy conservation for the 3 kg mass gives

$\dfrac{1}{2}m_2 v_2'^2 = \dfrac{1}{2}kx^2$

$\Rightarrow v_2' = \sqrt{\dfrac{k}{m_2}}x = \sqrt{\dfrac{185 \text{ N/m}}{3.0 \text{ kg}}} \times 0.80 \text{ m} = 6.28 \text{ m/s}$

Energy conservation before and after collision gives

$\dfrac{1}{2}m_1 v_1^2 = \dfrac{1}{2}m_1 v_1'^2 + \dfrac{1}{2}m_2 v_2'^2$

$\Rightarrow v_1^2 - v_1'^2 = \dfrac{m_2}{m_1}v_2'^2 = \dfrac{3.0 \text{ kg}}{1.6 \text{ kg}}(6.28 \text{ m/s})^2 = 74.0 \text{ m}^2/\text{s}^2 \qquad (1)$

Momentum conservation gives

$m_1 v_1 = m_1 v_1' + m_2 v_2'$

$\Rightarrow v_1 - v_1' = \dfrac{m_2}{m_1}v_2' = \dfrac{3.0 \text{ kg}}{1.6 \text{ kg}} \times 6.28 \text{ m/s} = 11.78 \text{ m/s} \qquad (2)$

Solving equations (1) and (2) simultaneously gives

$v_1 = 9.0 \text{ m/s}$

65. We will assume that the collision is perfectly collinear and that the incoming ball moves in the opposite direction after collision with twice the speed of the second ball.

$v_1' = 2v_2'$

Momentum conservation gives

$m_1 v_1 = -m_1 v_1' + m_2 v_2' = -2m_1 v_2' + m_2 v_2'$

$\Rightarrow \dfrac{v_1}{v_2'} = \dfrac{m_2}{m_1} - 2 \qquad (1)$

Energy conservation gives

$$\frac{1}{2}m_1v_1^2 = \frac{1}{2}m_1v_1'^2 + \frac{1}{2}m_2v_2'^2 = \frac{1}{2}m_1(2v_2')^2 + \frac{1}{2}m_2v_2'^2$$

$$\Rightarrow \left(\frac{v_1}{v_2'}\right)^2 = 4 + \frac{m_2}{m_1} \qquad (2)$$

Solving equations (1) and (2) simultaneously gives

$$m_2 = 5m_1$$

67. Momentum conservation gives

$$\left(\frac{3v}{2}\right)m_1 + m_2v = m_1v' + m_2(2v)$$

$$\Rightarrow \frac{v'}{v} = \frac{3}{2} - 3\frac{m_2}{m_1} \qquad (1)$$

Energy conservation gives

$$\frac{1}{2}m_1\left(\frac{3v}{2}\right)^2 + \frac{1}{2}m_2v^2 = \frac{1}{2}m_1v'^2 + \frac{1}{2}m_2(2v)^2$$

$$\Rightarrow \left(\frac{v'}{v}\right)^2 = \frac{9}{4} - 3\frac{m_2}{m_1} \qquad (2)$$

Solving (1) and (2) simultaneously yields

$$\frac{m_2}{m_1} = \frac{2}{3} \quad \text{and} \quad \frac{v'}{v} = \frac{1}{2}$$

69. Given: $R = 72.0$ kg/min $= 1.2$ kg/s

The standard rocket propulsion equation is

$$v_f - v_i = v_{er} \ln\left(\frac{M_i}{M_f}\right)$$

Because half the fuel is gone, and fuel is half the total mass:

$$M_f = \frac{3}{4}M_i \Rightarrow \frac{M_i}{M_f} = \frac{4}{3}$$

$$\Rightarrow 190 \text{ m/s} - 0 = v_{er} \ln\left(\frac{4}{3}\right)$$

$$\Rightarrow v_{er} = 660.5 \text{ m/s}$$

$$T = Rv_{er} = (1.2 \text{ kg/s})(660.5 \text{ m/s}) = 792.5 \text{ N}$$

71. (a) The water flow rate is given by

$$R = \frac{56 \text{ kg}}{6 \text{ s}} = 9.33 \text{ kg/s}$$

$$Q = \frac{R}{\rho} = \frac{9.33 \text{ kg/s}}{1000 \text{ kg/m}^3} = 9.33 \times 10^{-3} \text{ m}^3/\text{s}$$

$$\Rightarrow v_{er} = \frac{Q}{A} = \frac{9.33 \times 10^{-3} \text{ m}^3/\text{s}}{\pi (0.035 \text{ m})^2} = 2.43 \text{ m/s}$$

$$\Rightarrow a = \frac{R v_{er}}{M} = \frac{9.33 \text{ kg/s} \times 2.43 \text{ m/s}}{230 \text{ kg} + 56 \text{ kg}} = 0.0789 \text{ m/s}^2 = 7.9 \text{ cm/s}^2$$

(b) The mass remaining after 4 s is

$$M_f = (230 \text{ kg} + 56 \text{ kg}) - 9.33 \text{ kg/s} \times 4 \text{ s} = 248.68 \text{ kg}$$

We can now apply the standard rocket propulsion equation.

$$v_f - v_i = v_{er} \ln\left(\frac{M_i}{M_f}\right)$$

$$\Rightarrow v_f = 2.42 \ln\left(\frac{230 \text{ kg} + 56 \text{ kg}}{248.68 \text{ kg}}\right) = 1.15 \text{ m/s}$$

(c) Because it takes 6 s to expel all the water, it takes 1 s to expel the last 6th.

$$\Delta v = v_{er} \ln\left(\frac{M_i}{M_f}\right) = 2.42 \text{ m/s} \ln\left(\frac{230 \text{ kg} + \dfrac{56 \text{ kg}}{6}}{230 \text{ kg}}\right) = 0.096 \text{ m/s}$$

$$a_{avg} = \frac{\Delta v}{\Delta t} = \frac{0.096 \text{ m/s}}{1.0 \text{ s}} = 0.096 \text{ m/s}^2$$

73. Both the sun and star orbit the common centre of mass, which does not move. Defining the centre of mass as the origin in a coordinate system.

$$r_{cm} = \frac{m_{star}\vec{r}_{star} + m_{planet}\vec{r}_{planet}}{m_{star} + m_{planet}} = 0$$

$$\Rightarrow \vec{r}_{star} = -\frac{m_{planet}}{m_{star}}\vec{r}_{planet} = -\frac{\vec{r}_{planet}}{100}$$

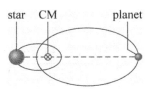

Therefore, the orbit of the star will mirror the orbit of the planet, and will be scaled down in size by a factor of 100. The size of the major and minor axes for the star are

long axis = 7 billion km / 100 = 70 million km

short axis = long axis / 1.4 = 50 million km

75. $x_{cm} = \dfrac{1.16 \text{ m}}{2} \sin(43°) = 0.396 \text{ m} = 39.6 \text{ cm}$

77. Conservation of momentum gives

$$m_b v_b = m_b v'_b + m_w v'_w$$

$$\Rightarrow v_w = \dfrac{m_b v_b - m_b v'}{m_w} = \dfrac{0.005 \text{ kg} \times 265 \text{ m/s} - 0.005 \text{ kg} \times 235 \text{ m/s}}{9.2 \text{ kg}} = 0.016 \text{ m/s} = 1.6 \text{ cm/s}$$

Since the watermelon will fall freely, its centre of mass will have

acceleration equal to $g = 9.81 \text{ m/s}^2$.

79. (a) First calculate the speed of the boxcar, assuming the elephant is moving forward. The total momentum of the elephant + boxcar is zero, because the elephant started from rest.

$$p = m_e(3.0 \text{ m/s} - v_b) - m_b v_b = 0$$

$$\Rightarrow v_b = 3.0 \text{ m/s} \dfrac{m_e}{m_e + m_b} = 3.0 \text{ m/s} \dfrac{2.4 \text{ t}}{2.4 \text{ t} + 7.4 \text{ t}} = 0.74 \text{ m/s}$$

$$\Rightarrow v_e = 3.0 \text{ m/s} - 0.74 \text{ m/s} = 2.26 \text{ m/s}$$

As the elephant throws herself against the spring, an inestatic collision occurs. When the spring is fully compressed, they move at the same speed. Because the system has zero momentum, that speed is zero. The loss of kinetic energy is

$$\Delta K = -\dfrac{1}{2} m_e v_e^2 - \dfrac{1}{2} m_b v_b^2 = -\dfrac{1}{2}\left(2400 \text{ kg} \times (2.26 \text{ m/s})^2 + 7400 \text{ kg} \times (0.74 \text{ m/s})^2\right) = -8200 \text{ J}$$

$$U_{spring} = \dfrac{1}{2} kx^2 = 8200 \text{ J}$$

$$\Rightarrow x = \sqrt{\dfrac{2 \times 8200 \text{ J}}{17500 \text{ N/m}}} = 0.97 \text{ m} = 97 \text{ cm}$$

(b) As stated above, the speed of the boxcar and elephant when the spring is completely compressed is zero.

81. Momentum conservation gives

$$m_c v_c = (m_c + m_w) v_w$$

$$\Rightarrow v_c = \left(\dfrac{19 \text{ kg} + 34 \text{ kg}}{19 \text{ kg}}\right)(2.11 \text{ m/s}) = 5.9 \text{ m/s}$$

83. The centre of mass cannot move, because no external force acts on the system, so

$$m_{bp} \Delta x_{bp} = m_c \Delta x_c \qquad (1)$$

The distance that the backpack moves in the stationary frame of reference is

$$\Delta x_{bp} = 3.0 \text{ m} - \Delta x_c \qquad (2)$$

Solving equations (1) and (2) simultaneously gives

$$\Delta x_c = 3.0 \text{ m} \times \frac{m_{bp}}{m_{bp} + m_c} = 3.0 \text{ m} \times \frac{15 \text{ kg}}{15 \text{ kg} + 19 \text{ kg} + 65 \text{ kg} + 72 \text{ kg}} = 0.26 \text{ m}$$

85. Given $v_{pod,rel}$ = 1400 km/h = 388.9 m/s

The speed of the ejected module relative to Earth is

$$v_{pod} = 388.9 \text{ m/s} - v'$$

Momentum conservation gives

$$11mv = 10mv' - mv_{pod} = 10mv' - m(388.9 \text{ m/s} - v')$$

$$\Rightarrow v' = \frac{11mv + m(388.9 \text{ m/s})}{11m} = v + \frac{388.9 \text{ m/s}}{11}$$

$$\Rightarrow v' - v = \frac{388.9 \text{ m/s}}{11} = 35.4 \text{ m/s}$$

87. Define the astronaut to be at the origin. The centre of mass is

$$r_{cm} = \frac{r_a m_a + r_m m}{m_a + m} = 20 \text{ m} \frac{40 \text{ kg}}{110 \text{ kg} + 40 \text{ kg}} = 5.33 \text{ m}$$

The astronaut and the mass rotate about the centre of mass. After 180° of rotation, she had moved a total distance of 10.66 m, twice the distance between her and the centre of mass.

89. Momentum is conserved in the horizontal direction. After the explosion, the lighter mass falls straight down. Therefore, according to conservation of momentum, we must have

$$m_{total} v = \left(\frac{2}{3} m_{total} \right) v_{heavy}$$

$$v_{heavy} = \frac{3}{2} v$$

Note that the kinetic energy of the heavy part is actually greater than the kinetic energy of the entire mass before the explosion. It is not necessary that energy is conserved, and the extra energy came from the explosion.

91. At the instant that the athlete begins running forward at speed v relative to the car, the spring is not exerting any force on the athlete/cart system. It can be shown that to conserve momentum, the initial speed of the cart is

$$v_{cart} = \frac{m}{m + M} v$$

This is also the maximum speed of the cart, because at any other position, the spring force slows the cart down.

Once running, the only force acting on the athlete/cart system is the spring force.
Although the athlete and cart have different speeds, they share the same acceleration, so

$$F = (m + M)a_{cart} = -kx_{cart}$$

$$\Rightarrow \ddot{x}_{cart} + \frac{k}{m+M}x_{cart} = 0$$

The solution to this ordinary differential equation is

$$x_{cart} = A\sin\left(\sqrt{\frac{k}{m+M}}t\right); A \text{ is maximum extension, } \phi \text{ a constant}$$

$$v_{cart} = \frac{dx_{cart}}{dt} = A\sqrt{\frac{k}{m+M}}\cos\left(\sqrt{\frac{k}{m+M}}t\right) = v_{max}\cos\left(\sqrt{\frac{k}{m+M}}t\right)$$

$$v_{max} = A\sqrt{\frac{k}{m+M}} = \frac{m}{m+M}v$$

$$\Rightarrow v = A\sqrt{k\left(1+\frac{M}{m}\right)}$$

93. (a) In the frame of reference where the arrow was originally moving in the positive y-direction, the arrow deflects 3° to the left of its original trajectory. The cart is moving at 40° to the right from the original trajectory of the arrow. Momentum is conserved in both directions, so

$$m_a v_a' \sin(3°) = m_c v_c' \sin(40°) \qquad (1)$$

$$m_a v_a' \cos(3°) + m_c v_c' \cos(40°) = m_a v_a \qquad (2)$$

Solving (1) and (2) simultaneously gives:

$$v_c' = \frac{m_a v_a}{m_c\left(\dfrac{\sin(40°)\cos(3°)}{\sin(3°)} + \cos(40°)\right)} = 0.201 \text{ m/s}$$

$$v_a' = \frac{m_c v_c' \sin(40°)}{m_a \sin(3°)} = 84.4 \text{ m/s}$$

$$\Delta K = \frac{1}{2}m_a v_a^2 - \frac{1}{2}m_a v_a'^2 - \frac{1}{2}m_c v_c'^2$$

$$\Delta K = \frac{1}{2}\left(0.110 \text{ kg} \times \left((90 \text{ m/s})^2 - (84.5 \text{ m/s})^2\right) - 3.78 \text{ kg} \times (0.201 \text{ m/s})^2\right) = 52.7 \text{ J}$$

(b) $v_c = 0.201$ m/s (as calculated above)

95. (a) The mass of the ball is given by

$$m_b = \left(\frac{4}{3}\pi r^3\right)(\rho) = \left(\frac{4}{3}\pi (0.2\ \text{m})^3\right)\left(4\times10^{17}\ \text{kg/m}^3\right) = 1.3\times10^{16}\ \text{kg}$$

With no external forces, the centre of mass of the system cannot change.

$$r_{cm} = \frac{m_E r_E + m_N r_N}{m_E + m_N} = 0$$

$$\Rightarrow m_E \Delta r_E = m_N \Delta r_N \qquad (1)$$

A change in Earth's position means moving toward the neutron ball and vice versa. The neutron ball moves 7000 m relative to Earth, so

$$\Delta r_N + \Delta r_E = 7000\ \text{m} \qquad (2)$$

Solving equations (1) and (2) gives

$$\Delta r_E = 7000\ \text{m}\times\frac{m_N}{m_E + m_N} = 7000\ \text{m}\times\frac{1.3\times10^{16}\ \text{kg}}{6\times10^{24}\ \text{kg}+1.3\times10^{16}\ \text{kg}} = 1.5\times10^{-5}\ \text{m} = 15\ \mu\text{m}$$

(b) The force on Earth is $m_N g$, so the acceleration of Earth up is

$$a_E = \frac{m_N}{m_E}g$$

The relative acceleration between the two is

$$a_E + a_N = g\left(1 + \frac{m_N}{m_E}\right) = \sim g$$

The time it takes for the gap of 7000 km to close is

$$t = \sqrt{\frac{2\times7000\ \text{m}}{9.81\ \text{m/s}^2}} = 37.8\ \text{s}$$

$$v_E = \frac{1}{2}a_E t^2 = \frac{1.3\times10^{16}\ \text{kg}}{6\times10^{24}\ \text{kg}}9.81\ \text{m/s}^2\times37.8\ \text{s} = 803\ \text{nm/s}$$

(c) In the centre of mass frame of reference, there is no momentum. After a collision, the velocity vectors of Earth and neutron star simply change direction while keeping the same magnitude in order to conserve energy and momentum. The rebound speed of Earth is therefore 803 nm/s.

97. First we will determine the expression for density with respect to distance.

$$m = \frac{5\mu_0 - \mu_0}{L - 0} = \frac{4\mu_0}{L}$$

$$\Rightarrow \mu = \frac{4\mu_0}{L}x + \mu_0$$

$$x_{cm} = \frac{\int_0^L \left(\frac{4\mu_0}{L} x + \mu_0 \right) x \, dx}{\int_0^L \left(\frac{4\mu_0}{L} x + \mu_0 \right) dx} = \frac{11}{18} L$$

99. We work in a frame of reference moving along side the masses at speed v before the spring is released. This way, the masses have no net momentum.

Momentum conservation gives

$$m v_1 = M v_2 \qquad (1)$$

Energy conservation gives

$$\frac{1}{2} k x^2 = \frac{1}{2} m v_1^2 + \frac{1}{2} M v_2^2 + \frac{1}{2} k x'^2 \qquad (2)$$

Solving equations (1) and (2) simultaneously gives

$$m v_1^2 + \frac{m^2}{M} v_1^2 = k \left(x^2 - x'^2 \right)$$

$$\Rightarrow v_1 = \sqrt{\frac{M k \left(x^2 - x'^2 \right)}{m \left(m + M \right)}} = \sqrt{\frac{1.4 \text{ kg} \times 430 \text{ N/m} \left((0.09 \text{ m})^2 - (0.07 \text{ m})^2 \right)}{(0.91 \text{ kg}) (0.91 \text{ kg} + 1.4 \text{ kg})}} = 0.96 \text{ m/s}$$

In the ground frame of reference, the velocity of the lighter mass is the vector sum $\vec{v}_1$ and $\vec{v}$.

101. The speed of the person just before he hits the platform is

$$v = \sqrt{2gh} = \sqrt{2 \times 9.81 \text{ m/s}^2 \times 7.0 \text{ m}} = 11.71 \text{ m/s}$$

The platform and man stick together. The speed immediately after the collision can be found using conservation of momentum.

$$mv = (m + M)v'$$

$$v' = v \frac{m}{m + M} = 11.71 \text{ m/s} \times \frac{67 \text{ kg}}{67 \text{ kg} + 300 \text{ kg}} = 2.14 \text{ m/s}$$

The platform will stop after moving down a distance x. Energy conservation gives:

$$\frac{1}{2} k_{eq} x^2 - (m + M) g x = \frac{1}{2} (m + M) v'^2$$

$$k_{eq} = 4 \times 410 \text{ N/m} = 1640 \text{ N/m}$$

$$\Rightarrow 820 \text{ N/m } x^2 - 3600 \text{ N } x - 840 \text{ Nm} = 0$$

$$\Rightarrow 4.61 \text{ m}$$

103. (a) When the masses stick together, the distance between them will oscillate. At moments when the spring has maximum or minimum compression (x), the masses have the same speed.

$$m_1 v = (m_1 + m_2)v' \qquad (1)$$

Conservation of total energy gives

$$\frac{1}{2}m_1 v^2 = \frac{1}{2}kx^2 + \frac{1}{2}(m_1 + m_2)v'^2 \qquad (2)$$

Solving equations (1) and (2) simultaneously gives

$$x = v\sqrt{\frac{m_1 m_2}{k(m_1 + m_2)}}$$

The maximum distance is therefore given by

$$d_{max} = L + v\sqrt{\frac{m_1 m_2}{k(m_1 + m_2)}}$$

(b) $d_{max} = 0.285\ \text{m} + 6.00\ \text{m/s}\sqrt{\dfrac{0.1\ \text{kg} \times 0.3\ \text{kg}}{50.0\ \text{N/m}\left(0.1\ \text{kg} + 0.3\ \text{kg}\right)}} = 0.517\ \text{m} = 52\ \text{cm}$

105. (a) When the spring is at maximum compression (x), the blocks are moving with the same speed, v'.

$$mv - Mu = (m + M)v' \qquad (1)$$

Energy conservation gives

$$\frac{1}{2}mv^2 + \frac{1}{2}Mu^2 = \frac{1}{2}(m + M)v'^2 + \frac{1}{2}kx^2 \qquad (2)$$

Solving (1) and (2) simultaneously gives

$$x = \sqrt{\frac{mM}{k(m + M)}}(v + u)$$

(b) $v' = \dfrac{mv - Mu}{m + M}$

(c) Once block M separates from the spring, the two move off as if they had a perfectly elastic collision.

$$mv - Mu = mv' + Mu' \qquad (3)$$

$$\frac{1}{2}mv^2 + \frac{1}{2}Mu^2 = \frac{1}{2}mv'^2 + \frac{1}{2}Mu'^2 \qquad (4)$$

Solving equations (1) and (3) simultaneously gives

$$v' = \frac{m-M}{m+M}v - \frac{2M}{m+M}u$$

$$u' = -\frac{M-m}{m+M}u + \frac{2m}{m+M}v$$

(d) At maximum compression, mass M has the same speed as mass m, answered in part (b).

107. In the frame of reference moving with the cars, each car moves away at the same speed. The potential energy is divided in two to make the kinetic energy of each car.

$$\frac{1}{2}mv'^2 = \frac{1}{2}\left(\frac{1}{2}kx^2\right)$$

$$\Rightarrow v' = \sqrt{\frac{k}{2m}}x$$

In the ground frame of reference, the speed of the car in the front is

$$v_f = v + \sqrt{\frac{k}{2m}}x$$

109. (a) The distance covered by the cart in 0.8 s is

$$x_w = 0.3 \text{ m/s} \times 0.8 \text{ s} = 0.24 \text{ m}$$

The distance covered by the cat in 0.8 s is

$$x_c = v_c \times 0.8 \text{ s}$$

We will assume that the cart is 1 m long.

$$\Rightarrow v_c \times 0.8 \text{ s} + 0.24 \text{ m} = 1.0 \text{ m}$$

$$\Rightarrow v_c = 0.95 \text{ m/s}$$

Momentum conservation gives

$$m_w v_w = m_c v_c$$

$$\Rightarrow m_w = \frac{3.15 \text{ kg} \times 0.95 \text{ m/s}}{0.3 \text{ m/s}} = 10.0 \text{ kg}$$

(b) When the cat lands, the speed of the cart will once again be zero because the total momentum of the system is zero.

111. (a) Momentum conservation gives

$$m_c v_c = (m_c + m_w)v_w$$

$$\Rightarrow v_w = \frac{m_c v_c}{m_c + m_w} = \frac{32 \text{ kg} \times 3.00 \text{ m/s}}{32 \text{ kg} + 167 \text{ kg}} = 0.48 \text{ m/s}$$

(b) The energy lost is

$$\Delta K = \frac{1}{2} m_c v_c^2 - \frac{1}{2}(m_c + m_w) v_w^2 = \frac{1}{2}\left(32 \text{ kg} \times (3.00 \text{ m/s})^2 - (32 \text{ kg} + 167 \text{ kg})(0.48 \text{ m/s})^2\right)$$

$$\Delta K = 121 \text{ J}$$

Assuming this is all lost to friction:

$$W_f = \mu_k m_c g d$$

$$\Rightarrow d = \frac{W_f}{\mu_k m_c g} = \frac{121 \text{ J}}{0.7 \times 32 \text{ kg} \times 9.81 \text{ m/s}^2} = 0.77 \text{ m}$$

Chapter 8—ROTATIONAL DYNAMICS

1. spherical shell, since its moment of inertia is larger

3. All three quantities for the point on the rim are larger since the radius is larger.

5. The smaller disk has the greater angular speed and radial acceleration on the circumference. The two disks have the same speed on their circumferences.

$$\omega_{small} = \frac{R}{r}\omega_{big}$$

$$a_{r,small} = \omega^2_{small}r = \left(\frac{R}{r}\omega_{big}\right)^2 r = \frac{R}{r}a_{r,big}$$

7. Because the medicine ball has no angular momentum (relative to the axis of the disk), your angular speed must increase to maintain angular momentum.

9. If one rod is longer by a factor of x, the torque on it will increase by x, and the moment of inertia will increase by x^2. Therefore the shorter rod will have a higher angular acceleration.

11. The equation is still valid if other units (such as degrees) are used for all of the angular quantities.

13. (a) no

    (b) The dot has radial acceleration but no tangential acceleration.

15. The total kinetic energy of the figure skater never changes. Therefore, she does zero work bringing her arms out, and zero work bringing her arms in. Because energy is conserved, her angular speed in the end is the same as her initial angular speed.

17. The moment of inertia of a ring about its axis is given by

$$I_{ring} = mr^2$$

A thin cylindrical shell can be considered as a number of rings stacked on top of one another. Now, since moment of inertia depends on how mass is distributed about the axis of rotation, the number of rings in the stack will not change the moment of inertia. That is,

$$I_{cylinder} = I_{ring} = mr^2$$

19. Assuming the ring was initially at rest, the total angular momentum of the system is zero. Therefore, the final speed of the insect is zero.

21. Tilting the axis of rotation changes the direction of the angular velocity vector, and hence changes the angular momentum vector.

23. Tension is the same when there is no acceleration.

25. Yes, at the moment when the angular acceleration reverses the direction of rotation of the wheel.

27. Yes, as long as they are not pushing the car straight toward the tree.

29. No, torque is defined about a pivot. Force is a vector quantity, so its direction depends on the frame of reference (although its magnitude is independent).

31. The 32nd parallels are the circles at $32°$ north and south of the equator.

$$r = r_{earth} \cos(32°) = (6\,371\,000 \text{ m}) \cos(32°) = 5.40 \times 10^6 \text{ m}$$

We know from question 30 that the angular velocity on any point on Earth's surface is given by

$$\omega = \frac{\Delta\theta}{\Delta t} = \frac{2\pi}{24 \times 3600 \text{ s}} = 7.27 \times 10^{-5} \text{ rad/s}$$

$$v = \omega r = (7.27 \times 10^{-5} \text{ rads/s})(5.40 \times 10^6 \text{ m}) = 393 \text{ m/s}$$

33. Given: $\omega = 7000 \text{ rev/min} = \dfrac{(7000)(2\pi)}{60 \text{ s}} = 733.0 \text{ rad/s}$ and $v = 340.0 \text{ m/s}$

$$r = \frac{v}{\omega} = \frac{340.0 \text{ m/s}}{733.0 \text{ rad/s}} = 0.464 \text{ m} = 46.4 \text{ cm}$$

35. $\omega = \dfrac{\Delta\theta}{\Delta t} = \dfrac{2\pi}{12 \times 3600 \text{ s}} = 1.45 \times 10^{-4} \text{ rad/s}$

37. (a) First find the tangential acceleration of a point on the large gear.

We will use $\Delta\theta = \omega_i t + \dfrac{1}{2}\alpha t^2$ with $\omega_i = 0$.

$$\Rightarrow \alpha_p = \frac{2\Delta\theta}{t^2} = \frac{2 \times 2 \times 2\pi}{(1.0 \text{ s})^2} = 25.13 \text{ rad/s}^2$$

$$\Rightarrow a_t = \alpha_p r_p = (25.13 \text{ rad/s}^2)(0.110 \text{ m}) = 2.76 \text{ m/s}^2$$

This will also be the acceleration of a point on the small gear's circumference.

$$\Rightarrow \alpha = \frac{a_t}{r_g} = \frac{2.76 \text{ m/s}^2}{0.040 \text{ m}} = 69 \text{ rad/s}^2$$

Since every point on a disk has the same angular acceleration, this is the angular acceleration of the real wheel.

(b) Since the initial velocity is zero, the velocity at time t is given by

$$v = a_t t = 2.76 \text{ m/s}^2 \times 1.5 \text{ s} = 4.14 \text{ m/s}$$

$$\Rightarrow a_r = \frac{v^2}{r_g} = \frac{(4.14 \text{ m/s})^2}{0.040 \text{ m}} = 428.5 \text{ m/s}^2$$

$$a = \sqrt{a_t^2 + a_r^2} = \sqrt{(2.76 \text{ m/s}^2)^2 + (428.5 \text{ m/s}^2)^2} = 428.5 \text{ m/s}^2 = 0.43 \text{ km/s}^2$$

(c) $a_t = 2.76 \text{ m/s}^2$ (as calculated in part a)

(d) The linear acceleration of the chain is equal to the tangential acceleration of a point on the gear's and pedal's circumference.

$$\Rightarrow a_t = 2.76 \text{ m/s}^2 \quad \text{(as calculated in part a)}$$

39. $\theta = \frac{1}{2}\alpha t^2$

$$\theta(7 \text{ s}) - \theta(5 \text{ s}) = 2 \times 2\pi = \frac{1}{2}\alpha\left((7 \text{ s})^2 - (5 \text{ s})^2\right)$$

$$\Rightarrow \alpha = 1.05 \text{ rad/s}^2$$
$$\omega = \alpha t = 1.05 \text{ rad/s}^2 \times 8 \text{ s} = 8.4 \text{ rad/s}$$

41. (a) Given: $\alpha = -1.3 \text{ rad/s}^2$, $\theta = 23° = 0.4 \text{ rad}$ in $t = 0.7 \text{ s}$

We will first calculate the initial angular velocity using the relation

$$\theta = \omega_i t + \frac{1}{2}\alpha t^2$$

$$\Rightarrow 0.4 \text{ rad} = 0.7 \text{ s} \times \omega_i - \frac{1.3 \text{ rad/s}^2 \times (0.7 \text{ s})^2}{2}$$

$$\Rightarrow \omega_i = 1.03 \text{ rad/s}$$

The total displacement can be calculated from

$$\omega_f^2 - \omega_i^2 = 2\alpha \Delta\theta_t$$

$$\Rightarrow \Delta\theta_t = \frac{0 - (1.03 \text{ rad/s})^2}{-2 \times 1.3 \text{ rad/s}^2} = 0.41 \text{ rad} = 23.3°$$

(b) $\omega_i = 1.03 \text{ rad/s}$ (as calculated above)

43. (a) $\omega = \omega_i + \alpha t = 0 + 3.5 \text{ rad/s}^2 \times 3.0 \text{ s} = 10.5 \text{ rad/s}$

$$a_r = \omega^2 r = (10.5 \text{ rad/s})^2 \times 0.42 \text{ m} = 46.3 \text{ m/s}^2$$

(b) $a_r(7\text{ s}) = (3.5\text{ rad/s}^2 \times 7\text{ s})^2 \times 0.42\text{ m} = 250\text{ m/s}^2$

The angular velocity is constant after $t = 7.0$ s; therefore, the radial acceleration at $t = 9.0$ s is 250 m/s^2.

45. $\vec{F} = \left(4\sin(133°)\cos(90°)\right)N\hat{i} + \left(4\sin(133°)\sin(90°)\right)N\hat{j} + \left(4\cos(133°)\right)N\hat{k}$

$\quad = (2.9\hat{j} - 2.7\hat{k})\text{ N}$

$\vec{r} \times \vec{F} = (-4 \times 2.7 - 0)\text{ Nm}\hat{i} - (-2 \times 2.7 - 0)\text{ Nm}\hat{j} + (2 \times 2.9 - 0)\text{ Nm}\hat{k}$

$\quad = \left(-10.8\hat{i} + 5.4\hat{j} + 5.8\hat{k}\right)\text{ Nm}$

47. $\vec{\tau} = \vec{r} \times \vec{F} = (14 \times 5 - 7 \times 6)\text{ Nm}\hat{i} - (-14 \times 2 - 7 \times 11)\text{ Nm}\hat{j} + (6 \times 2 + 5 \times 11)\text{ Nm}\hat{k}$

$\Rightarrow \vec{\tau} = \left(28\hat{i} + 105\hat{j} + 67\hat{k}\right)\text{ Nm}$

49. $|\vec{\tau}| = |\vec{r}||\vec{F}|\sin\theta = 2\text{ m} \times 670\text{ N} \times \sin(37°) = 806\text{ Nm}$

51. $\vec{r} \times \vec{F} = \left((-3 \times 1 - 5 \times 5)\hat{i} - (2 \times 1 - 5 \times 3)\hat{j} + (-2 \times 5 - 3 \times 3)\hat{k}\right)\text{ Nm} = \left(-28\hat{i} + 13\hat{j} - 19\hat{k}\right)\text{ Nm}$

$\left|\vec{r} \times \vec{F}\right| = \sqrt{(-28\text{ Nm})^2 + (13\text{ Nm})^2 + (-19\text{ Nm})^2} = 36.25\text{ Nm}$

$\theta = \cos^{-1}\left(\frac{-19}{36.25}\right) = 121.6°$

$\phi = \tan^{-1}\left(\frac{13}{-28}\right) = 150.1°$

$\vec{r} \times \vec{F} = \left(36\text{ N m}, 121.6°, 150.1°\right)$

53. Two of the masses lie on the axis of rotation, and two at a distance equal to half the diagonal.

$r = \frac{L}{2} = \frac{\sqrt{(1.1\text{ m})^2 + (1.1\text{ m})^2}}{2} = 0.78\text{ m}$

$I = 2mr^2 = 2 \times 0.4\text{ kg} \times (0.78\text{ m})^2 = 0.48\text{ kg m}^2$

55. The mass of each blade is

$m_b = \frac{1.32 - 0.9}{3} = 0.14\text{ kg}$

The moment of inertia of a blade about an axis passing through the centre of mass is

$I_b' = \frac{m_b}{12}\left(l^2 + w^2\right) = \frac{0.14\text{ kg}}{12}\left((1.1\text{ m})^2 + (0.17\text{ m})^2\right) = 0.0145\text{ kg m}^2$

The moment of inertia of a blade about the fan's axis of rotation is then given by

$I_b = I_b' + m_b r^2 = 0.0145\text{ kg m}^2 + 0.14\text{ kg} \times (0.11\text{ m} + 0.55\text{ m})^2 = 0.0754\text{ kg m}^2$

The total moment of inertia is then given by

$$I = I_c + 3I_b = \frac{0.900 \text{ kg} \times (0.11 \text{ m})^2}{2} + 3(0.0754 \text{ kg m}^2) = 0.232 \text{ kg m}^2$$

57. We will assume the sphere is centred at the origin with the axis of rotation as the z-axis. Using spherical coordinates, mass of a small differential piece of the sphere is

$dM = \rho r^2 \sin\theta \, dr \, d\theta \, d\varphi; \ \rho$ is the density

The distance from each element to the axis of rotation is

$r_{axis} = r \sin\theta$

$$I = \int dI = \int_0^R \int_0^\pi \int_0^{2\pi} r_{axis}^2 \, dM = \int_0^R \int_0^\pi \int_0^{2\pi} (r \sin\theta)^2 \, \rho r^2 \sin\theta \, dr \, d\theta \, d\varphi = \rho \int_0^R r^4 \, dr \int_0^\pi \sin^3\theta \, d\theta \int_0^{2\pi} d\varphi$$

$$I = \rho \left(\frac{r^5}{5} \right) \Big|_0^R \left(\frac{\cos 3\theta - 9\cos\theta}{12} \right) \Big|_0^\pi \, \varphi \Big|_0^{2\pi} = \frac{8\pi\rho}{15} R^5 = \frac{2MR^2}{5}$$

59. If x is the distance moved by the masses, then energy conservation gives

$$m_1 gx \sin\theta = \frac{1}{2}(m_1 + m_2)v^2 + \frac{1}{2}I\omega^2 = \frac{1}{2}\left(m_1 + m_2 + \frac{I}{r^2}\right)v^2$$

Applying kinematic equation $v^2 - v_0^2 = 2ax$, we get

$$a = \frac{m_1 g \sin\theta}{m_1 + m_2}$$

Substituting $a = \alpha r$ in this equation gives the required angular acceleration.

$$\alpha = \frac{m_1 g \sin\theta}{r(m_1 + m_2)}$$

61. Given: $v = 133 \text{ km/h} = 36.9 \text{ m/s}$

$$\left| \vec{L} \right| = \left| \vec{r} \times \vec{p} \right| = mvr \sin\theta$$

No matter what the position of the car, $r \sin\theta = 11.0 \text{ m}$

$L = mvr \sin\theta = 1250 \text{ kg} \times 36.9 \text{ m/s} \times 11.0 \text{ m} = 5.07 \times 10^5 \text{ kg m}^2/\text{s}$

Given that there is no net torque acting on the car about the officer, it makes sense that the angular momentum does not change.

63. The angle between the velocity and radius vectors is given by

$\theta = 90° - 42° = 48°$

The moment of inertia of the pendulum with the putty is

$$I = \frac{m_r L^2}{3} + m_s (L + R)^2 + \frac{2m_s R^2}{5} + m_p \left(\frac{4L}{5} \right)^2$$

$$I = \frac{0.310 \text{ kg} \times (0.51 \text{ m})^2}{3} + 0.190 \text{ kg} (0.51 \text{ m} + 0.17 \text{ m})^2$$

$$+ \frac{2 \times 0.190 \text{ kg} \times (0.17 \text{ m})^2}{5} + 0.053 \text{ kg} \times \left(\frac{4 \times 0.51 \text{ m}}{5} \right)^2 = 0.126 \text{ kg m}^2$$

Conservation of angular momentum gives

$$\omega = \frac{L_p}{I} = \frac{m_p v \sin \theta}{I} \frac{4L}{5} = \frac{0.053 \text{ kg} \times 14 \text{ m/s} \times \sin 48°}{0.126 \text{ kg m}^2} \frac{4 \times 0.51 \text{ m}}{5} = 1.79 \text{ rad/s}$$

The mechanical energy right after the collision is

$$K = \frac{1}{2} I \omega^2 = \frac{1}{2} \times 0.126 \text{ kg m}^2 \times (1.79 \text{ rad/s})^2 = 0.202 \text{ J}$$

The centre of mass relative to the pivot is

$$r_{cm} = \frac{53 \text{ g} \times (0.51 \text{ m} / 4) + 190 \text{ g} \times (0.51 \text{ m} + 0.17 \text{ m}) + 310 \text{ g} \times (0.51 \text{ m} / 2)}{53 \text{ g} + 190 \text{ g} + 310 \text{ g}} = 0.389 \text{ m}$$

The increase in potential energy of the centre of mass equals the original kinetic energy.

$$mgh = mgr_{cm} (1 - \cos \phi) = K$$

$$\Rightarrow \cos \phi = 1 - \frac{K}{mgr} = 1 - \frac{0.202 \text{ J}}{(0.053 \text{ kg} + 0.190 \text{ kg} + 0.310 \text{ kg}) \times 9.81 \text{ m/s}^2 \times 0.389 \text{ m}} = 0.096$$

$$\Rightarrow \phi = 25°$$

65. (a) The moment of inertia of the wheel with insect at the centre is given by

$$I = m_w r^2 + 12 \left(\frac{m_s r^2}{3} \right) = 0.200 \text{ kg} \times (0.45 \text{ m})^2 + 12 \left(\frac{0.021 \text{ kg} \times (0.45 \text{ m})^2}{3} \right)$$

$$= 0.0575 \text{ kg m}^2$$

The moment of inertia of the wheel with insect at the rim is given by

$$I' = I + m_i r^2 = 0.0575 \text{ kg m}^2 + 0.031 \text{ kg} \times (0.45 \text{ m})^2 = 0.0638 \text{ kg m}^2$$

Conservation of angular momentum gives

$$I' \omega' = I \omega$$

$$\Rightarrow \omega' = \frac{0.0575 \text{ kg m}^2 \times 0.30 \text{ rad/s}}{0.0638 \text{ kg m}^2} = 0.27 \text{ rad/s}$$

(b) $$W = \frac{1}{2} I' \omega'^2 - \frac{1}{2} I \omega^2$$

$$\Rightarrow W = \frac{1}{2} \left(0.0638 \text{ kg m}^2 \times (0.27 \text{ m/s})^2 - 0.0575 \text{ kg m}^2 \times (0.30 \text{ m/s})^2 \right)$$

$$= -2.6 \times 10^{-4} \text{ J}$$

67. $\vec{V} = \vec{V_1} \times \vec{V_2} = (-2 \times 4 - 5 \times 4)\hat{i} - (-2 \times 2 - 5 \times 7)\hat{j} + (4 \times 2 - 4 \times 7)\hat{k}$

$\Rightarrow \vec{V} = -28\hat{i} + 39\hat{j} - 20\hat{k}$

If $\vec{V}$ is perpendicular to the plane containing $\vec{V_1}$ and $\vec{V_2}$, then we must have

$\vec{V} \cdot \vec{V_1} = \vec{V} \cdot \vec{V_2} = 0$

$\vec{V} \cdot \vec{V_1} = -28 \times 2 + 39 \times 4 - 20 \times 5 = 0$

$\vec{V} \cdot \vec{V_2} = -28 \times 7 + 39 \times 4 + 20 \times 2 = 0$

$\vec{V'} = \vec{V_2} \times \vec{V_1} = (5 \times 4 + 2 \times 4)\hat{i} - (5 \times 7 + 2 \times 2)\hat{j} + (7 \times 4 - 4 \times 2)\hat{k}$

$\Rightarrow \vec{V'} = 28\hat{i} - 39\hat{j} + 20\hat{k} = -\vec{V}$

69. (a) The mass of the 19.9 m long cut part of the tree is reduced. Assuming uniform density along the length of the cylinder, the new mass is

$$m' = m\frac{l'}{l} = 4200 \text{ kg} \times \frac{19.9 \text{ m}}{21 \text{ m}} = 3980 \text{ kg}$$

The width is very small compared to the length, so the trunk can be treated as a long thin rod. Its moment of inertia about the pivot point is

$$I = \frac{1}{3}m'L^2 = \frac{1}{3} \times 3980 \text{ kg} \times (19.9 \text{ m})^2 = 5.25 \times 10^5 \text{ kg m}^2$$

When the trunk is parallel to the ground, the centre of mass has fallen a distance $\frac{l'}{2}$.

$$K = mg\frac{l'}{2} = \frac{1}{2}I\omega^2$$

$$\Rightarrow \omega = \sqrt{\frac{mgl'}{I}} = \sqrt{\frac{3980 \text{ kg} \times 9.81 \text{ m/s}^2 \times 19.9 \text{ m}}{5.25 \times 10^5 \text{ kg m}^2}} = 1.22 \text{ rad/s}$$

$$v_{cm} = \omega\frac{l'}{2} = 1.22 \text{ rad/s} \times \frac{19.9 \text{ m}}{2} = 12.1 \text{ m/s}$$

(b) The total kinetic energy of the centre of mass is

$$K = mg\frac{l'}{2} = 3980 \text{ kg} \times 9.81 \text{ m/s}^2 \times \frac{19.9 \text{ m}}{2} = 3.88 \times 10^5 \text{ J}$$

The speed of a car having this much kinetic energy can be calculated from

$$\frac{1}{2}m_c v^2 = K$$

$$\Rightarrow v = \sqrt{\frac{2 \times 3.88 \times 10^5 \text{ J}}{700 \text{ kg}}} = 33.3 \text{ m/s} = 120 \text{ km/h}$$

71. If we cause the astronaut to accelerate at $1.1g$, the normal force applied by the floor would be $1.1mg$, exactly as if the astronaut were on a planet with gravity 10% higher than Earth's.

$$a = r\omega^2$$

$$\Rightarrow \omega = \sqrt{\frac{1.1 \times 9.81 \text{ m/s}^2}{104 \text{ m}}} = 0.32 \text{ rad/s}$$

73. Given: $\alpha = \dfrac{d\omega}{dt} = -12t - 2.5t^3$

Integrating both sides gives

$$\int_{\omega_i}^{\omega_f} d\omega = -\int_0^t \left(12t + 2.5t^3\right)dt$$

$$\Rightarrow \omega_f - \omega_i = -6t^2 - 0.625t^4 \qquad (1)$$

For $\omega_f = \dfrac{\omega_i}{2}$ this becomes

$$0.625t^4 + 6t^2 - 31.5 = 0$$

The only positive real solution is $t = 1.94$ s.

For angular displacement we will integrate (1) on both sides with $\omega = \dfrac{d\theta}{dt}$.

$$\Rightarrow \int_0^\theta d\theta = \int_0^t \omega_i \, dt - \int_0^t \left(6t^2 + 0.625t^4\right)dt$$

$$\Rightarrow \theta = \omega_i t - 2t^3 - 0.125t^5$$

$$\Rightarrow \theta = (63)(1.94) - 2 \times 1.94^3 - 0.125 \times 1.94^5 = 104.2 \text{ rad}$$

75. When the two touch, they are slowed down by the same force of friction. The torque that slows each disk down is given as

$$\tau = I\alpha \Rightarrow fR = I\frac{d\omega}{dt}$$

$$\Rightarrow d\omega = \frac{R}{I}f\,dt \Rightarrow \Delta\omega = \frac{R}{I}\int f\,dt$$

Because both disks share the same force, we can write

$$\frac{\Delta\omega_1}{\Delta\omega_2} = \frac{R_1 I_2}{R_2 I_1} = \frac{1.42 \text{ m} \times 910 \text{ kg m}^2}{0.60 \text{ m} \times 1120 \text{ kg m}^2} = 1.923 \qquad (1)$$

Once they stop slipping, the speed at the edge of each disk is the same.

$$v_1 = v_2$$

$$\Rightarrow \omega_{1,f}R_1 = -\omega_{2,f}R_2$$

(the negative sign is necessary because the disk spin in opposite directions in the end)

$$\Rightarrow \left(\omega_{1,i} - \Delta\omega_{1}\right)R_{1} = -\left(\omega_{2,i} - \Delta\omega_{2}\right)R_{2} \qquad (2)$$

Solving (1) and (2) simultaneously we have

$$\Delta\omega_{1} = 6.87 \text{ rad/s} \Rightarrow \omega_{1,f} = 5 \text{ rad/s} - 6.87 \text{ rad/s} = -1.87 \text{ rad/s}$$

$$\Delta\omega_{2} = 3.57 \text{ rad/s} \Rightarrow \omega_{2,f} = 8 \text{ rad/s} - 3.57 \text{ rad/s} = 4.43 \text{ rad/s}$$

The energy lost to friction is therefore

$$E_{lost} = K_i - K_f = \frac{1}{2}I_1\left(\omega_{1,i}^2 - \omega_{1,f}^2\right) + \frac{1}{2}I_2\left(\omega_{2,i}^2 - \omega_{2,f}^2\right)$$

$$E_{lost} = \frac{1}{2}(1120 \text{ kg m}^2)\left((5 \text{ rad/s})^2 - (-1.87 \text{ rad/s})^2\right) + \frac{1}{2}(910 \text{ kg m}^2)\left((8 \text{ rad/s})^2 - (4.43 \text{ rad/s})^2\right)$$

$$E_{lost} = 32.2 \text{ kJ}$$

77. (a) The moment of inertia of the bar directly on the axis is zero, because there is zero distance between each mass and the axis. For the other bar parallel to the axis, the moment of inertia is the same as a point mass, because all mass is the same distance from the axis.

$$I = mL^2$$

(b) The parallel axis theorem can certainly be applied to each perpendicular bar.

(c) As explained in part (a), all of the mass is the exact same distance from the axis of rotation.

(d) The moment of inertia of the entire object about the axis a is

$$I_{total} = 3 \times \frac{1}{3}mL^2 + mL^2 = 2mL^2$$

79. Consider the sheet to be centred on the x-y plane, with length W parallel to the y-axis. Break the sheet into strips parallel to the y-axis.

The moment of inertia of each strip is r^2dM.

$$dM = W\rho dx \text{ where } \rho = \frac{M}{WL}$$

$$I = \int_{r_1}^{r_2} r^2 dM = W\rho \int_{-\frac{L}{2}}^{\frac{L}{2}} x^2 dx$$

$$\Rightarrow I = \frac{W\rho L^3}{12}$$

$$\Rightarrow I = \frac{ML^2}{12}$$

If the thickness is not negligible, the mass of a strip parallel to the axis of rotation is given by

$$dM = W\rho\, dx\, dy \quad \text{where } \rho = \frac{M}{WLT}$$

$$I = \int_{r_1}^{r_2} r^2\, dM = W\rho \int_{-\frac{L}{2}}^{\frac{L}{2}} \int_{-\frac{T}{2}}^{\frac{T}{2}} \left(x^2 + y^2 \right) dx\, dy$$

$$\Rightarrow I = \frac{W\rho}{12}\left(L^3 T + LT^3 \right)$$

$$\Rightarrow I = \frac{m}{12}\left(L^2 + T^2 \right)$$

81. (a) $v_s = r_s \omega_s = 0.17 \text{ m} \times 7.0 \text{ rad/s} = 1.2 \text{ m/s}$

(b) $v_l = r_l \omega_l = r_s \omega_s = 1.2 \text{ m/s}$

(c) $v_b = 1.2 \text{ m/s}$ (since there's no slippage)

(d) $\omega_l = \dfrac{v_l}{r_l} = \dfrac{1.2 \text{ m/s}}{0.52 \text{ m}} = 2.3 \text{ rad/s}$

(e) Since the tangential velocity is constant, the total acceleration is equal to the radial acceleration.

$$a_{rl} = \frac{v_l^2}{r_l}, \quad a_{rs} = \frac{v_s^2}{r_s}$$

$$\Rightarrow \frac{a_{rl}}{a_{rs}} = \frac{r_s}{r_l} = \frac{17 \text{ cm}}{52 \text{ cm}} = 0.33$$

83. (a) The moment of inertia of the disk about the pivot is

$$I = \frac{mr^2}{2} + mr^2 = \left(\frac{3}{2}\right)(12 \text{ kg})(0.91 \text{ m})^2 = 14.9 \text{ kg m}^2$$

(i) Here $h = r = 0.91 \text{ m}$

Energy conservation gives

$$\frac{1}{2}I\omega^2 = mgh$$

$$\Rightarrow \omega = \sqrt{\frac{2mgh}{I}} = \sqrt{\frac{2\times 12 \text{ kg} \times 9.81 \text{ m/s}^2 \times 0.91 \text{ m}}{14.9 \text{ kg m}^2}} = 3.8 \text{ rad/s}$$

$$\Rightarrow v = 2 \times 0.91 \text{ m} \times 3.79 \text{ rad/s} = 6.9 \text{ m/s}$$

(ii) Here $h = 2r = 1.82$ m

$$\Rightarrow \omega = \sqrt{\frac{2 \times 12 \text{ kg} \times 9.81 \text{ m/s}^2 \times 1.82 \text{ m}}{14.9 \text{ kg m}^2}} = 5.4 \text{ rad/s}$$

$$\Rightarrow v = 2 \times 0.91 \times 5.36 = 9.8 \text{ m/s}$$

(b) Here $h = r + r\sin(40°) = 1.49$ m

$$\Rightarrow \omega = \sqrt{\frac{2 \times 12 \text{ kg} \times 9.81 \text{ m/s}^2 \times 1.49 \text{ m}}{14.9 \text{ kg m}^2}} = 4.9 \text{ rad/s}$$

$$\Rightarrow v = 2 \times 0.91 \times 4.86 = 8.8 \text{ m/s}$$

$$I = I_2 + I_{rod} + I_1$$

$$I = \frac{m_2 r_2^2}{2} + \left(\frac{m_{rod} l_{rod}^2}{12} + m_{rod} \left(\frac{r_2 + l}{2} \right)^2 \right) + \left(\frac{m_1 r_1^2}{2} + m_1 (r_1 + l + r_2)^2 \right)$$

$$I = \frac{39 \text{ kg} \times (0.33 \text{ m})^2}{2} + \left(\frac{11 \text{ kg} \times (0.67 \text{ m})^2}{12} + 11 \text{ kg} \times \left(\frac{0.33 \text{ m} + 0.67 \text{ m}}{2} \right)^2 \right)$$

$$+ \left(\frac{23 \text{ kg} \times (0.17 \text{ m})^2}{2} + 23 \text{ kg} \times (0.33 \text{ m} + 0.67 \text{ m} + 0.17 \text{ m})^2 \right) = 37.1 \text{ kg m}^2$$

85.

$$r_{cm} = \frac{39 \text{ kg} \times 0 + 11 \text{ kg} \times \left(0.33 \text{ m} + \frac{0.67 \text{ m}}{2} \right) + 23 \text{ kg} \times (0.33 \text{ m} + 0.67 \text{ m} + 0.17 \text{ m})}{39 \text{ kg} + 11 \text{ kg} + 23 \text{ kg}} = 0.469 \text{ m}$$

Energy conservation gives

$$\frac{1}{2} I \omega^2 = mgr_{cm}$$

$$\Rightarrow \omega = \sqrt{\frac{2mgr_{cm}}{I}} = \sqrt{\frac{2(39 \text{ kg} + 11 \text{ kg} + 23 \text{ kg})(9.81 \text{ m/s}^2)(0.469 \text{ m})}{37.1 \text{ kg m}^2}} = 4.25 \text{ rad/s}$$

$$\Rightarrow v_{cm} = r_{cm} \omega = 0.47 \text{ m} \times 4.25 \text{ rad/s} = 2.00 \text{ m/s}$$

87. As you pull the crowbar up, you exert a torque on it about the pivot where it presses down on the lower piece of wood. The upper piece of wood exerts a torque pushing the crowbar down with equal and opposite torque. The vertical force that the upper block of wood must be just enough to counteract the force of the nails before the block starts moving.

$$\Sigma \tau = F \times 0.37 \text{ m} - 6 \times 110 \text{ N} \times 0.002 \text{ m} = 0$$

$$\Rightarrow F = \frac{6 \times 110 \text{ N} \times 0.002 \text{ m}}{0.37 \text{ m}} = 3.6 \text{ N}$$

89. The change in potential energy is converted into kinetic energy and heat lost to friction:

$$-\Delta U_g = \Delta K + W_f$$

$$m_2 g d - m_1 g d \sin\theta = \frac{1}{2}(m_1 + m_2)v^2 + \frac{1}{2}I\left(\frac{v}{r}\right)^2 + \mu_k m_1 g \cos\theta d$$

$$\Rightarrow v = \sqrt{\frac{2d(m_2 g - m_1 g \sin\theta - \mu_k m_1 g \cos\theta)}{m_1 + m_2 + \dfrac{I}{r^2}}}$$

91. (a) The moment of inertia or each disk is

$$I_1 = \frac{1}{2}m_1 r_1^2 = \frac{32\ \text{kg}\times(0.27\ \text{m})^2}{2} = 1.17\ \text{kg m}^2$$

$$I_2 = \frac{I_1}{8} = 0.146\ \text{kg m}^2$$

Conservation of angular momentum gives

$$I_1\omega_1 + I_2\omega_2 = (I_1 + I_2)\omega_f$$

$$\Rightarrow \omega_f = \frac{I_1\omega_1 + I_2\omega_2}{(I_1 + I_2)} = \frac{1.17\ \text{kg m}^2\times 17\ \text{rad/s} - 0.146\ \text{kg m}^2\times 34\ \text{rad/s}}{1.17\ \text{kg m}^2 + 0.146\ \text{kg m}^2} = 11.3\ \text{rad/s}$$

(b) The energy lost due to friction is

$$E_i = \frac{1}{2}\left(1.17\ \text{kg m}^2\times(17\ \text{rad/s})^2 + 0.146\ \text{kg m}^2\times(34\ \text{rad/s})^2\right) = 253\ \text{J}$$

$$E_f = \frac{1}{2}(I_1 + I_2)\omega_f^2 = \frac{1}{2}(1.17\ \text{kg m}^2 + 0.146\ \text{kg m}^2)(11.3\ \text{rad/s})^2 = 84\ \text{J}$$

$$E_{lost} = E_i - E_f = 253\ \text{J} - 84\ \text{J} = 169\ \text{J}$$

93. (a) Linear momentum within this system is conserved. If the child begins and ends stationary at the same point, the merry-go-round will be spinning at the same rate is was at the beginning, 6.0 rev/min.

(b) Angular momentum is conserved. As the child leaps away from the merry-go-round, he carries no angular momentum with respect to the pivot because he travels in the radial direction.

$$I_i\omega_i = I_f\omega_f$$

$$\Rightarrow \omega_f = \omega_i \frac{I_i}{I_f} = 6.0\ \text{rev/min}\times\frac{2700\ \text{kg m}^2 + 28\ \text{kg}\times(3.0\ \text{m})^2}{2700\ \text{kg m}^2} = 6.6\ \text{rev/min}$$

95. Given: $\alpha = \dfrac{d\omega}{dt} = -4e^{-2t}$

Integrating both sides we get

$$\int_{\omega_0}^{\omega} d\omega = -4\int_0^t e^{-2t}\, dt$$

$$\Rightarrow \omega = 35 \text{ rad/s} - 2\left(e^{-2t} - 1\right)$$

The smallest angular speed ever reached is 33 rad/s, so the flywheel will lever stop.

97. (a) Conservation of energy says that the gain in kinetic energy is the loss in gravitational potential energy.

$$mgd = \frac{1}{2}mv^2 + \frac{1}{2}I_p\omega_p^2 + \frac{1}{2}I_d\omega_d^2$$

$$mgd = \frac{1}{2}mv^2 + \frac{1}{2}I_p\left(\frac{v^2}{R_p^2}\right) + \frac{1}{2}\left(\frac{1}{2}m_d R_d^2\right)\left(\frac{v^2}{R_d^2}\right)$$

$$\Rightarrow v = \sqrt{\dfrac{2mgd}{m + \dfrac{I_p}{R_p^2} + \dfrac{m_d}{2}}} = \sqrt{\dfrac{2 \times 4.0 \text{ kg} \times 9.81 \text{ m/s}^2 \times 1.2 \text{ m}}{4.0 \text{ kg} + \dfrac{48 \text{ kg cm}^2}{(8.0 \text{ cm})^2} + \dfrac{2.9 \text{ kg}}{2}}} = 3.90 \text{ m/s}$$

$$\omega_p = \frac{v}{r_p} = \frac{3.90 \text{ m/s}}{0.080 \text{ m}} = 48.7 \text{ rad/s}$$

(b) Taking a derivative of the energy balance equation from part (a):

$$mgv = mva + I_p\left(\frac{v}{R_p^2}\right)a + \left(\frac{1}{2}m_d R_d^2\right)\left(\frac{v}{R_d^2}\right)a$$

$$\Rightarrow a = \dfrac{mg}{m + \dfrac{I_p}{R_p^2} + \dfrac{m_d}{2}} = \dfrac{4.0 \text{ kg} \times 9.81 \text{ m/s}^2}{4.0 \text{ kg} + \dfrac{48 \text{ kg cm}^2}{(8.0 \text{ cm})^2} + \dfrac{2.9 \text{ kg}}{2}} = 6.33 \text{ m/s}^2$$

99. (a) Positive. The friction force between the astronaut and the disk pulls the astronaut toward the centre, in the same direction that he walks. Otherwise, the astronaut would fly off the wheel.

(b) Angular speed will increase since the moment of inertia will decrease such that their product remains constant to preserve angular momentum.

(c) Conservation of angular momentum gives

$$I_1\omega_1 = I_2\omega_2$$

$$\Rightarrow \left(\frac{15000 \text{ kg} \times (17 \text{ m})^2}{2} + 76 \text{ kg} \times (17 \text{ m})^2 \right)(1.3 \text{ rad/s}) = \left(\frac{15000 \text{ kg} \times (17 \text{ m})^2}{2} \right)\omega_2$$

$$\Rightarrow \omega_2 = 1.31 \text{ rad/s}$$

(d) Kinetic energy will decrease. The astronauts moving to the edge of the disk is a spontaneous process that could happen without any effort from the astronauts.

(e) Conservation of angular momentum gives

$$I_1 = \frac{15000 \text{ kg} \times (17 \text{ m})^2}{2} = 2.17 \times 10^6 \text{ kg m}^2$$

$$I_2 = \frac{15000 \text{ kg} \times (17 \text{ m})^2}{2} + 76 \text{ kg} \times (17 \text{ m})^2 + 82 \text{ kg} \times (17 \text{ m})^2 = 2.21 \times 10^6 \text{ kg m}^2$$

$$I_1\omega_1 = I_2\omega_2$$

$$\Rightarrow \omega_2 = 1.3 \text{ rad/s} \frac{2.17 \times 10^6 \text{ kg m}^2}{2.21 \times 10^6 \text{ kg m}^2}$$

$$\Rightarrow \omega_2 = 1.27 \text{ rad/s}$$

The work done is given by

$$W = \frac{1}{2}I_2\omega_2^2 - \frac{1}{2}I_1\omega_1^2$$

$$\Rightarrow W = \frac{1}{2}\left[2.21 \times 10^6 \text{ kg m}^2 \left(1.27 \text{ rad/s}\right)^2 - 2.17 \times 10^6 \text{ kg m}^2 \left(1.3 \text{ rad/s}\right)^2 \right]$$

$$\Rightarrow W = -4.67 \times 10^4 \text{ J} = -47 \text{ kJ}$$

1. Because the rolling ring has a higher moment of inertia, it has more mechanical energy than the rolling disk. Therefore, it will reach a higher height when its kinetic energy is converted to gravitational potential energy.

3. (a) The ratio of rotational to translational kinetic energy is

$$\text{Cylinder: } \frac{K_{rot}}{K_{tr}} = \frac{I_{cm}\omega^2}{mv^2} = \frac{1}{2}mr^2\left(\frac{v}{r}\right)^2 \cdot \frac{1}{mv^2} = \frac{1}{2}$$

$$\text{Sphere: } \frac{K_{rot}}{K_{tr}} = \frac{I_{cm}\omega^2}{mv^2} = \frac{2}{5}mr^2\left(\frac{v}{r}\right)^2 \cdot \frac{1}{mv^2} = \frac{2}{5}$$

Because both objects have the same total kinetic energy at the bottom, the cylinder will have the higher rotational kinetic energy at the bottom.

(b) Consider the total kinetic energy as expressed by the momentary pivot approach. Because both objects have the same total kinetic energy, the sphere will have a higher angular speed because it has a lower moment of inertia about point b.

5. To prevent locking up of wheels that may cause skidding. With no skidding, the maximum force of friction to slow down the car is determined by the coefficient of static friction, which is in general much higher than the coefficient of kinetic friction.

7. Since the tires of the car are rotating without slipping, the friction is static.

9. All objects experience the same torque about point b because they have the same mass and radii. Equation (9-13) says that the lower the moment of inertia about b, the higher the acceleration. So the sphere will reach the bottom of the ramp first, followed by the disk, and then the ring.

11. Given: $K_{rot} = \dfrac{4}{6}K_{total} = \dfrac{4}{6}(K_{rot} + K_{tran})$

$\Rightarrow K_{rot} = 2K_{tran}$

$\Rightarrow \dfrac{1}{2}I\left(\dfrac{v}{r}\right)^2 = 2\left(\dfrac{1}{2}mv^2\right)$

$\Rightarrow I = 2mr^2$

13. Both disks will reach the bottom of the hill at the same time. Equation (9-13) shows that the acceleration of each object is independent of their mass:

$$\alpha = \frac{\tau_b}{I_b} = \frac{mgr\sin\theta}{mr^2 + \dfrac{1}{2}mr^2} = \frac{2g\sin\theta}{3r}$$

$$a = \alpha r = \frac{2}{3}g\sin\theta$$

15. The acceleration of a cylinder down a ramp is independent of mass and radius, as shown below. Both will reach the bottom at the same time.

$$\alpha = \frac{\tau_b}{I_b} = \frac{mgr\sin\theta}{mr^2 + \frac{1}{2}mr^2} = \frac{2g\sin\theta}{3r}$$

$$a = \alpha r = \frac{2}{3}g\sin\theta$$

17. It is shown below that the solid uniform cylinder has a higher acceleration than the spherical shell. If they both start from rest, the uniform cylinder must reach the bottom first.

spherical shell: $a = \alpha r = \frac{\tau_b}{I_b}r = \frac{mgr\sin\theta}{mr^2 + \frac{2}{3}mr^2}r = \frac{3}{5}g\sin\theta$

uniform cylinder: $a = \alpha r = \frac{\tau_b}{I_b}r = \frac{mgr\sin\theta}{mr^2 + \frac{1}{2}mr^2}r = \frac{2}{3}g\sin\theta$

19. It was shown in question 18 that for a rolling object, the friction/mass ratio is

$$\frac{f}{m} = g\sin\theta - a$$

Question 14 derived the acceleration of a rolling sphere. The acceleration of a rolling ring can easily be found with the same technique.

$$\frac{f_s}{m_s} = g\sin\theta - \frac{5}{7}g\sin\theta = \frac{2}{7}g\sin\theta$$

$$\frac{f_r}{m_r} = g\sin\theta - \frac{1}{2}g\sin\theta = \frac{1}{2}g\sin\theta$$

$$\Rightarrow \frac{f_s}{f_r}\frac{m_r}{m_s} = \frac{2}{7}\cdot\frac{2}{1} = \frac{4}{7}$$

$$\frac{m_r}{m_s} = \frac{4}{7}\frac{f_r}{f_s} = \frac{4}{7}\cdot10 = \frac{40}{7}$$

21. Air resistance or a sticky floor can possibly slow the object down. If either the wheel or surface are deformable, the normal force between the wheel and surface can exert a torque on the wheel's centre of mass to slow it down, as described in Section 9-7 (rolling friction, online supplements).

23. opposite to the direction of motion

25. The velocity of any point is the superposition of circular motion around the centre of mass plus the linear velocity of the centre of mass.

(a) $v_A = v_{cm} + v_{rel,cm} = 2v_{cm} = 2 \times 15$ m/s $= 30$ m/s

$$\vec{a}_A = \vec{a}_{cm} - \frac{v_{rel,cm}^2}{r}\hat{i} = -\frac{(15 \text{ m/s})^2}{0.37 \text{ m}}\hat{i} = -608 \text{ m/s}^2\hat{i}$$

(b) After a quarter period:

$$\vec{v}_A = \vec{v}_{cm} + \vec{v}_{rel,cm} = 15 \text{ m/s}\hat{i} - 15 \text{ m/s}\hat{j}$$

$$|\vec{v}_A| = \sqrt{(15 \text{ m/s})^2 + (15 \text{ m/s})^2} = 21.2 \text{ m/s}$$

$$\vec{a}_A = \vec{a}_{cm} - \frac{v_{rel,cm}^2}{r}\hat{j} = -\frac{(15 \text{ m/s})^2}{0.37 \text{ m}}\hat{j} = -608 \text{ m/s}^2\hat{j}$$

$$|\vec{a}_A| = 608 \text{ m/s}^2$$

After a half period:

$$\vec{v}_A = \vec{v}_{cm} + \vec{v}_{rel,cm} = 15 \text{ m/s}\hat{i} - 15 \text{ m/s}\hat{i} = 0$$

$$|\vec{v}_A| = 0$$

$$\vec{a}_A = \vec{a}_{cm} - \frac{v_{rel,cm}^2}{r}\hat{j} = \frac{(15 \text{ m/s})^2}{0.37 \text{ m}}\hat{i} = 608 \text{ m/s}^2\hat{i}$$

$$|\vec{a}_A| = 608 \text{ m/s}^2$$

$$\Delta\vec{x} = \frac{1}{2}\vec{a}_{linear}t^2$$

$$\Rightarrow \vec{a}_{linear} = \frac{2\Delta\vec{x}}{t^2} = \frac{2 \times 5.0 \text{ m}\hat{i}}{(7.0 \text{ s})^2} = 0.204 \text{ m/s}^2\hat{i}$$

$$\vec{a}_{total} = \vec{a}_t + \vec{a}_r + \vec{a}_{linear} = -1.07 \text{ m/s}^2\hat{i} + 64.6 \text{ m/s}^2\hat{j} + 0.204 \text{ m/s}^2\hat{i} = -0.87 \text{ m/s}^2\hat{i} + 65 \text{ m/s}^2\hat{j}$$

27. (a) The velocity of a point on the top of the tire is a superposition of the linear velocity and the velocity of the point relative to the centre of mass.

Given: $v_{rot} = 56$ km/h $= 15.56$ m/s, $v_c = 10$ km/h $= 2.78$ m/s

$v_{top} = 15.56$ m/s $+ 2.78$ m/s $= 18.3$ m/s

(b) The vector sum of linear and rotational velocity can never be zero. At certain angles the x component of rotational velocity will cancel the x component of linear velocity, but at this angle there will be non-zero rotational velocity in the y-direction.

29. (a) $a_t = r\alpha = 0.800 \text{ m} \times 7.10 \text{ rad/s} = 5.68 \text{ m/s}^2$

(b) $\vec{a}_t = -5.68 \text{ m/s}^2 \hat{i}$

$\omega_f = \omega_i + \alpha t = 0 + 7.10 \text{ rad/s}^2 \times 1.70 \text{ s} = 12.07 \text{ rad/s}$

$\vec{a}_r = \omega_f^2 r \hat{j} = (12.07 \text{ rad/s})^2 (0.800 \text{ m}) \hat{j} = 117 \text{ m/s}^2 \hat{j}$

$\vec{a}_{total} = \vec{a}_t + \vec{a}_r + \vec{a}_{linear} = -5.68 \text{ m/s}^2 \hat{i} + 117 \text{ m/s}^2 \hat{j} + 2.30 \text{ m/s}^2 \hat{i} = -3.4 \text{ m/s}^2 \hat{i} + 117 \text{ m/s}^2 \hat{j}$

$|\vec{a}_{total}| = \sqrt{(3.4 \text{ m/s}^2)^2 + (117 \text{ m/s}^2)^2} = 117 \text{ m/s}^2$

$\theta = \tan^{-1}\left(\dfrac{-3.4}{117}\right) = -1.7°$ with respect to the vertical

31. The apparent acceleration is due to change in frame of reference and not a real effect.

33. Considering motion at rotation about point b. Find the angle where the torque about point b is zero (i.e., the line of action passes through point b). Let φ be the angle measured from the vertical.

$\sin\varphi = \dfrac{r}{R} \Rightarrow \cos\theta = \dfrac{r}{R}$ (becasue $\varphi + \theta = 90°$)

$\Rightarrow \theta = \cos^{-1}\left(\dfrac{r}{R}\right)$

35. It is convenient to measure torque about point b, so that the tension in the string does not need to be found.

$a = \alpha r = \dfrac{\tau_b}{I_b} r = \dfrac{mgr}{mr^2 + mr^2} r = \dfrac{1}{2} g$

37. Consider torque about point b.

Directions are shown by the darker arrows.

F_1 will cause no motion.

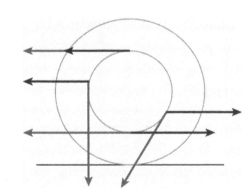

39. Equations (9-12) and (9-13) give equivalent expressions for angular acceleration:

$\alpha = \dfrac{\tau_{cm}}{I_{cm}} = \dfrac{\tau_b}{I_b}$

$\dfrac{fr}{\dfrac{2}{3}mr^2} = \dfrac{mgr\sin\theta}{mr^2 + \dfrac{2}{3}mr^2}$

$$\Rightarrow f = \frac{2}{5}mg\sin\theta = \frac{2}{5}\times 1.2 \text{ kg}\times 9.81 \text{ m/s}^2 \times \sin(32°) = 2.5 \text{ N}$$

The determination of the force of friction is independent of radius.

41. (a) $\tau_{cm} = I_{cm}\alpha$

$$\Rightarrow rf = (mr^2)\left(\frac{a}{r}\right)$$

$$\Rightarrow f = ma \qquad (1)$$

$$\Sigma F = mg\sin\theta - f = ma \qquad (2)$$

Solving equations (1) and (2) simultaneously gives

$$f = \frac{mg\sin\theta}{2} = \frac{1.3 \text{ kg}\times 9.81 \text{ m/s}^2 \times \sin(37°)}{2} = 3.84 \text{ N}$$

(b) The equation of motion is given by

$$\tau_b = I_b\alpha$$

$$\Rightarrow mg\sin\theta = (2mr^2)\left(\frac{a}{r}\right)$$

$$\Rightarrow ma = \frac{mg\sin\theta}{2} \qquad (3)$$

Solving equations (2) (from part a) and (3) simultaneously gives

$$f = \frac{mg\sin\theta}{2} = \frac{1.3 \text{ kg}\times 9.81 \text{ m/s}^2 \times \sin(37°)}{2} = 3.84 \text{ N}$$

43. If f is the force of friction between a log and the stone, the acceleration of the stone is

$$a_s = \frac{\Sigma F}{M} = \frac{F - 2f}{M}$$

The acceleration of a log is

$$a_l = \alpha_l R = \frac{\tau_b}{I_b}R = \frac{2fR}{mR^2 + \frac{1}{2}mR^2}R = \frac{4}{3}\frac{f}{m}$$

Because the stone moves without slipping, the acceleration of a log's centre of mass is half the acceleration of the stone:

$$a_l = \frac{4}{3}\frac{f}{m} = \frac{a_s}{2} \Rightarrow f = \frac{3}{8}a_s m$$

Substituting this value into the expression for a_s gives

$$a_s = \frac{4F}{4M - 3m}$$

45. (a) The equation of motion is

$$\frac{\tau_b}{I_b} = \frac{\tau_{cm}}{I_{cm}} \Rightarrow \frac{mg\sin\theta r}{0.5mr^2 + mr^2} = \frac{fr}{0.5mr^2}$$

$$\Rightarrow f = \frac{1}{3}mg\sin\theta = \frac{1}{3}0.340 \text{ kg}\times9.81 \text{ m/s}^2 \sin(39°) = 0.70 \text{ N}$$

For rolling without slipping the maximum force of friction is given by

$$f_{max} = \mu_s mg\cos(39°) = 0.17\times0.340 \text{ kg}\times9.81 \text{ m/s}^2 \times\cos(39°) = 0.44 \text{ N}$$

Since $f > f_{max}$, the cylinder cannot roll without slipping.

(b) $$fr = I\alpha = \left(\frac{mr^2}{2}\right)\alpha$$

For rolling with slipping:

$$f = \mu_k mg\cos(39°) = 0.12\times0.340 \text{ kg}\times9.81 \text{ m/s}^2 \times\cos(39°) = 0.31 \text{ N}.$$

$$\Rightarrow \alpha = \frac{2f}{mr} = \frac{(2)(0.31 \text{ N})}{(0.340 \text{ kg})(0.030 \text{ m})} = 60.8 \text{ rad/s}^2$$

(c) $$a = g\left(\sin 39° - \mu_k \cos 39°\right) = 5.26 \text{ m/s}^2$$

The time taken by the cylinder to reach the bottom is given by

$$t = \sqrt{\frac{2d}{a}} = \sqrt{\frac{2h}{a\sin 39°}} = \sqrt{\frac{2\times4}{5.26\sin 39°}} = 1.55 \text{ s}$$

We can now calculate linear and angular velocities at the bottom as follows.

$$v_f = at = 5.26 \text{ m/s}^2 \times 1.55 \text{ s} = 8.15 \text{ m/s}$$

$$\omega = \alpha t = 60.8 \text{ rad/s}^2 \times 1.55 \text{ s} = 94.2 \text{ rad/s}$$

The energy lost to friction is the difference in mechanical energy:

$$E_{lost} = mgh - \frac{1}{2}mv^2 - \frac{1}{2}I\omega^2$$

$$E_{lost} = 0.340 \text{ kg}\times\left(9.81 \text{ m/s}^2 \times 4 \text{ m} - \frac{1}{2}(8.15 \text{ m/s})^2 - \frac{1}{4}(0.03 \text{ m}\times94.2 \text{ rad/s})^2\right)$$

$$E_{lost} = 1.37 \text{ J}$$

47. $$\frac{\tau_{cm}}{I_{cm}} = \frac{\tau_b}{I_b} \Rightarrow \frac{mg\sin\theta r}{\frac{2}{5}mr^2} = \frac{\mu_s mgr}{mr^2 + \frac{2}{5}mr^2}$$

$$\Rightarrow \sin\theta = \frac{7}{4}\mu_s \Rightarrow \theta = \sin^{-1}\left(\frac{7}{5}\times0.30\right) = 24.8°$$

49. If there is no friction from the tires, the only force pulling the bicycle down the hill is the force of gravity.

$$a = mg \sin \theta$$

The torque on the wheel is caused only by the brake pads. The net force of friction is $2\mu_s N$, because the brake pads act on both sides of the wheel.

The deceleration of the wheels must match the bicycle:

$$\alpha = \frac{g \sin \theta}{r} = \frac{\tau_{cm}}{I} = \frac{2\mu_k Nr}{m_w r^2}$$

$$\Rightarrow N = \frac{m_w g \sin \theta}{2\mu_k} = \frac{0.900 \text{ kg} \times 9.81 \text{ m/s}^2 \times \sin(17°)}{2 \times 0.20} = 6.45 \text{ N}$$

51. (a) For a rolling cylinder, the ratio of translational to total kinetic energy is

$$\frac{K_{tr}}{K_{total}} = \frac{mv^2}{mv^2 + I\omega^2} = \frac{mv^2}{mv^2 + \left(\frac{mr^2}{2}\right)\left(\frac{v}{r}\right)^2} = \frac{2}{3}$$

$$\Rightarrow K_{tr} = \frac{2}{3}(mgh) = \frac{2}{3}1.2 \text{ kg} \times 9.81 \text{ m/s}^2 \times 3.4 \text{ m} = 26.7 \text{ J}$$

$$\Rightarrow K_{rot} = \frac{1}{3}(mgh) = 13.3 \text{ J}$$

(b) For a rolling ring, the ratio of translational to total kinetic energy is

$$\frac{K_{tr}}{K_{total}} = \frac{mv^2}{mv^2 + I\omega^2} = \frac{mv^2}{mv^2 + (mr^2)\left(\frac{v}{r}\right)^2} = \frac{1}{2}$$

$$\Rightarrow K_{tr} = K_{rot} = \frac{1}{2}(mgh) = \frac{1}{2}1.2 \text{ kg} \times 9.81 \text{ m/s}^2 \times 3.4 \text{ m} = 20.0 \text{ J}$$

53. (a) Energy conservation gives

$$\frac{K_{tr}}{K_{total}} = \frac{mv^2}{mv^2 + I\omega^2} = \frac{mv^2}{mv^2 + \left(\frac{2mr^2}{5}\right)\left(\frac{v}{r}\right)^2} = \frac{5}{7}$$

$$\Rightarrow K_{tran} = \frac{5}{7}(mgh)$$

On the frictionless ramp, only translational kinetic energy contributes to movement up the ramp.

$$\Rightarrow mgh' = \frac{5}{7}(mgh) \Rightarrow h' = \frac{5}{7}h = 5.0 \text{ m}$$

(b) Yes, because there is no friction as the ball skids up the ramp.

(c) The ball keeps on spinning at the same angular velocity it had at the bottom of the ramp.

$$K_{rot} = \frac{2}{7}mgh = \frac{1}{2}\left(\frac{2mr^2}{5}\right)\omega^2$$

$$\Rightarrow \omega = \sqrt{\frac{10gh}{7r^2}} = \sqrt{\frac{10 \times 9.81 \text{ m/s}^2 \times 7.0 \text{ m}}{7 \times (0.127 \text{ m})^2}} =$$

$$\Rightarrow \omega = 78.0 \text{ rad/s}$$

55. (a) It was shown in question 53 that for a rolling sphere, 5/7ths of its kinetic energy is translational energy. Because the ball does not stop spinning once it passes the bottom, only 5/7th of its total mechanical energy is converted to gravitational potential energy.

$$\Rightarrow K_{tran} = \frac{5}{7}(mgh) = mgh'$$

$$\Rightarrow h' = \frac{5}{7}h$$

Energy is conserved. Although the ball attains a lower height, it reserves some energy as rotational kinetic energy.

(b) At the highest point the translational kinetic energy will be zero.

(c) $K_{rot} = K_{total} - K_{tr} = \left(\frac{2}{7}\right)(mgh)$

(d) A ring of the same mass has a higher moment of inertia, so more of the total energy will go toward rotational motion as it moves down. Less translational energy means that the ring will move up to a lesser height on the right side as compared to the solid sphere.

57. (a) The energy and momentum of the car are

$$K = \frac{1}{2}Mv^2 + 4 \times \frac{1}{2}I\left(\frac{v}{R}\right)^2$$

$$p = Mv$$

If the brakes are just on the verge of slipping, the external force slowing the car down is four times the friction force between a wheel and the road. Because the wheels are not slipping with respect to the road, this removes no energy from the system. The friction force between the brakes and the wheel *does* remove energy from the system. The rate of energy loss (power) caused by the brakes is

$$F_{ext} = -4 \times \mu_r \frac{Mg}{4} = -\mu_r Mg$$

Applying Newton's second law:

$$F_{ext} = M\frac{dv}{dt} \Rightarrow \frac{dv}{dt} = -\mu_r g \quad (1)$$

107

Power is the rate of energy loss:

$$P = F_{brake} v_{wheel/brakepad} = -\mu_b N r_b \left(\frac{v}{R}\right)$$

$$P = \frac{dK}{dt} = Mv\frac{dv}{dt} + 4\frac{I}{R^2}v\frac{dv}{dt} = -\mu_b N r_b \left(\frac{v}{R}\right) \quad (2)$$

Substituting equation (1) into (2) gives

$$N = \frac{\mu_r g\left(MR^2 + 4I_{axel}\right)}{4\mu_b r_b R} = 30.4 \text{ kN}$$

(b) The presence of the force of gravity will both remove momentum *and* energy

$$F_{ext} = -\mu_r Mg - Mg\sin\theta = -Mg(\mu_r + \sin\theta)$$

$$F_{ext} = M\frac{dv}{dt} \Rightarrow \frac{dv}{dt} = -(\mu_r + \sin\theta)g \qquad (1)$$

$$P = -F_{brake}v_{wheel/brakepad} - Mgv = -\mu_b N r_b \left(\frac{v}{R}\right) - Mgv$$

$$P = \frac{dK}{dt} = Mv\frac{dv}{dt} + 4\frac{I}{R^2}v\frac{dv}{dt} = -\mu_b N r_b \left(\frac{v}{R}\right) - Mgv \quad (2)$$

Substituting (1) into (2) gives:

$$N = \frac{g\mu_r\left(MR^2 + 4I_{axel}\right) + g\sin\theta MR^2}{4\mu_b r_b R} = 41.1 \text{ kN}$$

59. (a) The points of contact between each string and the pulleys, the block on the table, and the cylinder all have the same acceleration. This acceleration is *twice* the acceleration of the centre of mass of the sphere. Here, a will represent the acceleration of the sphere centre of mass, all other objects have acceleration $2a$.

$$a = \frac{\Sigma F_{sphere}}{M_s} = \frac{T_1 - f_s - M_s g\sin\theta_1}{M_s}$$

Summing torques about the sphere's axis of rotation gives

$$T_1 R_s + f_s R_s = I_s\alpha = \left(\frac{2M_s R_s^2}{5}\right)\left(\frac{a}{R_s}\right) \Rightarrow f_s = \frac{2M_s a}{5} - T_1$$

$$\Rightarrow T_1 = \frac{7}{10}M_s a + \frac{M_s g\sin\theta_1}{2}$$

$$\alpha_{p1} = \frac{\Sigma\tau_{p1}}{I_1} = \frac{2a}{r_1} = \frac{T_2 r_1 - T_1 r_1}{I_1}$$

$$\Rightarrow T_2 = \left(\frac{2I_1}{r_1^2} + \frac{7}{10}M_s\right)a + \frac{M_s g\sin\theta_1}{2}$$

$$a_{blk} = 2a = \frac{\Sigma F_{blk}}{m} = \frac{T_3 - T_2}{m}$$

$$\Rightarrow T_3 = \left(2m + \frac{2I_1}{r_1^2} + \frac{7}{10}M_s\right)a + \frac{M_s g \sin\theta_1}{2}$$

$$\alpha_{p2} = \frac{\Sigma \tau_{p2}}{I_2} = \frac{2a}{r_2} = \frac{T_4 r_2 - T_3 r_2}{I_2}$$

$$\Rightarrow T_4 = \left(\frac{2I_2}{r_2^2} + 2m + \frac{2I_1}{r_1^2} + \frac{7}{10}M_s\right)a + \frac{M_s g \sin\theta_1}{2}$$

$$a_{cyl} = 2a = \frac{\Sigma F_{cyl}}{M_c} = \frac{M_c g \sin\theta_2 - f_c - T_4}{M_c}$$

Summing torques about the cylinder's axis of rotation gives

$$f_c R_c = I_c \alpha = \left(\frac{M_c R_c^2}{2}\right)\left(\frac{2a}{R_c}\right) \Rightarrow f_c = M_c a$$

$$\Rightarrow 3M_c a + T_4 = M_c g \sin\theta_2$$

$$\Rightarrow \left(3M_c + \frac{2I_2}{r_2^2} + 2m + \frac{2I_1}{r_1^2} + \frac{7}{10}M_s\right)a + \frac{M_s g \sin\theta_1}{2} = M_c g \sin\theta_2$$

$$\Rightarrow a = \frac{2M_c g \sin\theta_2 - M_s g \sin\theta_1}{\left(6M_c + \frac{4I_2}{r_2^2} + 4m + \frac{4I_1}{r_1^2} + \frac{7}{5}M_s\right)}$$

(b) At the point in time where the sphere has moved forward a distance d, and the cylinder and mass have moved forward a distance $2d$, the gravitational potential energy released by the cylinder moving down, minus gravitational potential energy stored by the sphere moving up, is converted to kinetic energy.

$$\Delta K = -\Delta U = 2M_c g d \sin\theta_2 - M_s g d \sin\theta_1 \qquad (1)$$

The mass, cylinder, and tops of each pulley move at speed $2v_s$.

The kinetic energy held by the 5 objects is

$$\Delta K = \frac{7}{10}M_s v_s^2 + \frac{1}{2}I_1\left(\frac{2v_s}{r_1}\right)^2 + \frac{1}{2}m(2v_s)^2 + \frac{1}{2}I_2\left(\frac{2v_s}{r_2}\right)^2 + \frac{3}{4}M_c(2v_s)^2$$

$$\Delta K = \left(\frac{7}{10}M_s + 2\frac{I_1}{r_1^2} + 2m + 2\frac{I_2}{r_2^2} + 3M_c\right)v^2 \qquad (2)$$

Taking derivatives of equations (1) and (2) with respect to time and equating, we get:

$$2M_cgv_s\sin\theta_2 - M_sgv_s\sin\theta = 2\left(\frac{7}{10}M_s + 2\frac{I_1}{r_1^2} + 2m + 2\frac{I_2}{r_2^2} + 3M_c\right)v_sa$$

$$\Rightarrow a = \frac{2M_cgv_s\sin\theta_2 - M_sgv_s\sin\theta}{\left(\frac{7}{5}M_s + 4\frac{I_1}{r_1^2} + 4m + 4\frac{I_2}{r_2^2} + 6M_c\right)}$$

61. (a) The object will roll in the direction of the force.

   (b) We can find acceleration by summing the torques about either the centre of mass or the pivot and get the same result. Assuming the force of friction is zero, then:

   $$\alpha = \frac{\tau_{cm}}{I_{cm}} = \frac{\tau_b}{I_b} = \frac{Fr}{\frac{1}{3}mr^2} = \frac{F(r+R)}{\frac{4}{3}mr^2}$$

   $$\Rightarrow r = \frac{R}{3}$$

63. Conservation of energy can be used to determine the speed of the rolling where right before the collision.

   $$K = \frac{1}{2}mv_1^2 + \frac{1}{2}I\omega^2 = \frac{1}{2}mv_1^2 + \frac{1}{2}\frac{2}{5}mr^2\left(\frac{v_1^2}{r^2}\right) = \frac{7}{10}mv_1^2 = mgh$$

   $$\Rightarrow v_1 = \sqrt{\frac{10}{7}gh} = \sqrt{\frac{10}{7}9.81 \text{ m/s}^2 \times 4.3 \text{ m}} = 7.76 \text{ m/s}$$

   After the collision, momentum and energy are conserved. For a non-slipping ball:

   Momentum: $mv_1 = mv_1' + Mv_2'$ (1)

   Energy: $mgh = \frac{7}{10}mv_1'^2 + \frac{1}{2}Mv_2'^2$ (2)

   Combining (1) and (2):

   $$v_1'^2\left(\frac{7}{10}m + \frac{1}{2}\frac{m^2}{M}\right) - v_1'\left(\frac{m^2}{M}v_1\right) + \frac{1}{2}\frac{m^2}{M}v_1^2 - mgh = 0$$

   $$\Rightarrow v_1'^2 (1.26 \text{ kg}) - v_1'(2.21 \text{ kg m/s}) - 58.9 \text{ J} = 0$$

   $$\Rightarrow v_1' = -6.02 \text{ m/s}$$

65. (a) The height covered by the centre of mass of the ball is given by

   $$h = (R-r) - (R-r)\cos(72°) = (1 - \cos(72°))(R-r)$$

   Energy conservation gives

   $$\frac{1}{2}mv^2 + \frac{1}{2}I\omega^2 = mgh$$

Copyright © 2015 by Nelson Education Ltd.

$$\Rightarrow \frac{1}{2}mv^2 + \frac{1}{2}\left(\frac{2mr^2}{3}\right)\left(\frac{v}{r}\right)^2 = \frac{5}{6}mv^2 = mg\left(1-\cos(72°)\right)\left(R-r\right)$$

$$F_c = \frac{mv^2}{(R-r)} = \frac{6}{5}mg\left(1-\cos(72°)\right)$$

The centripetal force is the difference between the normal and the gravitational forces:

$$N = F_c + mg = \frac{mg}{5}\left(11-6\cos(72°)\right)$$

(b) If the object slides, all kinetic energy is translational kinetic energy.

$$\frac{1}{2}mv^2 = mg\left(1-\cos(72°)\right)\left(R-r\right)$$

$$\Rightarrow F_c = \frac{mv^2}{(R-r)} = 2mg\left(1-\cos(72°)\right)$$

The centripetal force is the difference between the normal and the gravitational forces

$$N = F_c + mg = mg\left(3-2\cos(72°)\right)$$

(c) With rolling, the normal force is

$$N = \frac{mg}{5}\left(11-6\cos(72°)\right) = \frac{0.67 \text{ kg}\times 9.81 \text{ m/s}^2}{5}\left(11-6\cos(72°)\right) = 12.0 \text{ N}$$

Without rolling, the normal force is

$$N = mg\left(3-2\cos(72°)\right) = 0.67 \text{ kg}\times 9.81 \text{ m/s}^2 \times\left(3-2\cos(72°)\right) = 15.7 \text{ N}$$

67. (a) Once a ball starts rolling without slipping, we can view rolling as rotation about the pivot. There is no torque about the pivot, so the sphere does not accelerate once it is rolling on a horizontal surface.

(b) The angular acceleration can be found by considering the torque provided by friction:

$$\tau_f = I\alpha \Rightarrow r\mu_k mg = \left(\frac{2mr^2}{5}\right)\alpha$$

$$\Rightarrow \alpha = \frac{5\mu_k g}{2r} = 31.2 \text{ rad/s}^2$$

Linear deceleration is caused by friction as well:

$$a = \frac{\Sigma F}{m} = -\mu_k g = -1.37 \text{ m/s}^2$$

The linear speed will drop and angular speed will increase until $v_f = r\omega_f$

$$v_f = v_i + at$$

$$\Rightarrow r\omega_f = 9.0 \text{ m/s} - 1.37 \text{ m/s}^2 t$$

$$\Rightarrow \omega_f = 81.80 \text{ rad/s} - 12.45 \text{ rad/s } t \qquad (1)$$

$$\omega_f = \omega_i + \alpha t = \omega_f = 31.2 \text{ rad/s } t \qquad (2)$$

Solving equations (1) and (2):

$$t = 1.87 \text{ s}$$

(c) $\omega_f = 31.2 \text{ rad/s} \times 1.87 \text{ s} = 58.5 \text{ rad/s}$

$$v_f = r\omega_f = 0.11 \text{ m} \times 58.5 \text{ rad/s} = 6.42 \text{ m/s}$$

$$\Delta K = \frac{1}{2}mv_i^2 - \frac{1}{2}mv_f^2 - \frac{1}{2}I\omega_f^2 = \frac{1}{2}mv_i^2 - \frac{7}{10}mv_f^2 = m\left(\frac{v_i^2}{2} - \frac{v_2^2}{2}\right)$$

$$\Delta K = 2.1 \text{ kg}\left(\frac{(9.0 \text{ m/s})^2}{2} - \frac{7(6.42 \text{ m/s})^2}{10}\right) = 24.5 \text{ J}$$

69. The equation of motion for the hanging mass is

$$Mg - T_1 = Ma \qquad (1)$$

Summing torques about the rotation axis of the pulley gives

$$rT_1 - rT_2 = I\alpha$$

$$\Rightarrow rT_1 - rT_2 = I\left(\frac{a}{r}\right)$$

$$\Rightarrow T_1 - T_2 = \frac{Ia}{r^2} \qquad (2)$$

Summing torques about the cylinder's axis of rotation gives

$$RT_2 + Rf = I_c\alpha_c = \left(\frac{m_cR^2}{2}\right)\left(\frac{a_c}{R}\right)$$

$$\Rightarrow T_2 + f = \frac{m_c a_c}{2} \qquad (3)$$

The equation of motion for the cylinder is

$$T_2 - f = m_c a_c$$

Solving this with equation (3) gives

$$T_2 = \frac{3m_c a_c}{4}$$

But $a_c = \frac{a}{2}$

$$\Rightarrow T_2 = \frac{3m_c a}{8}$$

Substituting this into equation (2) gives

$$T_1 = \frac{Ia}{r^2} + \frac{3m_c a}{8}$$

Substituting this into equation (1) gives

$$Mg = Ma + \frac{Ia}{r^2} + \frac{3m_c a}{8}$$

$$\Rightarrow a = \frac{Mg}{M + \dfrac{I}{r^2} + \dfrac{3}{8}m_c}$$

71. Only if the object is already rolling without slipping. Otherwise, the object will slip because there is no frictional torque to change the angular speed.

73. If the marble is a perfectly rigid sphere and the horizontal surface is totally frictionless, the marble will not lose any energy to the surface. However, air resistance will cause the marble to gradually come to a stop.

75. Equation (9-18) is

$$K = \frac{1}{2}I_b\omega^2$$

Here I_b is the moment of inertia about point b on the surface. According to the parallel axis theorem, the moment of inertia about the central axis is given by

$$I_b = I_{cm} + mr^2$$

Substituting this into the above equation gives

$$K = \frac{1}{2}\left(I_{cm} + mr^2\right)\omega^2 = \frac{1}{2}I_{cm}\omega^2 + \frac{1}{2}mr^2\left(\frac{v_{cm}}{r}\right)^2$$

This gives equation 9-17; that is

$$K = \frac{1}{2}I_{cm}\omega^2 + \frac{1}{2}mv_{cm}^2$$

77. (a) It is not immediately known which force of friction will reach it's maximum value first. Assume that both forces act in the forward direction. If the log is moving without slipping, the angular acceleration can be found by summing torques about either the centre of mass or the pivot. The force acting on the top and bottom of each log will be indicated by subscript t and b respectively.

$$\alpha = \frac{\Sigma\tau_{cm}}{I_{cm}} = \frac{\Sigma\tau_b}{I_b} = \frac{R(f_t - f_b)}{0.5\,mR^2} = \frac{2Rf_t}{0.5\,mR^2 + mR^2}$$

$$\Rightarrow f_t = 3f_b$$

The maximum value of the bottom force of friction is much higher than the top because the normal force is higher, and also the coefficient of friction is higher. At the same time, the top force of friction is required to be three times as high as the bottom force, so it is guaranteed to be the first one to reach its maximum value. So consider the case where both top and bottom friction forces take the maximum value of the top friction force. The acceleration of a log's centre of mass is

$$a_l = \frac{\Sigma F_l}{m} = \frac{f_t^{max} + \frac{1}{3}f_t^{max}}{m} = \frac{\frac{4}{3} \times 0.35 \times \frac{M}{2} g}{m} = 0.35\frac{2}{3}\frac{M}{m}g$$

The acceleration of the plank is twice the acceleration of a log. Summing all forces acting on the plank gives

$$a_p = \frac{\Sigma F_p}{m} = \frac{F - 2f_t^{max}}{m} = \frac{F - 2 \times 0.35 \times \frac{M}{2} g}{M} = 0.35\frac{4}{3}\frac{M}{m}g$$

$$\Rightarrow F = 0.35Mg\left(1 + \frac{4}{3}\frac{M}{m}\right)$$

(b) As shown in part (a) $a_p = 0.35\frac{4}{3}\frac{M}{m}g$

79. Assume force pushes in the direction opposite to motion.

$$\frac{\tau_b}{I_b} = \frac{\tau_{cm}}{I_{cm}} \Rightarrow \frac{2Fr}{\beta mr^2 + mr^2} = \frac{(F+f)r}{\beta mr^2}$$

$$\Rightarrow f = F\left(\frac{\beta - 1}{\beta + 1}\right)$$

If the moment of inertia is less than mr^2 ($\beta < 1$), the force of friction will negative. Because we assumed positive f meant force opposite to the direction of motion, the force will be in the direction of motion, pushing the object forward.

81. (a) The vertical component of the normal force must balance gravity, while the horizontal component must provide the centripetal acceleration of the ball.

$$\tan\theta = \frac{mg}{F_c} \Rightarrow F_c = \frac{mg}{\tan\theta} = \frac{mv^2}{R - r}$$

$$\tan\theta = \frac{R - h}{\sqrt{R^2 - (R-h)^2}} = \frac{R - h}{\sqrt{2Rh - h^2}}$$

$$\Rightarrow v = \sqrt{g\frac{R-r}{R-h}\sqrt{2Rh - h^2}}$$

(b) The height h is where the marble will be stable. If it enters above h, gravity will overpower the vertical component of the normal force, and it will move down toward its stable height. When it gets there, its vertical momentum will carry it further below its stable height. At this point, the normal force will overcome gravity, and it will be pushed back toward its stable point. The height of the marble will oscillate indefinitely because with no friction, energy is conserved.

83. (a) Using the work-energy approach, the gravitational potential energy released when mass m moves a distance d up the ramp, and mass M drops a distance $2d$ is converted into the kinetic energy of both masses and also the spinning pulley.

$$\Delta K = -\Delta U = 2Mgd - mgd\sin\theta \quad (1)$$

The speeds of the falling mass and a point on the rim of the pulley are twice the speed of the sphere, v. The kinetic energy is

$$\Delta K = \frac{1}{2}mv^2 + \frac{1}{2}I_s\omega_s^2 + \frac{1}{2}I_p\omega_p^2 + \frac{1}{2}Mv_M^2$$

$$\Delta K = \frac{1}{2}mv^2 + \frac{1}{2}\left(\frac{2mR^2}{5}\right)\left(\frac{v}{R}\right)^2 + \frac{1}{2}I_p\left(\frac{2v}{r}\right)^2 + \frac{1}{2}M(2v)^2 = \left(\frac{7}{10}m + 2\frac{I}{r^2} + 2M\right)v^2 \quad (2)$$

Equating (1) and (2) and differentiating with respect to time gives

$$2Mgv - mgv\sin\theta = 2\left(\frac{7}{10}m + 2\frac{I}{r^2} + 2M\right)va$$

$$\alpha = \frac{a}{R} = \frac{g(2M - m\sin\theta)}{R\left(\frac{7}{5}m + 4\frac{I}{r^2} + 4M\right)} = \frac{9.81\text{ m/s}^2(2\times5.6\text{ kg} - 1.1\text{ kg}\times\sin32°)}{\left(\frac{7}{5}1.1\text{ kg} + 4\frac{0.40\text{ kg m}^2}{(0.11\text{ m})^2} + 4\times5.6\text{ kg}\right)} = 0.67\text{ rad/s}^2$$

Below is a Newton's second law approach. We note that the falling mass and a point on the top of the pulley have acceleration of $2a$.

$$Mg - T_1 = M2a$$

$$\Rightarrow T_1 = M(g - 2a)$$

Summing torques about the pulley's axis of rotation gives

$$T_1 r - T_2 r = I_p\left(\frac{2a}{r}\right)$$

$$\Rightarrow T_2 = T_1 - \frac{2I_p a}{r^2} = Mg - \left(2M + 2\frac{I_p}{r^2}\right)a$$

Summing torque on the sphere about the pivot gives

$$2T_2 R - mg\sin\theta R = \left(mR^2 + \frac{2}{5}mR^2\right)\left(\frac{a}{R}\right) \Rightarrow 2T_2 = \frac{7}{5}ma + mg\sin\theta$$

$$\Rightarrow 2Mg - \left(4M + 4\frac{I_p}{r^2}\right)a = \frac{7}{5}ma + mg\sin\theta$$

$$\Rightarrow \alpha = \frac{a}{R} = \frac{g(2M - m\sin\theta)}{R\left(\frac{7}{5}m + 4\frac{I}{r^2} + 4M\right)} = 0.67\text{ rad/s}^2$$

(b) An expression for T_2 was found in the previous work:

$$T_2 = Mg - \left(2M + 2\frac{I_p}{r^2}\right)\alpha R = 43.1 \text{ N}$$

Fiction can be found by summing torques about the centre of mass:

$$\alpha = \frac{\tau_{cm}}{I} = \frac{(T_2 + f)R}{\frac{2}{5}mR^2} = \frac{5(T_2 - f)}{2mR}$$

$$f = T_2 - \frac{2}{5}m\alpha R = 43.0 \text{ N}$$

Chapter 10—EQUILIBRIUM AND ELASTICITY

1. No, even when the object is at its highest point an unbalanced force (gravity) acts on it.

3. c

5. c

7. d

9. No, an unbalanced centripetal force is acting on the object.

11. Yes; for example, the centre of gravity of a V-shaped object lies outside the object.

13. no

15. If the gravitational field strength due to the star at the farther planet is g, the gravitational field strength at the closer planet is $4g$ (see Equation 11-1).

$$r_g = \frac{\sum r_i m_i g_i}{\sum m_i g_i} = \frac{rm(4g) + (2r)mg}{m(4g) + mg} = \frac{6}{5}r \text{ from the centre of the star}$$

17. No. Young's modulus is a material property.

19. If toughness is defined as resistance to deformation, Young's modulus quantifies toughness. According to Table 10-1, steel is tougher than human hair.

21. compression

23. c

25. a and c

27. b

29. a

31. Tension is the same everywhere in the rope because the pulley is not accelerating.

$T = mg$ to balance the force acting on the hanging mass.

This must be equal to the force of friction on 12 kg mass on the verge of slipping.

$$\Rightarrow f = \mu Mg = T$$

$$\Rightarrow \mu = \frac{T}{Mg} = \frac{m}{M} = \frac{5 \text{ kg}}{12 \text{ kg}} = 0.42$$

33. Each end of the bar acts against the ice with force N. Balancing all forces acting on the bar:

$2\mu N = mg$

$\Rightarrow N = \dfrac{mg}{2\mu} = \dfrac{65 \text{ kg} \times 9.81 \text{ m/s}^2}{2 \times 0.12} = 2656 \text{ N}$

35. By balancing torque about the centre of mass, we see that both forces are equal.

$F_1 + F_2 = 2F_1 = F_g = mg$

$\Rightarrow F_1 = F_2 = \dfrac{mg}{2} = \dfrac{14.0 \times 10^3 \text{ kg} \times 9.81 \text{ m/s}^2}{2} = 6.87 \times 10^4 \text{ N}$

37. The angle that the long branch makes with the ground is $180° - 67° - 105° = 8°$. Because the ends of the branches are modelled as circular, the force of friction acts along a line dividing the $105°$ between the branches in half (shown below). We assume the force of friction acts up on the long branch and down on the short branch. The contact force between the two branches (N) acts $90°$ to the force of friction.

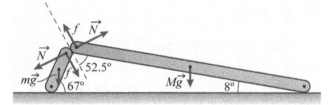

Summing torques on each branch about the point of contact with the ground:

$\Sigma \tau_{long} = Mg \cos(8°) \dfrac{R}{2} - N \sin(37.5°)R - f \sin(52.5°)R = 0$

$\Sigma \tau_{short} = -mg \cos(67°) \dfrac{r}{2} + N \sin(37.5°)r - f \sin(52.5°)r = 0$

$\Rightarrow 2N \sin(37.5°) + 2f \sin(52.5°) = Mg \cos(8°)$ (1)

$\Rightarrow 2N \sin(37.5°) - 2f \sin(52.5°) = mg \cos(67°)$ (2)

Solving equations (1) and (2) simultaneously gives

$N = g \dfrac{M \cos(8°) + m \cos(67°)}{4 \sin(37.5°)}$

$f = g \dfrac{M \cos(8°) - m \cos(67°)}{4 \sin(52.5°)}$

$\mu = \dfrac{f}{N} = \dfrac{4.0 \text{ kg} \times \cos(8°) - 2.7 \text{ kg} \times \cos(67°)}{4.0 \text{ kg} \times \cos(8°) + 2.7 \text{ kg} \times \cos(67°)} \times \dfrac{\sin(37.5°)}{\sin(52.5°)} = 0.44$

39. Balancing forces gives

$T_1 \cos(21°) = T_2 \cos(37°)$ (1)

$T_1 \sin(21°) + T_2 \sin(37°) = mg$ (2)

Solving equations (1) and (2) simultaneously gives

$$T_2 = \frac{300 \text{ kg} \times 9.81 \text{ m/s}^2}{\cos(37°)\tan(21°)+\sin(37°)} = 3240 \text{ N}$$

$$T_1 = 3240 \text{ N} \times \frac{\cos(37°)}{\cos(21°)} = 2770 \text{ N}$$

41. If θ is the angle that the force of gravity makes with the leaning person, then

$$\sin\theta = \frac{24}{110} \Rightarrow \theta = 12.6°$$

Balancing torques about the passenger's feet, and assuming that the net force of the wind acts at the passenger's centre of mass:

$$mgh_{cm}\sin\theta = F_w h_{cm}\cos\theta$$

$$\Rightarrow F_w = mg\tan\theta = 75 \text{ kg} \times 9.81 \text{ m/s}^2 \tan(12.6°) = 164 \text{ N}$$

43. Define λ as linear mass density.

$$r_g = \frac{\int_0^L rg(r)\lambda dz}{\int_0^L g(r)\lambda dr} = \frac{\int_0^L 1400re^{-r/120}\lambda dr}{\int_0^L 1400e^{-r/120}\lambda dr} = \frac{\int_0^L re^{-r/120}dr}{\int_0^L e^{-r/120}dr}$$

$$r_g = \frac{\left(-120e^{-r/120}(r+120)\right)\Big|_0^L}{\left(-120e^{-r/120}\right)\Big|_0^L} = \frac{120-e^{-L/120}(L+120)}{1-e^{-L/120}}$$

$$r_{cm}-r_g = \frac{L}{2} - \frac{120-e^{-L/120}(L+120)}{1-e^{-L/120}}$$

45. If the direction of the gravitational field does not vary, calculation of the centre of gravity is simplified. We need to find the radius of the neutron star. In Chapter 11, we learn that gravitational feed strength is inversely proportional to the square of the radius from a circular body, and points toward the centre.

$$\left(\frac{r+200 \text{ m}}{r}\right)^2 = \frac{110g}{107g} \Rightarrow r = 14.4 \text{ km}$$

The change in the direction of the field moving 200 m over the surface is

$$\Delta\theta = \frac{100 \text{ m}}{14.4 \text{ km}} = 6.96 \times 10^{-3} \text{ rad} = 0.40°$$

∴ we can ignore the curvature of the neutron star and assume the field points down. The centre of gravity is defined as any point at which the force of gravity will exert zero net torque on an object. If we define the origin to be the position of the object at the surface of the star, then

$$110mgx_g = 107mg(100 \text{ m} - x_g)$$

$$\Rightarrow x_g = \frac{107}{107 + 110} \times 100 \text{ m} = 49.3 \text{ m}$$

The centre of gravity is not uniquely defined. If we constrain it to be along the line joining the two objects, then the y-coordinate is

$$y_g = 2x_g = 98.6 \text{ m}$$

$$\Rightarrow r_g = (49.3\hat{i} + 98.6\hat{j}) \text{ m}$$

47. Balancing torque at the hinge, we get

$$mg\frac{r}{2} = T\sin(46°)r$$

$$\Rightarrow T = \frac{mg}{2\sin(46°)} = \frac{210 \text{ kg} \times 9.81 \text{ m/s}^2}{2\sin(46°)} = 1432 \text{ N}$$

49. Balancing forces gives

$$m_1 g = 2T$$
$$m_2 g = T$$
$$\Rightarrow m_1 = 2m_2$$

51. Balance torques about the pivot point:

$$F \times 0.50 \text{ m} = mg\sqrt{(0.40 \text{ m})^2 - (0.20 \text{ m})^2}$$

$$\Rightarrow F = \frac{120 \text{ kg} \times 9.81 \text{ m/s}^2 \sqrt{(0.40 \text{ m})^2 - (0.20 \text{ m})^2}}{0.50 \text{ m}} = 816 \text{ N}$$

53. First we will balance torques on hinges A and B.

Assume that C_x and C_y act in the positive directions on the upper bar. Netwon's third law says they will act in the opposite directions on the lower bar.

$$\tau_A = 0 \Rightarrow 2100.0 \text{ kg} \times 9.81 \text{ m/s}^2 \times \frac{4.50 \text{ m}}{2} + C_x \times 1.00 \text{ m} = C_y \times 2.00 \text{ m} \qquad (1)$$

$$\tau_B = 0 \Rightarrow 700 \text{ kg} \times 9.81 \text{ m/s}^2 \times \frac{2.00 \text{ m}}{2} + C_y \times 2.00 \text{ m} = C_x \times 2.50 \text{ m} \qquad (2)$$

Solving equations (1) and (2) simultaneously gives

$$C_x = 35480 \text{ N} \quad \text{and} \quad C_y = 40916 \text{ N}$$

Balancing forces on the top bar gives

$A_x = -C_x = -35480$ N

$A_y = 2100$ kg $\times 9.81$ m/s$^2 - C_y = -20315$ N [negative sign means it acts downward]

Balancing forces on the bottom bar gives

$B_x = C_x = 35480$ N

$B_y = C_y + 700$ kg $\times 9.81$ m/s$^2 = 47783$ N

55. First find the angles at points A, B, and C:

$$\sin(\angle B) = \frac{2.0 \text{ m}}{4.0 \text{ m}} = 0.5 \Rightarrow \angle B = 30°$$

Because the triangle is isosceles, $\angle A = \angle C = 75°$

Length of the larger beam is therefore given by

$$L_l = \frac{6.0 \text{ m}}{\sin(75°)} = 6.21 \text{ m}$$

For the contact forces between the beam, we assume they act in the positive x and y directions for the shorter beam, and the negative x and y directions for the longer beam. Considering only longer beam, summing torques about A gives

$$mg\cos(75°)\left(\frac{6.2 \text{ m}}{2}\right) + F_y\left(4.0 \text{ m} - 4.0 \text{ m}\cos(30°)\right) = F_x\left(2.0 \text{ m}\right)$$

$$\Rightarrow 2.0\, F_x - 0.536 \text{ m}\, F_y = 8658 \text{ Nm} \qquad (1)$$

Considering only shorter beam, summing torques about B gives

$$mg\cos(30°)\left(\frac{4.0 \text{ m}}{2}\right) = F_y\cos(30°)\left(4.0 \text{ m}\right) + F_x\left(2.0 \text{ m}\right)$$

$$\Rightarrow 2.0 \text{ m}\, F_x + 3.46 \text{ m}\, F_y = 18691 \text{ Nm} \qquad (2)$$

Solving equations (1) and (2) simultaneously gives

$F_x = 5002$ N, and $F_y = 2511$ N

The friction forces must be equal and opposite because they are the only external forces acting in the horizontal direction. f will act to the right on A and to the left on B. Considering that the only two forces that act on each beam in the horizontal direction are the force of friction and the horizontal contact force:

$f = F_x = 5002$ N

57. Since $\dfrac{F}{A} = G\dfrac{\Delta L}{L}$

$\Rightarrow \dfrac{mg}{A} = G\dfrac{\Delta L}{L}$

$\Rightarrow m = \dfrac{190 \times 10^6 \text{ Pa} \times 0.13 \text{ m} \times 25 \times 10^{-4} \text{ m}^2}{9.81 \text{ m/s}^2 \times 4.0 \text{ m}} = 1574 \text{ kg}$

59. The mass supported by the shear force at the edge of the limestone beam is the mass of the load plus the mass of the beam. The density of limestone is 2500 kg/m^3.

$\dfrac{(m_{max} + m_{beam})g}{A} = \tau_{max}$

$\Rightarrow m_{max} = \dfrac{25 \times 10^6 \text{Pa} \times 100 \times 10^{-4} \text{ m}^2}{9.81 \text{ m/s}^2} - 2500 \text{ kg/m}^3 \times 100 \times 10^{-4} \text{ m}^2 \times 2.3 \text{ m} = 25\,440 \text{ kg}$

61. $\dfrac{F}{A} = Y\dfrac{\Delta L}{L} \Rightarrow Y = \dfrac{F}{A}\dfrac{L}{\Delta L} = \dfrac{2000 \text{ N}}{\pi(0.21 \text{ m})^2}\dfrac{3.0 \text{ m}}{0.015 \text{ m}} = 2.89 \times 10^6 \text{ Pa} = 2.89 \text{ MPa}$

63. (a) $\sigma_{ys} = \dfrac{F}{A} = \dfrac{mg}{\pi r^2}$

$\Rightarrow m = \dfrac{\pi r^2 \sigma_{ys}}{g} = \dfrac{\pi(0.030 \text{ m})^2 (250 \times 10^6 \text{ Pa})}{9.81 \text{ m/s}^2} = 7.2 \times 10^4 \text{ kg}$

(b) $m = \dfrac{\pi r^2 \sigma_{uts}}{g} = \dfrac{\pi(0.030 \text{ m})^2 (450 \times 10^6 \text{ Pa})}{9.81 \text{ m/s}^2} = 1.3 \times 10^5 \text{ kg}$

(c) $D = 2r = 2\sqrt{\dfrac{mg}{\pi\sigma_{uts}}} = 2\sqrt{\dfrac{1.3 \times 10^5 \text{ kg} \times 9.81 \text{ m/s}^2}{\pi(1000 \times 10^6 \text{ Pa})}} = 0.040 \text{ m} = 4.0 \text{ cm}$

65. The yield strength of femur is approximately given by $\sigma_{ys} = 100 \text{ MPa}$.

The average radius of a female femur bone is 12.7 mm.

$\sigma = \dfrac{F}{A} = \dfrac{82 \text{ kg} \times 9.81 \text{ m/s}^2}{\pi(0.0127 \text{ m})^2} = 1.6 \times 10^6 \text{ Pa} = 1.6 \text{ MPa}$

$\Rightarrow \dfrac{\sigma}{\sigma_{ys}} = 0.016 = 1.6\%$

67. The bottom block can withstand a force of

$$F = \left(250 \times 10^6 \text{ Pa}\right)\left(0.300 \text{ m}\right)^2 = 2.25 \times 10^7 \text{ N}$$

This is equivalent to a mass of

$$M = \frac{F}{g} = \frac{2.25 \times 10^7 \text{ N}}{9.81 \text{ m/s}^2} = 2.30 \times 10^6 \text{ kg}$$

The mass of one block is

$$m = \left(0.300 \text{ m}\right)^3 \left(2700 \text{ kg/m}^3\right) = 72.9 \text{ kg}$$

Therefore, the number of blocks that can be stacked is

$$N = \frac{M}{m} = \frac{2.30 \times 10^6 \text{ kg}}{72.9 \text{ kg}} = 31550$$

69. (a) Change in length before breaking is given by

$$\Delta L = \frac{\sigma L}{Y} = \frac{300 \times 10^6 \text{ Pa} \times 2.0 \text{ m}}{73 \times 10^9 \text{ Pa}} = 0.0082 \text{ m}$$

In question 56, it was shown that the effective amount of work it takes to compress a rod a distance ΔL is

$$W = \frac{YA}{2L} \Delta L^2 = \frac{73 \times 10^9 \text{ Pa} \times \pi \times (0.245 \text{ m})^2}{2 \times 2.0 \text{ m}} (0.0082 \text{ m})^2 = 2.31 \times 10^5 \text{ J}$$

(b) The work done against the falling mass is equal to the work done to break the rod:

$$W = mgh \Rightarrow m = \frac{W}{gh} = \frac{2.31 \times 10^5 \text{ J}}{9.81 \text{ m/s}^2 \times 20 \text{ m}} = 1180 \text{ kg}$$

71. Balancing torques at the hinge gives

$$m_{beam} g r_{cm} \cos(11°) + mgr_m \cos(11°) = Nr_N \sin(11° + 30°)$$

$$\Rightarrow N = 9.81 \text{ m/s}^2 \times \cos(11°) \frac{900 \text{ kg} \times \dfrac{3.7 \text{ m}}{2} + 200 \text{ kg} \times 3.0 \text{ m}}{3.7 \text{ m} \times \sin(11° + 30°)} = 8985 \text{ N}$$

For reaction forces, we balance horizontal and vertical forces.

$$R_x = N \cos(30°) = 7782 \text{ N}$$

$$R_y = (200 \text{ kg} + 900 \text{ kg}) \times 9.81 \text{ m/s}^2 - N \sin(30°) = 6298 \text{ N}$$

73. The normal force nor the force of gravity can exert a torque on the sphere's centre of mass. There will be no net torque when the line of action of the force of tension passes through the sphere's centre of mass.

$$\sin \theta = \frac{R}{7R} \Rightarrow \theta = \sin^{-1}\left(\frac{1}{7}\right) = 8.2°$$

75. Net torque on the pulley is zero, so the horizontal and vertical tensions are related as

$$T_h = 2T_v$$

T_h balances the force of friction, and T_v balances the hanging mass, so

$$\mu Mg = 2(mg) \Rightarrow m = \frac{\mu M}{2} = \frac{0.35 \times 3.0 \text{ kg}}{2} = 0.525 \text{ kg}$$

77. Balancing torques at the hinge gives

$$T\frac{L}{3}\sin(85°) = Mg\frac{L}{2}\cos(30°) + mgL\cos(30°)$$

$$\Rightarrow T = 3g\cos(30°)\frac{M+2m}{2\sin(85°)} = 15351 \text{ N}$$

Balancing forces gives

$$R_x = T\cos(85° - 30°) = 8805 \text{ N}$$

$$R_y = (m+M)g - T\sin(85° - 30°) = -2961 \text{ N [acting down]}$$

79. Balancing torques about the centre of the sphere gives

$$F(d-r) = fr \qquad (1)$$

Since $f = \mu N$ and $N = F$

$$\Rightarrow f = \mu F \qquad (2)$$

Solving equations (1) and (2) simultaneously, we get

$$d = r(1+\mu)$$

81. We note that both contact points are on the verge of slipping (one cannot slip without the other one slipping).

Here subscripts 1 and 2 refer to inclined and horzontal surfaces respectively.

$$\mu_1 N_1 \times 0.50 \text{ m} + \mu_2 N_2 \times 0.50 \text{ m} = 120 \text{ N} \times 0.20 \text{ m}$$

$$\Rightarrow 0.2N_1 + \mu_2 N_2 = 48 \text{ N} \qquad (1)$$

Summing forces in the x- and y-directions:

$$F + \mu_2 N_2 = \mu_1 N_1 \sin(32°) + N_1 \cos(32°)$$

$$\Rightarrow 0.954N_1 - \mu_2 N_2 = 120 \text{ N} \qquad (2)$$

$$N_2 + \mu_1 N_1 \sin(32°) = N_1 \cos(32°)$$

$$\Rightarrow 0.742N_1 - N_2 = 0 \qquad (3)$$

Solving equations (1), (2), and (3) simultaneously gives

$$N_2 = 108.0 \text{ N}$$

$$\mu_2 N_2 = 18.88 \text{ N}$$

$$\Rightarrow \mu_2 = \frac{18.88 \text{ N}}{108.0 \text{ N}} = 0.175$$

83. Balancing torques about the point where the bar is attached to the string gives

$$mg\cos(42°)\left(\frac{L}{2}\right) = N\cos(42°)(L) + f\sin(42°)(L)$$

Balancing forces gives

$$f = T\sin(21°)$$

$$T\cos(21°) + N = mg$$

$$0.743\frac{N}{m} + 0.669\frac{f}{m} = 3.65 \text{ m/s}^2 \qquad (1)$$

$$0.375\frac{T}{m} - \frac{f}{m} = 0 \qquad (2)$$

$$0.933\frac{T}{m} + \frac{N}{m} = 9.81 \text{ m/s}^2 \qquad (3)$$

Solving (1), (2), and (3) simultaneously gives

$$\frac{N}{m} = 2.135 \text{ m/s}^2 \text{ and } \frac{f}{m} = 3.085 \text{ m/s}^2$$

$$\mu = \frac{f}{N} = 1.44$$

85. (a) The minimum force needed to move the fridge forward (assuming that the force does not cause enough torque to tip the fridge) is

$$F = f = \mu_s mg = 0.34 \times 92 \text{ kg} \times 9.81 \text{ m/s}^2 = 306.9 \text{ N}$$

(b) If on the verge of tipping, the pedestal closest to the applied force would come off the ground, so there would be no normal force applied to that pedestal. Therefore, the maximum torque you can apply about the pedestal is

$$\tau_F = Fr_F = 306.9 \text{ N} \times 1.10 \text{ m} = 337.5 \text{ Nm}$$

The counter-clockwise torque caused by the force of gravity is

$$\tau_G = mgr_g = 92 \text{ kg} \times 9.81 \text{ m/s}^2 \times \frac{0.76 \text{ m}}{2} = 343.0 \text{ Nm}$$

Since $\tau_F < \tau_G$, the refrigerator will not tip first.

(c) The torques about the pedestal farthest from the applied force must balance

$$Fy = \tau_G$$

$$\Rightarrow y = \frac{343.0 \text{ Nm}}{306.9 \text{ N}} = 1.12 \text{ m}$$

87. Balancing torques about the front edge of the shelf gives

$$F \sin \theta h_F = mgx_{cm}$$

$$\Rightarrow F = \frac{mgx_{cm}}{\sin \theta h_F} = \frac{15 \text{ kg} \times 9.81 \text{ m/s}^2}{\sin(30°) \times 1.4 \text{ m}} \times \frac{0.27 \text{ m}}{2} = 28.4 \text{ N}$$

89. (a) The slope of the graph is Young's modulus. $Y = 1.3$ GPa.

(b)

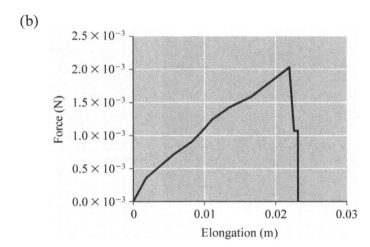

(c) Question 56 derived the amount of work needed to stretch a material by a distance x.

$$W = \frac{YA}{2L}x^2 = \frac{YAL}{2}\left(\frac{x}{L}\right)^2 = \frac{1.3 \times 10^9 \text{ Pa} \times \pi \times (0.75 \times 10^{-6} \text{ m})^2 \times 0.11 \text{ m}}{2}\left(\frac{1}{100}\right)^2$$

$$= 1.26 \times 10^{-8} \text{ J}$$

(d) Around 1.5%, 200 MPa

(e) The work needed is the area under the curve produced in part (b).

$$W = 2.2 \times 10^{-5} \text{ J}$$

If a fly were to transfer all of its kinetic energy to the silk to break it:

$$K = \frac{1}{2}mv^2 \Rightarrow v = \sqrt{\frac{2K}{m}} = \sqrt{\frac{2 \times 2.2 \times 10^{-5} \text{ J}}{50 \times 10^{-6} \text{ kg}}} = 0.94 \text{ m/s}$$

It is reasonable to believe that a fly could fly this fast.

91. From graph: $Y = \frac{(160 - 20) \times 10^6 \text{ Pa}}{(0.8 - 0.2) \times 0.01} = 23.3$ GPa

$$k = Y\frac{A}{L} = (23.3 \times 10^9 \text{ Pa})\frac{\pi(0.50 \times 10^{-6} \text{ m})^2}{0.015 \text{ m}} = 1.22 \text{ N/m}$$

93. Weight will be distributed equally in all four bars, so all bars feel the same stress.

Stress in each bar is given by

$$\sigma = \frac{mg}{4A} = \frac{(120 \text{ kg} + 4500 \text{ kg}) \times 9.81 \text{ m/s}^2}{4 \times \pi \times (0.050 \text{ m})^2} = 1.44 \times 10^6 \text{ P} = 1.44 \text{ MPa}$$

Strain in each bar is given by

$$\varepsilon = \frac{\sigma}{Y} = \frac{1.44 \times 10^6 \text{ Pa}}{200 \times 10^9 \text{ Pa}} = 7.2 \times 10^{-6}$$

95. Kinetic energy of the bullet is

$$E = \frac{1}{2}mv^2 = \frac{1}{2}(0.0075 \text{ kg})(390 \text{ m/s})^2 = 570.4 \text{ J}$$

The point at which the steel plate will reach its yield point is give by

$$\frac{\Delta L}{L} = \frac{0.58 \times \text{Yield Strength}}{G} = \frac{0.58 \times 250 \times 10^6 \text{ Pa}}{G} = \frac{1.45 \times 10^8 \text{ Pa}}{G}$$

If we assume the centre of the plate has moved a distance ΔL from the edge, $L = 2.0$
Assuming the steel plate is held firmly in place within the vest, the area over which
the sheer force is spread can be estimated as the perimeter of the plate times the thickness.

$$A = 4 \times 0.040 \text{ m} \times T = 0.16 \text{ m} \times T$$

Similar to the approach in question 94, the work needed to deform a material by sheering is

$$W = \frac{GAL}{2}\left(\frac{\Delta L}{L}\right) = \frac{GL(0.16 \text{ m} \times T)}{2}\left(\frac{\Delta L}{L}\right) = K_{bullet}$$

$$\Rightarrow T = \frac{2K_{bullet}}{LG\left(\frac{\Delta L}{L}\right) \times 0.16 \text{ m}} = \frac{2 \times 570.4 \text{ J}}{0.020 \text{ m} \times 1.45 \times 10^8 \text{ Pa} \times 0.16 \text{ m}} = 2.46 \times 10^{-3} \text{ m} = 2.5 \text{ mm}$$

97. (a) Velocity of athlete just before hitting ground is $v = \sqrt{2gh}$.

The change in momentum when impacting is therefore given by $\Delta p = m\sqrt{2gh}$.
This is equal to the impulse $2Ft$, where F is the force provided by one femur.

$$\Rightarrow 2Ft = m\sqrt{2gh}$$

But $F = \sigma_{Ult}A$

$$\Rightarrow h = \frac{1}{2g}\left(\frac{2\sigma_{Ult}At}{m}\right)^2 = \frac{1}{2 \times 9.81 \text{ m/s}^2}\left(\frac{2 \times 190 \times 10^6 \text{ Pa} \times \pi(0.017 \text{ m})^2 \times 0.220 \text{ s}}{70 \text{ kg}}\right)^2$$

$$= 5.99 \times 10^4 \text{ m}$$

(b) $\dfrac{F}{A} = Y\dfrac{\Delta L}{L} = \sigma_{Ult}$

$\Rightarrow \Delta L = L\dfrac{\sigma_{Ult}}{Y} = 52 \text{ cm} \times \dfrac{190 \times 10^6 \text{ Pa}}{17 \times 10^9 \text{ Pa}} = 0.58 \text{ cm}$

99. Longitudinal compression: $F = 200 \times 10^6 \text{ Pa} \times \pi (0.015 \text{ m})^2 = 1.4 \times 10^5 \text{ N}$

Longitudinal tension: $F = 135 \times 10^6 \text{ Pa} \times \pi (0.015 \text{ m})^2 = 9.5 \times 10^4 \text{ N}$

Transverse compression: $F = 131 \times 10^6 \text{ Pa} \times 0.53 \text{ m} \times 2 \times 0.015 \text{ m} = 2.1 \times 10^6 \text{ N}$

Transverse shear: $F_{low} = 65 \times 10^6 \text{ Pa} \times \pi (0.015 \text{ m})^2 = 4.6 \times 10^4 \text{ N}$

$$F_{high} = 71 \times 10^6 \text{ Pa} \times \pi (0.015 \text{ m})^2 = 5.0 \times 10^4 \text{ N}$$

101. (a) The ultimate tensile strength increases as the steel passes through subsequent cycles. From the graph we see that the UTS of the steel was about 850 MPa. The UTS increased to about 950 MPa after first pass and to about 1150 MPa after four passes.

(b) The pressing process increases the ultimate tensile strength of the steel, and hence increases the energy required to break the steel.

1. b

3. Weight is a force and should be measured in units such as newtons.

5. b. Acceleration due to gravity is given by

    $$g = \frac{GM}{R^2}$$

    Because the density of the two planets is the same, it follows that

    $$\frac{M_2}{(4/3)\pi R_2^{\,3}} = \frac{M_1}{(4/3)\pi R_1^{\,3}}$$

    $$\frac{M_2}{R_2^{\,3}} = \frac{M_1}{R_1^{\,3}}$$

    It follows that

    $$\frac{g_2}{R_2} = \frac{g_1}{R_1}$$

    and therefore

    $$\frac{g_2}{g_1} = \frac{R_2}{R_1}$$

    Thus, if $R_2 = 2R_1$

    $$\Rightarrow \frac{g_2}{g_1} = \frac{2R_1}{R_1}$$

    $$\Rightarrow \frac{g_2}{g_1} = 2$$

    Hence, the larger planet has greater acceleration due to gravity at its surface.

7. c

    $$g = \frac{GM}{R^2}$$

    $$\Rightarrow g \propto \frac{1}{R^2}$$

9. a

    $$g = \frac{GM}{R^2}$$

    G is the universal gravitational constant.

11. The satellite speeds up as it loses altitude due to atmospheric drag. As shown, the magnitude of the satellite's kinetic energy can be shown for a circular orbit to be half of the magnitude of the satellite's potential energy, but the potential energy is negative and the kinetic energy is positive. Therefore, when the r is less, v is more.

From $F = ma$ with the gravitational force and with centripetal acceleration:

$$F_g = \frac{GMm}{r^2} = \frac{mv^2}{r} \text{ which can be written as}$$

$$\frac{GMm}{2r} = \frac{1}{2}mv^2$$

13. b

$$T^2 \propto a^3$$

$$\Rightarrow T \propto a^{\frac{3}{2}}$$

$$\Rightarrow T' \propto (4a)^{\frac{3}{2}}$$

$$\Rightarrow T' = 8T$$

15. c. Because the star has no component of velocity toward or away from us, there is no redshift or blueshift.

17. c. The situation described here is called a Hohmann transfer and is widely used in interplanetary spaceflight. When you provide a speed boost to an object in an elliptical orbit applied at perihelion, the extra energy will result in the object being able to go to a greater distance (it now has more kinetic energy). Therefore, both the aphelion distance and the semi-major axis will increase. However, the motion will still bring it back to the same point (i.e., the orbit is continuous), so the perihelion distance will not change.

19. $r = 2.3 \times 10^6 \text{ ly} = (2.3 \times 10^6)(3.00 \times 10^8 \times 3.16 \times 10^7) = 2.2 \times 10^{22} \text{ m}$

$M = (10^{12})(1.989 \times 10^{30}) = 1.989 \times 10^{42} \text{ kg}$

$F = G\frac{Mm}{r^2} = 6.672 \times 10^{-11} \frac{(1.989 \times 10^{42})(50)}{(2.2 \times 10^{22})^2} = 1.4 \times 10^{-11} \text{ N}$

21. Nearest point: $r_n = 3.84 \times 10^8 - 6.378 \times 10^6 = 3.776 \times 10^8 \text{ m}$

Farthest point: $r_f = 3.84 \times 10^8 + 6.378 \times 10^6 = 3.904 \times 10^8 \text{ m}$

$F = G\frac{Mm}{r^2}$

$\Rightarrow \frac{F_n}{m} - \frac{F_f}{m} = GM\left(\frac{1}{r_n^2} - \frac{1}{r_f^2}\right) = (6.674 \times 10^{-11})(7.348 \times 10^{22})\left(\frac{1}{(3.776 \times 10^8)^2} - \frac{1}{(3.904 \times 10^8)^2}\right)$

$$\Rightarrow \frac{F_n - F_f}{m} = 2.21 \times 10^{-6} \text{ N/kg}$$

As a percentage of the force per kg at the centre of Earth, the difference between the forces at the nearest point and farthest point is

$$\frac{\dfrac{F_n - F_f}{m}}{G\dfrac{M}{r_c^2}} = \frac{GM\left(\dfrac{1}{r_n^2} - \dfrac{1}{r_f^2}\right)}{\dfrac{GM}{r_c^2}}$$

$$= r_c^2 \left(\frac{1}{r_n^2} - \frac{1}{r_f^2}\right)$$

$$= \left(3.84 \times 10^8\right)^2 \left(\frac{1}{\left(3.776 \times 10^8\right)^2} - \frac{1}{\left(3.904 \times 10^8\right)^2}\right)$$

$$= 0.067$$

$$= 6.7\%$$

23. $g = \dfrac{GM}{r^2} = \dfrac{\left(6.672 \times 10^{-11}\right)\left(6.42 \times 10^{23}\right)}{\left(3.40 \times 10^6\right)^2} = 3.71 \text{ m/s}^2$

25. $\rho = \dfrac{M_d}{\dfrac{4}{3}\pi r^3}$

$$\Rightarrow M_d = \frac{4}{3}\pi r^3 \rho = \frac{4}{3}\pi (500)^3 (4800) = 2.513 \times 10^{12} \text{ kg}$$

$$g = \frac{GM}{R^2} + \frac{GM_d}{r^2} = 9.80 + \frac{\left(6.672 \times 10^{-11}\right)\left(2.513 \times 10^{12}\right)}{(500)^2} = 9.801 \text{ m/s}^2$$

27. First, we will calculate the mass of the planet:

$$\frac{GMm}{r^2} = \frac{mv^2}{r}$$

$$\Rightarrow M = \frac{v^2 r}{G} = \frac{(8000)^2 (8000 \times 10^3)}{6.672 \times 10^{-11}} = 7.67 \times 10^{24} \text{ kg}$$

$$g = \frac{GM}{R^2}$$

$$\Rightarrow g_{sat} = \frac{(6.672 \times 10^{-11})(7.67 \times 10^{24})}{(8000 \times 10^3)^2} = 8.00 \text{ m/s}^2$$

$$g_{planet} = \frac{(6.672 \times 10^{-11})(7.67 \times 10^{24})}{(5000 \times 10^3)^2} = 20.5 \text{ m/s}^2$$

29. The work done against gravity is equal to the change in potential energy:

$$W = -\frac{GMm}{R_2} - \left(-\frac{GMm}{R_1}\right)$$

$$W = GMm\left(\frac{1}{R_1} - \frac{1}{R_2}\right)$$

$$W = (6.672 \times 10^{-11})(6.42 \times 10^{23})(25)\left(\frac{1}{3.40 \times 10^6} - \frac{1}{3.422 \times 10^6}\right)$$

$$W = 2.02 \times 10^6 \text{ J}$$

31. $F = -\nabla U$

$$\Rightarrow F_x = -\frac{\partial U}{\partial x} = -\frac{\partial ky^2}{\partial x} = 0$$

$$F_y = -\frac{\partial U}{\partial y} = -\frac{\partial ky^2}{\partial y} = -2ky$$

$$F_z = -\frac{\partial U}{\partial z} = -\frac{\partial ky^2}{\partial z} = 0$$

33. $v_e = \sqrt{\dfrac{2GM}{r}}$

$$\Rightarrow v_e = \sqrt{\frac{2(6.672 \times 10^{-11})(7.348 \times 10^{22})}{1.737 \times 10^6}} = 2.4 \times 10^3 \text{ m/s} = 2.4 \text{ km/s}$$

35. $M_{bh} = (4 \times 10^6)(1.989 \times 10^{30}) = 7.956 \times 10^{36} \text{ kg}$

$a = 3.5 \times 24 \times 3600 \times 3.00 \times 10^8 = 9.072 \times 10^{13} \text{ m}$

(a) $T^2 = \dfrac{4\pi^2}{G(M_{bh} + M_s)} a^3$

Since $M_{bh} \gg M_s$

$\Rightarrow T^2 \approx \dfrac{4\pi^2}{GM_{bh}} a^3 = \dfrac{4\pi^2}{(6.672 \times 10^{-11})(7.956 \times 10^{36})} (9.072 \times 10^{13})^3$

$\Rightarrow T \approx 2.36 \times 10^8$ s $= 7.5$ y

(b) $v = \dfrac{2\pi a}{T} = \dfrac{2\pi(9.07 \times 10^{13})}{2.36 \times 10^8} = 2.4 \times 10^6$ m/s $= 2400$ km/s

37. $\dfrac{mv^2}{r} = \dfrac{GMm}{r^2}$

$\Rightarrow v = \sqrt{\dfrac{GM}{r}} = \sqrt{\dfrac{(6.672 \times 10^{-11})(7.35 \times 10^{22})}{1.737 \times 10^6}} = 1680$ m/s

39. (a) $r_{cm} = \dfrac{m_{sun} r_{sun} + m_{earth} r_{earth}}{m_{sun} + m_{earth}}$

With origin of the coordinate axes at the centre of the Sun, we have $r_{sun} = 0$.

$\Rightarrow r_{cm} = \dfrac{0 + (5.972 \times 10^{24})(1.496 \times 10^{11})}{1.989 \times 10^{30} + 5.972 \times 10^{24}} = 4.492 \times 10^5$ m

$= 449$ km from the centre of the Sun

(b) $r_{sun} = 6.955 \times 10^8$ m

$\Rightarrow \dfrac{r_{cm}}{r_{sun}} = \dfrac{4.492 \times 10^5}{6.955 \times 10^8}$

$\Rightarrow r_{cm} = 0.00065 r_{sun}$

41. (a) $\rho = \dfrac{M}{\dfrac{4}{3}\pi r^3}$

$\Rightarrow M = \dfrac{4}{3}\pi r^3 \rho = \left(\dfrac{4}{3}\pi\right)(10.5 \times 10^3)^3 (1880) = 9.12 \times 10^{15}$ kg

$g = \dfrac{GM}{R^2}$

$\Rightarrow g = \dfrac{(6.672 \times 10^{-11})(9.12 \times 10^{15})}{(10.5 \times 10^3)^2} = 5.5 \times 10^{-3}$ m/s^2

Since the object is dropped, the initial velocity is zero.

$$\Rightarrow y = \frac{1}{2}gt^2$$

$$\Rightarrow t = \sqrt{\frac{2y}{g}} = \sqrt{\frac{2 \times 1.5}{5.5 \times 10^{-3}}} = 23.3 \text{ s}$$

(b) $T^2 = \frac{4\pi^2}{G(M_{Phobos} + M_{Mars})}a^3$

Here, the mass of Phobos is much less than that of Mars and can be ignored. We solve for the mass of Mars from Phobos' period and semi-major axis using Kepler's third law.

$$\Rightarrow M_{Mars} = \frac{4\pi^2}{GT^2}a^3 = \frac{4\pi^2}{(6.672 \times 10^{-11})(27562)^2}(9.4 \times 10^6)^3 = 6.47 \times 10^{23} \text{ kg}$$

This is close to the accepted value for the mass of Mars.

43. (a) The moment of inertia of a point mass a distance r from axis of rotation is given by

$$I = mr^2$$

Thus, the total moment of inertia of the two masses is

$$I = 2(0.73)(0.9)^2 = 1.18 \text{ kg m}^2$$

(b) $T = 2\pi\sqrt{\frac{I}{\kappa}}$

$$\Rightarrow \kappa = \frac{4\pi^2 I}{T^2} = \frac{4\pi^2(1.18)}{(420)^2} = 2.65 \times 10^{-4} \text{ N} \cdot \text{m}$$

(c) $F = G\frac{Mm}{r^2} = (6.672 \times 10^{-11})\frac{(158)(0.73)}{(0.230)^2} = 1.45 \times 10^{-7} \text{ N}$

(d) $\tau = rF = (0.9)(1.45 \times 10^{-7}) = 1.31 \times 10^{-7} \text{ N} \cdot \text{m}$

(e) $\tau_t = -\kappa\theta$

$$\Rightarrow |\theta| = \frac{\tau_t}{\kappa} = \frac{(2)(1.31 \times 10^{-7})}{2.65 \times 10^{-4}} = 9.89 \times 10^{-4} \text{ rad}$$

45. (a) $g = \dfrac{GM}{R^2}$

$\Rightarrow M = \dfrac{gR^2}{G} = \dfrac{(1.8)(4300 \times 10^3)^2}{6.672 \times 10^{-11}} = 5.0 \times 10^{23}$ kg

(b) Escape speed is given by

$v_e = \sqrt{\dfrac{2GM}{R}} = \sqrt{\dfrac{(2)(6.672 \times 10^{-11})(5,0 \times 10^{23})}{4300 \times 10^3}} = 3.9 \times 10^3$ m/s $= 3.9$ km/s

47. To be gravitationally bound to the Sun, the kinetic energy of the object must be less than the magnitude of the gravitational potential energy; that is

$\dfrac{1}{2}mv^2 < \dfrac{GMm}{r}$

$\Rightarrow v < \sqrt{\dfrac{2GM}{r}} = \sqrt{\dfrac{2(6.672 \times 10^{-11})(1.989 \times 10^{30})}{1.49 \times 10^{11}}} = 42.2 \times 10^3$ m/s

$\Rightarrow v < 42.2$ km/s

49. The total angle subtended by Sun is $0.53°$. The radius of the Sun can be estimated from half of this angle (see figure) and the Sun–Earth mean distance D.

$R = D \tan(0.265°) = (1.50 \times 10^{11}) \tan(0.265°) = 6.938 \times 10^8$ m

Hence, the density is

$\rho = \dfrac{M}{\dfrac{4}{3}\pi R^3} = \dfrac{1.989 \times 10^{30}}{\dfrac{4}{3}\pi(6.938 \times 10^8)^3} = 1400$ kg/m^3

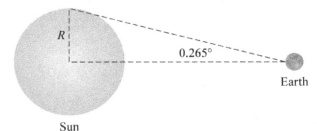

Sun ... 0.265° ... Earth

51. For an object in circular orbit about another object, the centripetal force is provided by, and therefore equal to, the gravitational force; that is

$\dfrac{mv^2}{r} = \dfrac{GMm}{r^2}$

$\Rightarrow M = \dfrac{rv^2}{G}$

53. (a) For an object in circular orbit about another object, the centripetal force is provided by, and therefore equal to, the gravitational force; that is

$$\frac{mv^2}{r} = \frac{GMm}{r^2}$$

$$\Rightarrow v = \sqrt{\frac{GM}{r}} = \sqrt{\frac{(6.672\times10^{-11})(4.4\times10^{30})}{55000}} = 7.3\times10^7 \text{ m/s}$$

(b) $T = \dfrac{2\pi r}{v} = \dfrac{(2\pi)(55\,000)}{7.3\times10^7} = 0.0047 \text{ s} = 4.7 \text{ ms}$

55. The simulation shows increased numbers of asteroids that have orbital radii and periods similar to Jupiter, but travel approximately 60° ahead of or behind Jupiter. These locations are two of the Lagrange points (L4 and L5).

57. (a) According to Kepler's third law

$$T^2 = \frac{4\pi^2}{G(M_{Star} + M_{Planet})}a^3$$

$$\Rightarrow a = \left[\frac{T^2 G(M_{Star} + M_{Planet})}{4\pi^2}\right]^{1/3}$$

Assuming the masses of both planets to be much smaller than the mass of the star, we get

$$a = \left[\frac{T^2 G M_{Star}}{4\pi^2}\right]^{1/3}$$

This can be used to calculate the ratio of semi-major axes for both planet.

$$\Rightarrow \frac{a_1}{a_2} = \left(\frac{T_1}{T_2}\right)^{2/3}$$

Given: $T_1 = 10$ days and $T_1 = 25$ days

$$\Rightarrow a_1 < a_2$$

Hence, the planet with 10-day period is closer to the star.

(b) Momentum balance for the planet with the 10-day period is

$$m_1 v_1 = m_{Star} v_{Star}$$

Similarly, for the second planet we have

$$m_2 v_2 = m_{Star} v'_{Star}$$

$$\Rightarrow \frac{m_1 v_1}{m_2 v_2} = \frac{v_{Star}}{v'_{Star}}$$

$$\Rightarrow \frac{m_1}{m_2} = \frac{v_{Star} v_2}{v'_{Star} v_1} \qquad (1)$$

Let us assume that both orbits are circular.

$$\Rightarrow v_1 = \frac{2\pi a_1}{T_1} \text{ and } v_2 = \frac{2\pi a_2}{T_2}$$

Substituting these in equation (1) gives

$$\frac{m_1}{m_2} = \frac{v_{Star}}{v'_{Star}} \frac{a_2}{a_1} \frac{T_1}{T_2}$$

But we saw earlier that $\dfrac{a_1}{a_2} = \left(\dfrac{T_1}{T_2}\right)^{2/3}$

$$\Rightarrow \frac{m_1}{m_2} = \frac{v_{Star}}{v'_{Star}} \left(\frac{T_1}{T_2}\right)^{-2/3} \frac{T_1}{T_2} = \frac{v_{Star}}{v'_{Star}} \left(\frac{T_1}{T_2}\right)^{1/3}$$

$$\Rightarrow \frac{m_1}{m_2} = \frac{40}{20} \left(\frac{10}{50}\right)^{1/3} = 1.17$$

Hence, the planet with a 10-day period is slightly more massive.

59. Suppose x is the distance from the centre of the Moon where the Moon's gravitational force balances Earth's gravitational force. The force balance for an object of mass m at that distance gives

$$\frac{GmM_{Moon}}{x^2} = \frac{GmM_{Earth}}{(r-x)^2} \qquad \text{(Here } r \text{ is the distance between Earth and Moon)}$$

$$\Rightarrow x = \frac{r}{1+\sqrt{\dfrac{M_{Earth}}{M_{Moon}}}} = \frac{3.84\times10^8}{1+\sqrt{\dfrac{5.97\times10^{24}}{7.34\times10^{22}}}} = 3.83\times10^7 \text{ m}$$

This is the minimum distance the object has to travel to reach the point at which it could fall into Earth. The minimum initial speed needed to take the object to this point is calculated by comparing the potential energy per unit mass on the Moon's surface compared to that at this point.

$$\frac{U_{Moon\ surface}}{m} = -\frac{GM_{Moon}}{R_{Moon}} - \frac{GM_{Earth}}{r} = -3.86\times10^6 \text{ J/kg}$$

Here, r represents the distance from the centre of the Moon to Earth as before.

Now calculate the potential energy per unit mass at the point distance x from the centre of the Moon.

$$\frac{U_x}{m} = -\frac{GM_{Moon}}{x} - \frac{GM_{Earth}}{r-x} = -1.28\times10^6 \text{ J/kg}$$

The difference in these can be equated to the kinetic energy per unit mass:

$$\frac{v^2}{2} = 2.58 \times 10^6 \text{ J/kg}$$

$$v = 2.3 \text{ km/s}$$

61. We know that force is equal to the negative gradient of potential; that is

$$F = -\frac{dU}{dr}$$

$$\Rightarrow F = -\frac{d}{dr}\left(-\frac{kMm}{r^3}\right) = kMm\frac{d}{dr}\left(\frac{1}{r^3}\right)$$

$$\Rightarrow F = -\frac{3kMm}{r^4}$$

$$\Rightarrow |F| = \frac{3kMm}{r^4}$$

63. (a) Given $r_0 = \frac{3}{4}\left(6.955 \times 10^8\right) = 5.22 \times 10^8$ m

Since the mass of the Sun was concentrated in a shell extremely far away from its centre, its potential energy can be assumed to be zero.

$$U_1 \approx 0$$

The potential energy of the Sun after contracting is given by

$$U_2 = -\frac{GM^2}{r_0}$$

The difference in these potential energies is given by

$$\Delta U = U_1 - U_2 = \frac{GM^2}{r_0}$$

Hence, the energy available to be radiated away by the Sun after contracting is

$$E = \frac{GM^2}{r_0}$$

$$\Rightarrow E = \frac{\left(6.672 \times 10^{-11}\right)\left(1.989 \times 10^{30}\right)^2}{5.22 \times 10^8} = 5.1 \times 10^{41} \text{ J}$$

(b) $t = \frac{5.1 \times 10^{41}}{3.85 \times 10^{26}} = 1.3 \times 10^{15}$ s $= 4.2 \times 10^7$ y $= 42$ million years

65. (a) According to Kepler's third law:

$$T^2 = \frac{4\pi^2}{G(M_{Star} + M_{Planet})} a^3$$

$$\Rightarrow a = \left[\frac{T^2 G(M_{Star} + M_{Planet})}{4\pi^2} \right]^{\frac{1}{3}}$$

$$= \left[\frac{(12 \times 24 \times 3600)^2 (6.672 \times 10^{-11})(1.8 \times 10^{30} + 4.5 \times 10^{26})}{4\pi^2} \right]^{\frac{1}{3}}$$

$$\Rightarrow a = 1.5 \times 10^{10} \text{ m}$$

Assuming spherical orbit, the velocity of the planet is

$$v_{Planet} = \frac{2\pi a}{T} = \frac{2\pi (1.5 \times 10^{10})}{12 \times 24 \times 3600} = 9.0 \times 10^4 \text{ m/s}$$

Momentum balance gives

$$M_{Star} v_{Star} = M_{Planet} v_{Planet}$$

$$\Rightarrow v_{Star} = \frac{M_{Planet}}{M_{Star}} v_{Planet} = \frac{4.5 \times 10^{26}}{1.8 \times 10^{30}} 9.0 \times 10^4 = 22.5 \text{ m/s}$$

(b) The spectral shift due to velocity is given by

$$T^2 = \frac{4\pi^2}{G(M_{Star} + M_{Planet})} a^3$$

$$\Rightarrow a = \left[\frac{T^2 G(M_{Star} + M_{Planet})}{4\pi^2} \right]^{\frac{1}{3}}$$

$$= \left[\frac{(12 \times 24 \times 3600)^2 (6.672 \times 10^{-11})(1.8 \times 10^{30} + 4.5 \times 10^{26})}{4\pi^2} \right]^{\frac{1}{3}}$$

$$\Rightarrow a = 1.5 \times 10^{10} \text{ m}$$

$$\frac{\Delta\lambda}{\lambda} = \frac{v_{Star}}{c}$$

$$\Rightarrow \Delta\lambda = \frac{v_{Star}}{c} \lambda = \frac{22.5}{3.0 \times 10^8} 396.78100 = 0.00003 \text{ nm}$$

$$\Rightarrow \lambda_{obs} = (396.78100 - 0.00003) \text{ to } (396.78100 + 0.00003) \text{ nm}$$

Or $\lambda_{obs} = 396.78097$ to 396.78103 nm

67. (a) We know that the escape velocity is inversely proportional to the square root of distance. Let us assume a comet in an orbit very close to the centre of the Sun, say $r = 5 \times 10^{10}$ m. If this comet has a velocity higher than the escape velocity, it can break out of Sun's gravitational pull and enter Earth's atmosphere. The escape velocity is given by

$$v_e = \sqrt{\frac{2GM}{r}} = \sqrt{\frac{2\left(6.672\times10^{-11}\right)\left(1.989\times10^{30}\right)}{5\times10^{10}}} = 72.9\times10^3 \text{ m/s} \approx 73 \text{ km/s}$$

(b) The mean radius of Halley's comet is approximately $r = 4000$ m with a mean density of $\rho = 600 \text{ kg/m}^3$.

$$\Rightarrow M = \rho V = (600)\frac{4}{3}\pi(4000)^3 = 1.6\times10^{14} \text{ kg}$$

(c) $KE = \frac{1}{2}Mv^2 = \frac{1}{2}\left(1.6\times10^{14}\right)(73\,000)^2 = 4.27\times10^{23}$ J

This is equivalent to

$KE = \left(4.27\times10^{23}\right)/\left(4.18\times10^9\right) = 1.02\times10^{14}$ tonnes of TNT

or about 100 million Megatonnes of TNT

Reasonable answers will of course vary.

1. c. Density is inversely proportional to volume.

3. d. The net force at the bottom of the cylinder is proportional to the area of the bottom.

5. ABC. The density corresponds to the slope of the graphed line.

7. a. When the iron cube is in the boat, the volume of water displaced has the same weight as the weight of the cube. However, when the cube is dropped in the water, the volume of water displaced is equal to the volume of the cube. The first volume of water is greater than the second volume of water. Therefore, the water level will fall.

9. c. The magnitude of the buoyant force on a floating object is equal to the weight of the object. In other words, the floating object displaces an amount of water equal to its weight. When the ice cube melts, the amount of extra water in the glass is exactly the same as the amount of water displaced, so the water level remains the same.

11. b. The scale measures the weight of the container and its contents. If instead of placing the ball in the container, an amount of water with weight equal to the ball's weight were placed in the container, the scale would read the same in each case.

13. Yes, the pressure at the bottom depends on the height of the water, not on the surface area at the bottom.

15. a. The tension on the string is given by

$$T = Mg - F_B$$
$$\Rightarrow T = \rho_b V_b g - \rho_f V_b g$$

Given: $\rho_b = 4\rho_f$

$$\Rightarrow T = \rho_b V_b g - \frac{\rho_b}{4} V_b g = \frac{3}{4} \rho_b V_b g = \frac{3}{4} Mg$$

17. Number of water molecules in 1 cm³ is

$$N = \frac{6.02 \times 10^{23}}{18}(1.0) = 3.344 \times 10^{22} \text{ molecules}$$

Assume each molecule to be at the centre of a cube of length x. The volume of each of these cubes is given by

$$x^3 = \frac{1}{3.346 \times 10^{22}} = 2.99 \times 10^{-23} \text{ cm}^3$$
$$x = 3.1 \times 10^{-8} \text{ cm}$$

Since the molecules are at the centre of these cubes, the distance between the molecules is also 3.1×10^{-8} cm $= 0.31$ nm.

19. $P = \dfrac{F}{A} = \dfrac{(70)(9.81)}{0.010} = 68670 \text{ N/m}^2 = 69 \text{ kN/m}^2$

21. (a) $\rho = \dfrac{M}{\dfrac{4}{3}\pi R^3} = \dfrac{3.00 \times 10^{30}}{\dfrac{4}{3}\pi \left(1.20 \times 10^3\right)^3} = 4.14 \times 10^{20} \text{ kg/m}^3$

    (b) $W = mg = \rho V g$

    $\Rightarrow W = \left(4.14 \times 10^{20}\right)\left(1.0 \times 10^{-6}\right)(9.81) = 4.1 \times 10^{15} \text{ N}$

    It would be impossible for one person to lift such a heavy weight.

23. The volume of the solid portion of the shell is

    $V = \dfrac{4}{3}\pi r_2^3 - \dfrac{4}{3}\pi r_1^3$

    $\Rightarrow V = \dfrac{4}{3}\pi \left(r_2^3 - r_1^3\right) = \dfrac{4}{3}\pi\left(0.15^3 - 0.14^3\right) = 0.00264 \text{ m}^3$

    Therefore, the mass of the solid portion is

    $M = \rho V = (6000)(0.00264) = 15.86 \text{ kg}$

    The average density of the shell, including the hollow portion (assuming the hollow is evacuated), is given by

    $\rho_{ave} = \dfrac{M}{\dfrac{4}{3}\pi r_2^3} = \dfrac{15.86}{\dfrac{4}{3}\pi 0.15^3} = 1120 \text{ kg/m}^3$

    Since the average density is greater than the density of water, the shell will not float in water.

25. The force on the window from inside the submarine is

    $F_{in} = P_0 A = \left(1.013 \times 10^5\right)\left(\pi \times 0.2^2\right) = 12730.0 \text{ N}$

    The pressure on the window from outside the submarine is

    $P = \rho g h + P_0 = (1030)(9.81)(45) + 101300 = 5.560 \times 10^5 \text{ Pa}$

    Therefore, the force on the window from outside the submarine is

    $F_{out} = PA = \left(5.560 \times 10^5\right)\left(\pi \times 0.2^2\right) = 69870.0 \text{ N}$

    Hence, the net force on the window is

    $F_{net} = F_{out} - F_{in} = 69870.0 - 12730.0 = 57140.0 \text{ N} = 57 \text{ kN}$

27. $P = P_0 + \rho g h$

    $= (1.013 \times 10^5 \text{ Pa}) + \left(1000 \text{ kg/m}^3\right)\left(9.81 \text{ m/s}^2\right)(0.9 \text{ m} - 0.3 \text{ m}) = 107 \text{ kPa}$

29. $\dfrac{F_s}{A_s} = \dfrac{F_l}{A_l}$

$\Rightarrow F_s = F_l \dfrac{A_s}{A_l} = (2000 \times 9.81) \dfrac{\pi \times 0.0267^2}{\pi \times 0.2^2} = 350 \text{ N}$

31. a (See question 8)

33. When empty, the volume of the ship submerged in water is given by

$V_{in} = (A)(9) \text{ m}^3$

If h and A are the height and area of the base respectively, then

$\dfrac{V_{in}}{V_o} = \dfrac{\rho_{ship}}{\rho_{fl}} = \dfrac{M_{ship}}{Ah\rho_{fl}}$

But $V_o = Ah$

$\Rightarrow \dfrac{9A}{Ah} = \dfrac{M_{ship}}{Ah\rho_{fl}}$

$\Rightarrow A = \dfrac{M_{ship}}{9\rho_{fl}} = \dfrac{2 \times 10^6}{(9)(1020)} = 217.86 \text{ m}^2$

Now, with oil the ship will sink according to

$\dfrac{V_{in}}{V_o} = \dfrac{\rho_{ship+oil}}{\rho_{fl}} = \dfrac{M_{ship+oil}}{Ah\rho_{fl}}$

If d is the depth to which the ship sinks, then $V_{in} = dA$.

$\Rightarrow \dfrac{dA}{hA} = \dfrac{M_{ship+oil}}{Ah\rho_{fl}}$

$\Rightarrow d = \dfrac{M_{ship+oil}}{A\rho_{fl}} = \dfrac{2 \times 10^6 + 4 \times 10^6}{(217.86)(1020)} = 27 \text{ m}$

35. The apparent weight of wood in alcohol is given by

$W_{app} = W - F_B$

$\Rightarrow F_B = W - W_{app} = (0.48)(9.81) - 0.46 = 4.25 \text{ N}$

But $F_B = \rho_{alc}gV$

$\Rightarrow V = \dfrac{F_B}{\rho_{alc}g} = \dfrac{4.25}{(790)(9.81)} = 5.48 \times 10^{-4} \text{ m}^3$

Hence, the density of wood is given by

$\rho = \dfrac{M}{V} = \dfrac{0.48}{5.48 \times 10^{-4}} = 880 \text{ kg/m}^3$

37. The volume of the cork is

$$V = \frac{m}{\rho} = \frac{10}{0.33} = 30.3 \text{ cm}^3 = 30.3 \times 10^{-6} \text{ m}^3$$

The buoyant force on the cork is given by

$$F_B = \rho_w g V = (1000)(9.81)(30.3 \times 10^{-6}) = 0.297 \text{ N}$$

The weight of the cork is

$$W = mg = (0.01)(9.81) = 0.0981 \text{ N}$$

Hence, the tension in the string is

$$T = F_B - W = 0.297 - 0.0981 = 0.20 \text{ N}$$

The tension in the string is acting downwards (in the direction of the weight).

39. The total mass when the beaker is half full is

$$m_t = 390 + \frac{500}{2}(1.0) = 640.0 \text{ g}$$

When the beaker is half full it is neutrally buoyant, that is

$$\rho_w V_t g = m_t g$$

$$\Rightarrow V_t = \frac{m_t}{\rho_w} = 640 \text{ cm}^3$$

Therefore, the volume of the glass is

$$V_g = V_t - 500 = 140 \text{ cm}^3$$

And the density of the glass is

$$\rho_g = \frac{390}{140} = 2.79 \text{ g/cm}^3$$

41. (a) $\dfrac{V_{sub}}{V_0} = \dfrac{\rho_{boat}}{\rho_{water}} = \dfrac{700}{1020} = 0.69$

(b) For the boat not to sink, we must have

$$\frac{V_{sub}}{V_0} = \frac{\rho_{boat}}{\rho_{water}} \le 1$$

$$\Rightarrow \rho_{boat} \le \rho_{water}$$

Mass of the boat is $(700)(5.5) = 3850$ kg

If there are N survivors, then the total mass of the boat is

$$m = 3850 + 65N$$

$$\Rightarrow \frac{3850 + 65N}{5.5} \le 1020$$

$$\Rightarrow N \le 27$$

(c) Yes, some of the weight of the survivors would be offset by the buoyant force corresponding to the volume of water displaced by their legs.

43. The acceleration depends on the net force acting on the sphere, which is equal to the apparent weight of the sphere when fully submerged.

$$F = mg - F_B = mg - \rho_w gV = \rho_0 gV - \rho_w gV$$

But $F = ma = \rho_0 aV$

$$\Rightarrow \rho_0 aV = \rho_0 gV - \rho_w gV$$

$$\Rightarrow a = \frac{\rho_0 - \rho_w}{\rho_0} g = \left(\frac{2.7 - 1.03}{2.7} \right) 9.81 = 6.1 \text{ m/s}^2$$

The acceleration will decrease as the speed of the ball increases, due to drag forces from the water. Although drag is the dominant factor, the density of the water also increases slightly with depth, so the buoyant force will also increase slightly with depth, which will also act to slightly decrease the acceleration as the ball's depth increases.

45. Force balance gives

$$\rho_{amb} gV = mg + \rho_{hot} gV$$

$$\Rightarrow V = \frac{m}{\rho_{amb} - \rho_{hot}} = \frac{280}{1.28 - 0.85} = 650 \text{ m}^3$$

47. The volume of the lake in terms of area and height is given by

$$V = Ah$$

Differentiating with respect to time gives

$$\frac{dV}{dt} = A\frac{dh}{dt}$$

$$\Rightarrow \frac{dh}{dt} = \frac{1}{A}\frac{dV}{dt}$$

$$\Rightarrow \Delta h \approx \frac{1}{A}\frac{dV}{dt}\Delta t$$

$$\Rightarrow \Delta h \approx \frac{1}{1.00 \times 10^8}\left(2.5 \times 10^4 - 1.9 \times 10^4\right)\left(6 \times 3600\right) = 1.3 \text{ m}$$

49. (a) Flow rate is given by

$$Q = \frac{\Delta V}{\Delta t} = \frac{500 \times 10^{-6}}{15} = 3.3 \times 10^{-5} \text{ m}^3/\text{s}$$

(b) Since $Q = vA$

$$\Rightarrow v = \frac{Q}{A} = \frac{3.33 \times 10^{-5}}{3.0 \times 10^{-4}} = 0.11 \text{ m/s}$$

(c) $v_f^2 - v_i^2 = 2gh$

$\Rightarrow v_f = \sqrt{v_i^2 + 2gh} = \sqrt{0.11^2 + 2(9.81)(0.2)} = 2.0$ m/s

(d) $A_1 v_1 = A_2 v_2$

$\Rightarrow A_2 = \dfrac{A_1 v_1}{v_2} = \dfrac{(3.0 \times 10^{-4})(0.11)}{1.98} = 1.7 \times 10^{-5}$ m^2 = 0.17 cm^2

51. (The numbers for this problem have been changed. In the wide section, the flow speed is 2.0 m/s and the water pressure is two times the atmospheric pressure.)

(a) $A_1 v_1 = A_2 v_2$

$\Rightarrow v_2 = \dfrac{A_1 v_1}{A_2} = \dfrac{\pi r_1^2 v_1}{\pi r_2^2} = \left(\dfrac{r_1}{r_2}\right)^2 v_1$

Given: $\dfrac{d_1}{d_2} = \dfrac{r_1}{r_2} = 2$

$\Rightarrow v_2 = (2)^2 (2.0) = 8.0$ m/s

(b) Since height does not change, by Bernoulli's principle,

$P_1 + \dfrac{1}{2} \rho v_1^2 = P_2 + \dfrac{1}{2} \rho v_2^2$

$P_2 = P_1 + \dfrac{1}{2} \rho \left(v_1^2 - v_2^2\right)$

$P_2 = (2)(1.01 \times 10^5) + \dfrac{1}{2}(1000)(2.0^2 - 8.0^2)$

$= 2.02 \times 10^5 - 30.0 \times 10^3 = 172.0 \times 10^3$ kPa $= 172$ kPa

53. (a) Assuming height does not change, Bernoulli's principle gives

$P_1 + \dfrac{1}{2} \rho v_1^2 = P_2 + \dfrac{1}{2} \rho v_2^2$

Using $A_1 v_1 = A_2 v_2$ or $v_2 = \dfrac{A_1 v_1}{A_2}$ in the above equation we get

$\Rightarrow P_2 = P_1 - \dfrac{1}{2} \rho v_1^2 \left[\left(\dfrac{A_1}{A_2}\right)^2 - 1\right]$

$\Rightarrow P_2 = 13.6 \times 10^3 - \dfrac{1}{2}(1060)(0.12)^2 \left[\left(\dfrac{1}{0.2}\right)^2 - 1\right] = 13.4 \times 10^3$ Pa

(b) The percentage decrease in the blood pressure is

$\dfrac{P_2 - P_1}{P_1} \times 100 = \dfrac{13.6 \times 10^3 - 13.4 \times 10^3}{13.6 \times 10^3} \times 100 = 1.47\%$

55. (a) Given: $A_2 v_2 = 4 \times 10^{-3}$ m^3/s

$$\Rightarrow v_2 = \frac{4 \times 10^{-3}}{\pi 0.02^2} = 3.18 \text{ m/s}$$

Also, since $A_1 v_1 = A_2 v_2$

$$\Rightarrow v_1 = \frac{A_2 v_2}{A_1} = \frac{4 \times 10^{-3}}{\pi 2^2} = 3.18 \times 10^{-4} \text{ m/s}$$

According to Bernoulli's principle

$$P_1 + \rho g y_1 + \frac{1}{2} \rho v_1^2 = P_2 + \rho g y_2 + \frac{1}{2} \rho v_2^2$$

$$\Rightarrow P_1 - P_2 = \rho g (y_2 - y_1) + \frac{1}{2} \rho (v_2^2 - v_1^2)$$

$$\Rightarrow P_1 - P_2 = (1000)(9.81)(12 - 7) + \frac{1}{2}(1000)\left[3.18^2 - \left(3.18 \times 10^{-4}\right)^2\right]$$

$$\Rightarrow P_1 - P_2 = 5.41 \times 10^4 \text{ Pa}$$

$$\Rightarrow P_2 = P_1 - 5.41 \times 10^4 = 1.01 \times 10^5 - 5.41 \times 10^4 = 4.7 \times 10^4 \text{ Pa} = 47 \text{ kPa}$$

(b) For maximum height, we will assume that the water barely reaches the top. In this case, the Bernoulli's equation becomes

$$P_1 + \rho g y_1 = P_2 + \rho g y_{\max}$$

$$\Rightarrow y_{\max} - y_1 = \frac{P_1 - P_2}{\rho g} = \frac{5.41 \times 10^4}{(1000)(9.81)} = 5.5 \text{ m}$$

57. Bernoulli's equation indicates that the pressure along a flow tube increases when the speed of the fluid is reduced. So the pressure will increase.

59. Conservation of mass gives

$$A_1 v_1 = A_2 v_2$$

$$\Rightarrow v_2 = \frac{A_1}{A_2} v_1$$

According to Bernoulli's equation

$$P_0 + \rho g y_1 + \frac{1}{2} \rho v_1^2 = P_t + \rho g y_2 + \frac{1}{2} \rho v_2^2$$

$$\Rightarrow P_0 + \rho g y_1 + \frac{1}{2} \rho v_1^2 = P_t + \rho g y_2 + \frac{1}{2} \rho \left(\frac{A_1}{A_2}\right)^2 v_1^2$$

$$\Rightarrow v_1 = A_2 \sqrt{\frac{2(P_t - P_0) + 2\rho g (y_2 - y_1)}{\rho (A_2^2 - A_1^2)}}$$

61. (a) Pressure in radial direction is given by

$$p = \frac{1}{2}\rho v^2$$

Using $v = r\omega$ we get

$$p = \frac{1}{2}\rho r^2 \omega^2$$

Differentiating with respect to r gives

$$\frac{dp}{dr} = \rho r \omega^2$$

(b) $\dfrac{dp}{dr} = \rho r \omega^2$

$$\Rightarrow dp = \rho r \omega^2 \, dr$$

Integrating both sides we get

$$\int_{P_0}^{P_R} dp = \int_0^R \rho r \omega^2 \, dr$$

$$\Rightarrow P_R - P_0 = \rho \omega^2 \frac{R^2}{2}$$

$$\Rightarrow P_R = P_0 + \frac{1}{2}\rho \omega^2 R^2$$

(c) Consider a particle of mass m in the rotating frame of reference of the fluid. In this frame, the fluid is in static equilibrium. For simplicity, we will look at the particle in xy frame only and then extend the result to the third dimension. The centripetal force acting on the particle is in x-direction and is given by

$$F_c = m\omega^2 x$$

The gravitational force is in y-direction and is given by

$$F_g = mg$$

The angle the resultant force makes with horizontal is given by

$$\tan\theta = \frac{F_c}{F_g} = \frac{\omega^2 x}{g}$$

But $\tan\theta = \dfrac{dy}{dx}$

$$\Rightarrow dy = \frac{\omega^2 x}{g} \, dx$$

Integrating both sides we get

$$y = \frac{\omega^2 x^2}{2g} + C$$

This is the equation of a parabola. In three dimensions, this will form a paraboloid.

63. Given: $A = 10^{-4} \text{ m}^2$

$$\Rightarrow R^4 = \left(\frac{A}{\pi}\right)^2 = \left(\frac{10^{-4}}{\pi}\right)^2 = 1.013 \times 10^{-9} \text{ m}^2$$

The pressure difference is given by

$$\Delta P = \frac{8 \mu L Q}{\pi R^4} = \frac{8(10^{-3})(5)(5 \times 10^{-4})}{\pi(1.013 \times 10^{-9})} = 6.28 \times 10^3 \text{ Pa} = 6.3 \text{ kPa}$$

65. (a) $P_{dynamic} = P_{static} + \frac{\rho_{air} v^2}{2}$

$$\Rightarrow v = \sqrt{\frac{2(P_{dynamic} - P_{static})}{\rho_{air}}}$$

But $P_{dynamic} - P_{static} = \rho_f g h$

$$\Rightarrow v = \sqrt{\frac{2\rho_f g h}{\rho_{air}}}$$

(b) $P_{dynamic} = P_{static} + \frac{\rho_{air} v^2}{2}$

$$\Rightarrow v = \sqrt{\frac{2(P_{dynamic} - P_{static})}{\rho_{air}}}$$

But $P_{dynamic} - P_{static} = \Delta P$

$$\Rightarrow v = \sqrt{\frac{2(\Delta P)}{\rho_{air}}}$$

67. Pressure due to flow is given by

$$P = \frac{1}{2}\rho v^2$$

Since $v = \frac{Q}{A}$, where Q is the flow rate through area A, therefore

$$P = \frac{1}{2}\rho\left(\frac{Q}{A}\right)^2 = \frac{1}{2}(1000)\left(\frac{2.0 \times 10^{-3}}{\pi 0.01^2}\right)^2 = 20264.2 \text{ Pa}$$

Since $F = PA$

$$\Rightarrow F = (20264.2)(\pi 0.01^2) = 6.37 \text{ N}$$

Total torque due to water flowing out of both ends is given by

$$\tau = 2rF = (2)(0.15)(6.37) = 1.9 \text{ N} \cdot \text{m}$$

69. b. As the elevator accelerates up, the effective downward force on the block increases to $m(g + a)$ and therefore the block sinks deeper into water.

Chapter 13—OSCILLATIONS

1. The total distance travelled in one cycle by a simple harmonic oscillator is four times the amplitude, or $4A$. The oscillator reaches maximum positive amplitude of A and maximum negative amplitude of $-A$. Since the oscillator returns to the same position after one cycle, the net displacement after one cycle is zero.

3. For a mass–spring system, the period T is independent of the amplitude A because the period depends only on the angular frequency of the motion. This is evident by examining the equation of motion:

$$x(t) = A\cos(\omega t + \phi)$$

The period of a sinusoidal function is

$$T = \frac{2\pi}{\omega}$$

$$T = 2\pi\sqrt{\frac{m}{k}}$$

which is independent of the amplitude of oscillations.

5. If the mass of the spring in a mass–spring oscillator were not ignored, the period would be greater, because

$$T = 2\pi\sqrt{\frac{m}{k}}$$

If more mass is oscillating, the formula for period indicates a greater period.

7. The period of the pendulum is greater if the mass of the string is not negligible, because then the pendulum is no longer a "simple" pendulum. The formula for the period of a physical pendulum includes the mass of the pendulum, and greater masses mean greater periods.

9. a. When the elevator accelerates upward the effective value of g in the elevator increases. Since $T = 2\pi\sqrt{\dfrac{L}{g}}$ for a pendulum, the period will decrease.

11. The period of the swing decreases when the child stands up. If we approximate the standing child as a rod of length L, then the new period is $T = 2\pi\sqrt{\dfrac{\frac{1}{3}ML^2}{Mg\frac{L}{2}}} = 2\pi\sqrt{\dfrac{2}{3}}\sqrt{\dfrac{L}{g}}$, which is less than the period when the child sits and can be approximated as a point mass.

13. c. The period of a physical pendulum in air is greater than in vacuum because air resistance slows the pendulum.

15. All of (a), (b), and (c) will not result in simple harmonic motion. For simple harmonic motion, the force must be restoring and a linear function of position.

17. d. At the maximum amplitude of oscillation, the total energy E_1 is equal to the potential energy U, $E_1 = \frac{1}{2}kA^2$. If the amplitude is doubled without "kicking" the oscillator, then the new energy is

$$E_2 = \frac{1}{2}k(2A)^2$$

$$E_2 = 4 \cdot \left[\frac{1}{2}kA^2\right]$$

$$E_2 = 4 \cdot E_1$$

If the amplitude is doubled, the total energy increases by a factor of 4.

19. Given quantities are $m = 2.0$ kg, $k = 5.0$ N/m, $x(0) = 0.3$ m, $v(0) = 1.0$ m/s

 (a) The total energy of the mass–spring system is

$$E = \frac{1}{2}mv^2 + \frac{1}{2}kx^2 = \frac{1}{2} \times 2.0 \times (1.0)^2 + \frac{1}{2} \times 5.0 \times (0.3)^2 = 1.225 \text{ J}$$

Since, $E = \frac{1}{2}kA^2$, therefore

$$A = \sqrt{\frac{2E}{k}} = \sqrt{\frac{2 \times 1.225}{5.0}} = 0.7 \text{ m}$$

 (b) At the maximum speed the kinetic energy is equal to the total energy of the mass–spring system. Therefore,

$$v_{max} = \sqrt{\frac{2E}{m}} = 1.1 \text{ m/s}$$

 (c) Since $\omega = \sqrt{\frac{k}{m}} = \sqrt{\frac{5.0}{2.0}} = 1.6 \text{ rad/s}$

$$a_{max} = \omega^2 A = 2.5 \text{ (rad/s)}^2 \times 0.7 \text{ m} = 1.7 \text{ m/s}^2$$

 (d) Knowing the position and velocity at $t = 0$, we can determine the phase constant (see Example 13-3, equation 5):

$$\phi = \tan^{-1}\left(-\frac{v(0)}{\omega x(0)}\right) = \tan^{-1}(-2.1) = -1.1 \text{ rad}$$

Therefore, displacement as a function of time for the oscillator is given by

$$x(t) = (0.7 \text{ m}) \cos(1.6t - 1.1)$$

21. Because the mass is pulled to the left and then released from rest, we can take the phase angle to be π radians. Therefore, we can write its displacement from the equilibrium position as a function of time as

$$x(t) = A\cos(\omega t + \pi) - A\cos(\omega t)$$

where A is the amplitude of oscillations. Differentiating with respect to time, we obtain

$$v(t) = \omega A \sin(\omega t)$$

(a) Substituting the given information into the position and velocity equations, we obtain

$$0.10 = -A\cos(0.3\omega)$$
$$0.857 = \omega A \sin(0.3\omega)$$

Dividing the second equation by the first, we obtain

$$8.57 \cos(0.3\omega) = -\omega \sin(0.3\omega)$$

By plotting the two sides of the equation on the same graph as a function of ω, we can look for the solutions where the curves intersect. The lowest value of ω at which the two curves intersect is

$$\omega = 7.7 \text{ rad/s}$$

and therefore the frequency is

$$f = \frac{\omega}{2\pi} = 1.2 \text{ s}^{-1}$$

(b) $x(t) = -A\cos(0.3\omega)$

$$0.10 = -A\cos(0.3(7.7))$$
$$A = -\frac{0.10}{\cos(2.3)}$$
$$A = -0.15 \text{ m}$$
$$A = 15 \text{ cm}$$

(c) The phase constant is π radians.

(d) $x(t) = -(0.15 \text{ m})\cos(7.7t + \pi) = -(0.15 \text{ m})\cos(7.7t)$

23. Because the period is 2.0 s, the angular frequency is

$$\omega = \frac{2\pi}{T} = \frac{2\pi}{2} = \pi$$

(a) An equation for the position of the oscillator is of the form

$$x = A\cos(\pi t + \phi)$$

Because the position at $t = 0$ is $x = 0$, then

$$0 = A\cos(\phi)$$

and therefore $\phi = \dfrac{\pi}{2}$. (You can obtain an equivalent equation for the position by choosing other values of ϕ.) Thus, an equation for the position of the oscillator is

$$x = A\cos\left(\pi t + \frac{\pi}{2}\right)$$

To determine the value of A, substitute $x = -0.4$ and $t = 0.5$ into the position function to obtain

$$-0.4 = A\cos\left(0.5\pi + \frac{\pi}{2}\right)$$

$$-0.4 = A\cos(\pi) = -A$$

$$A = 0.4$$

Thus, an equation for the position of the oscillator is

$$x = 0.4\cos\left(\pi t + \frac{\pi}{2}\right)$$

(b) Differentiating the position function in part (a), we obtain

$$v = -0.4\pi\sin\left(\pi t + \frac{\pi}{2}\right)$$

and

$$a = -0.4\pi^2\cos\left(\pi t + \frac{\pi}{2}\right)$$

Thus, the maximum velocity is $0.4\pi = 1.2$ m/s and the maximum acceleration is $0.4\pi^2 = 3.9$ m/s^2.

(c) The total energy of the oscillator is equal to the maximum kinetic energy, which is

$$E = \frac{1}{2}mv_{max}^2 = \frac{1}{2}(0.1\text{ kg})(0.4\pi\text{ m/s})^2 = 0.008\pi^2\text{ J} = 0.08\text{ J}$$

25. No. Applying the given information to a position function $x(t) = A\cos(\omega t + \phi)$, we get

$$x(0) = A\cos(\phi) \qquad (1)$$

and

$$x\left(\frac{T}{2}\right) = A\cos\left(\omega\frac{T}{2} + \phi\right)$$

$$x\left(\frac{T}{2}\right) = A\cos\left(\frac{2\pi}{T} \times \frac{T}{2} + \phi\right)$$

$$x\left(\frac{T}{2}\right) = A\cos(\pi + \phi)$$

$$x\left(\frac{T}{2}\right) = -A\cos(\phi) \qquad (2)$$

Equations (1) and (2) are not independent, so there is no way to use only these two equations to solve for the values A and ϕ, given the values of $x(0)$ and $x(T/2)$.

27. (a) The lower the angular frequency (the factor of t), the higher the period; therefore, the periods, from high to low, are $3 > 2 > 4 > 1 > 5$.

 (b) The maximum speed is ωA , so the maximum speeds from high to low are $5 > 1 > 4 > 2 > 3$.

 (c) The phase angles, from high to low, are $2 > 4 > 1 > 3 = 5$.

 (d) The maximum acceleration is $\omega^2 A$, so the maximum accelerations from high to low are $5 > 1 > 4 > 2 > 3$.

29. Considering only the rotational motion of Earth, the period of a point on the equator is 1 day. The angular speed is 1 revolution per day, which is

$$\frac{2\pi \text{ rad}}{\text{day}} = \frac{2\pi \text{ rad}}{24 \times 3600 \text{ s}} = \frac{\pi}{43200} = 7.27 \times 10^{-5} \text{ rad/s}$$

$$T = 86\ 400 \text{ s}$$

31. Because the period is

$$T = 2\pi\sqrt{\frac{m}{k}}$$

the rank of the periods, from small to large, is $D < A < B < E < F < C$

33. (a) $\omega = \sqrt{\frac{k}{m}} = \sqrt{\frac{100}{0.1}} = \sqrt{1000} = 32 \text{ rad/s}$

 $T = \frac{2\pi}{\omega} = \frac{2\pi}{\sqrt{1000}} = 0.20 \text{ s}$

(b) $v_{max} = \omega A = \sqrt{1000} \times 0.20 = 6.3$ m/s

$a_{max} = \omega^2 A = \left(\sqrt{1000}\right)^2 (0.20) = 200$ m/s^2

(c) $E = \dfrac{1}{2}kA^2 = \dfrac{1}{2}(100)(0.20)^2 = 2.0$ J

35. (a) $k = \dfrac{F}{x}$

$= \dfrac{10.0 \text{ N}}{5.0 \text{ cm}}$

$= \dfrac{10.0 \text{ N}}{5.0 \times 10^{-2} \text{ m}}$

$= 200$ N/m

(b) $f = \dfrac{1}{2\pi}\sqrt{\dfrac{k}{m}}$

$= \dfrac{1}{2\pi}\sqrt{\dfrac{200}{2.0}}$

$= 1.6$ s^{-1}

(c) From the result of part (b), the angular frequency is

$$\omega = 2\pi f = 2\pi \times \dfrac{10}{2\pi} = 10 \text{ rad/s}$$

The amplitude is given as 5.0 cm, which is equivalent to 0.050 m. Thus, a formula for the position function is

$$x(t) = 0.050\cos(10t + \phi)$$

There are a number of reasonable choices for the phase angle. The maximum displacement occurs at $t = 0.050$ m, and at this time the argument of the cosine function is either 0 or an integer multiple of 2π; taking an argument of 2π, we can determine one of the reasonable values for the phase angle:

$10t + \phi = 2\pi$

$10(0.50) + \phi = 2\pi$

$\phi = 2\pi - 5$

$\phi = 1.28$

Thus, a reasonable formula for the position function is

$$x(t) = 0.050\cos(10t + 1.28)$$

(d) The period of the oscillator is

$$T = \frac{1}{f}$$

$$= \frac{\pi}{5}$$

$$= 0.6283 \text{ s}$$

In the time interval from 0.5 s to 1.76 s, the oscillator makes two complete cycles, travelling a total of 40 cm. In the remainder of the interval, from 1.76 s to 2 s, the displacement is

$$\left| x(2) - x(1.76) \right| = \left| 0.05 \cos\left(10(2) + 1.28\right) - 0.05 \cos\left(10(1.76) + 1.28\right) \right|$$

$$= 0.088 \text{ m}$$

Thus, the total distance travelled is 0.40 m + 0.088 m = 0.49 m.

37. (a) By symmetry, the block remains in contact with the spring for one half of its cycle of oscillation. The period of oscillation is

$$T = 2\pi\sqrt{\frac{m}{k}}$$

$$= 2\pi\sqrt{\frac{0.5}{50}}$$

$$= \frac{\pi}{5}$$

Thus, the block remains in contact with the spring for $\pi/10 = 0.31$ s.

(b) The initial speed of the block played no role in the calculation of part (a), so changes in the initial speed of the block do not change the time of contact. (The amplitude of the oscillation depends on the initial speed of the block, but not the period of the oscillation.)

39. (a) Each spring supports one-fourth of the car's mass, which is 300 kg. The spring constant of each spring is

$$k = m\omega^2$$

$$k = 300\left(2\pi \times 2.0\right)^2$$

$$k = 47 \text{ kN/m}$$

(b) Each passenger is supported by one spring, so each spring now supports 370 kg. Thus, the oscillation frequency of each spring is

$$f = \frac{1}{2\pi}\sqrt{\frac{k}{m}}$$

$$f = \frac{1}{2\pi}\sqrt{\frac{47374}{370}}$$

$$f = 1.8 \text{ s}^{-1}$$

It makes sense that the frequency is smaller, because more mass has been added to the system.

41. (a) Consider the point joining the two springs. The right spring exerts a force F_2 on the left spring, and the left spring exerts a force F_1 on the right spring. By Newton's third law, $F_1 = -F_2$.

Thus $F_2 = -F_1$

$$k_2 x_2 = k_1 x_1$$

$$x = x_1 + x_2 = x_1 + \frac{k_1}{k_2} x_1 = \left(\frac{k_1 + k_2}{k_2}\right) x_1$$

$$x = \left(\frac{k_1 + k_2}{k_1}\right) x_2$$

$$F = -k_2 x_2$$

$$F = -\frac{k_1 k_2}{k_1 + k_2} x$$

$$F = -\left(\frac{k_1 k_2}{k_1 + k_2}\right) x$$

This is indeed simple harmonic motion with an effective spring constant of $\left(\frac{k_1 k_2}{k_1 + k_2}\right)$

(b) $\omega_1^2 = \frac{k_1}{m}$ $\qquad \omega_2^2 = \frac{k_2}{m}$

$$\omega^2 = \frac{1}{m}\left(\frac{k_1 k_2}{k_1 + k_2}\right) = \frac{\omega_1^2 \omega_2^2}{\omega_1^2 + \omega_2^2} \Rightarrow f^2 = \frac{f_1^2 f_2^2}{f_1^2 + f_2^2}$$

(c) $U_1 = \frac{1}{2} k_1 x_1^2$ $\qquad U_2 = \frac{1}{2} k_2 x_2^2$ $\qquad \Rightarrow k_1 U_1 = k_2 U_2$

(d) No; because $k_1 x_1 = k_2 x_2$, if one of the springs is in its equilibrium position then the other one is also in its equilibrium position.

43. $T = 2\pi\sqrt{\dfrac{m}{k}} \quad \Rightarrow \quad \dfrac{4\pi^2 m}{T^2} = k \quad \Rightarrow \quad \dfrac{m}{k} = \dfrac{T^2}{4\pi^2}$

$k\Delta x = mg \Rightarrow \Delta x = \left(\dfrac{m}{k}\right)g$

$\Delta x = \dfrac{gT^2}{4\pi^2}$

$\Delta x = \dfrac{(9.81)(0.5)^2}{4\pi^2}$

$\Delta x = 0.062$ m

$\Delta x = 6.2$ cm

45. (a) $T = 2\pi\sqrt{\dfrac{m}{k}} \rightarrow k = \dfrac{4\pi^2 m}{T^2} = \dfrac{4\pi^2 (1.0)}{(2.0)^2} = 9.9$ N/m

(b) $\dfrac{1}{2}kA^2 = \dfrac{1}{2}(9.9)(0.05)^2 = 0.012$ J

(c) $U = \dfrac{1}{2}k(0.03)^2 \Rightarrow K = \dfrac{1}{2}kA^2 - \dfrac{1}{2}k(0.03)^2$

$\dfrac{K}{E} = \dfrac{\dfrac{1}{2}kA^2 - \dfrac{1}{2}k(0.03)^2}{\dfrac{1}{2}kA^2} = 1 - \dfrac{(0.03)^2}{(0.05)^2} = 1 - \left(\dfrac{3}{5}\right)^2 = 1 - \dfrac{9}{25} = 64\%$

(d) $x = 0.050\cos(\pi t)$ because $\omega = \sqrt{\dfrac{k}{m}} = \dfrac{2\pi}{T} = \dfrac{2\pi}{2.0} = \pi$

47. (a) $U = \dfrac{1}{2}kx^2 = \dfrac{1}{2}(500)(0.50)^2 = 63$ J

(b) $\dfrac{1}{2}mv^2 = \dfrac{1}{2}kx^2$

$v = \sqrt{\dfrac{62.5 \text{ J}}{\dfrac{1}{2}m}} = \sqrt{\dfrac{125}{0.030}} = 65$ m/s

49. $U + K = E$ and $U = K$ so $K = \dfrac{1}{2}E$

$$\dfrac{1}{2}mv^2 = \dfrac{1}{2}\left(\dfrac{1}{2}kA^2\right)$$

$$v^2 = \dfrac{1}{2}\dfrac{k}{m}A^2$$

$$v = \left(\dfrac{1}{\sqrt{2}}\sqrt{\dfrac{k}{m}}\right)A$$

$$v = \dfrac{0.10}{\sqrt{2}}\cdot\sqrt{\dfrac{k}{m}}$$

$$v = \dfrac{1}{10\sqrt{2}}\cdot\sqrt{\dfrac{k}{m}}$$

$$v = \dfrac{1}{14.1}\cdot\sqrt{\dfrac{k}{m}}$$

51. $T = 2\pi\sqrt{\dfrac{L}{g}}$

Thus, the periods are in the same rank order as the lengths: A, D, B, C, E, F

53. (a) $T = 2\pi\sqrt{\dfrac{L}{g}}\cdot 100 = 200\pi\sqrt{\dfrac{1.000}{9.81}} = 200.6$ s

(b) $T = 2\pi\sqrt{\dfrac{1.000}{.9972g}}\cdot 100 = 200\pi\sqrt{\dfrac{1.000}{(0.9972)(9.81)}} = 200.9$ s

Yes, the true difference is 0.3 s and can be measured with a stop watch accurate to 0.1 s.

55. (a) $T = 2\pi\sqrt{\dfrac{L}{g}} = \dfrac{2\pi}{\sqrt{g}}L^{1/2}$

$$\dfrac{dT}{dL} = \dfrac{2\pi}{\sqrt{g}}\cdot\dfrac{1}{2}L^{-1/2}$$

$$dT = \dfrac{2\pi}{\sqrt{g}}\dfrac{1}{\sqrt{L}}\cdot\dfrac{1}{2}dL$$

$$\dfrac{dT}{T} = \dfrac{2\pi g^{-1/2}L^{-1/2}2^{-1}dL}{2\pi L^{1/2}g^{-1/2}}$$

$$\frac{dT}{T} = \frac{dL}{2L}$$

$$dT = \left(\frac{T}{2L}\right) dL$$

(b) $\dfrac{dT}{T} = \dfrac{dL}{2L} \Rightarrow \dfrac{dL}{L} = 2\dfrac{dT}{T}$

$\dfrac{dT}{T} = +\dfrac{5 \text{ s}}{1 \text{ h}} = +\dfrac{5 \text{ s}}{3600 \text{ s}} = +\dfrac{5}{3600}$

The positive sign is because we wish to increase the period.

$\dfrac{dL}{L} = 2\dfrac{dT}{T} = +\dfrac{10}{3600} = +\dfrac{1}{360} = 0.28\%$

Thus, the length of the pendulum must be increased by 0.28 percent.

57. $I_p = \dfrac{2}{5}mR^2 + mL^2$

$T = 2\pi\sqrt{\dfrac{I_p}{mgL}}$

$T = 2\pi\sqrt{\dfrac{\dfrac{2}{5}mR^2 + mL^2}{mgL}}$

$T = 2\pi\sqrt{\dfrac{2R^2 + 5L^2}{5gL}}$

$T = 2\pi\sqrt{\dfrac{L}{g} + \dfrac{2R^2}{5gL}}$

$T = 2\pi\sqrt{\dfrac{L}{g}\left[1 + \dfrac{2R^2}{5L^2}\right]}$

$T = 2\pi\sqrt{\dfrac{L}{g}}\sqrt{1 + \dfrac{2R^2}{5L^2}}$

$T = T_\circ\sqrt{1 + \dfrac{2R^2}{5L^2}}$

59. (a) Reading from the graph, the amplitude is 0.6 m and the period is 3 s.

(b) If we use a cosine function to model the position function, then the phase constant is

$$\frac{0.75 \text{ s}}{3 \text{ s}} \times 2\pi = \frac{\pi}{2}$$

(c) The phase at $t = 1$ s is

$$\frac{2\pi}{3}(1) + \frac{\pi}{2} = \frac{7\pi}{6}$$

The phase at $t = 2$ s is

$$\frac{2\pi}{3}(2) + \frac{\pi}{2} = \frac{11\pi}{6}$$

(d) $x(t) = 0.6\cos\left(\frac{2\pi}{3}t + \frac{\pi}{2}\right)$

(e) Differentiating the formula for the position function, we obtain

$$v(t) = -0.4\pi\sin\left(\frac{2\pi}{3}t + \frac{\pi}{2}\right)$$

Thus,

$$v(1) = -0.4\pi\sin\left(\frac{2\pi}{3} + \frac{\pi}{2}\right)$$

$$v(1) = -0.4\pi\sin\left(\frac{7\pi}{6}\right)$$

$$v(1) = 0.63 \text{ m/s}$$

(f) Differentiating the formula for the velocity function, we obtain

$$a(t) = -\frac{0.8\pi^2}{3}\cos\left(\frac{2\pi}{3}(2) + \frac{\pi}{2}\right)$$

Thus,

$$a(2) = -\frac{0.8\pi^2}{3}\cos\left(\frac{4\pi}{3} + \frac{\pi}{2}\right)$$

$$a(2) = -\frac{0.8\pi^2}{3}\cos\left(\frac{11\pi}{6}\right)$$

$$a(2) = -2.3 \text{ m/s}^2$$

61. (a) Reading from the graph, the amplitude of the oscillation is 0.2 m, and the period is 3 s.

(b) The equilibrium position is 1 m.

(c) Reading from the graph, the phase constant is $\frac{0.5 \text{ s}}{3 \text{ s}} \times 2\pi = \frac{\pi}{3}$.

(d) $x(t) = A\cos(\omega t + \phi) + 1$

$$x(t) = 0.2\cos\left(\frac{2\pi}{3}t + \frac{\pi}{3}\right) + 1$$

(e) $x(t) = 0.2\cos\left(\frac{2\pi}{3}t + \frac{\pi}{3} \pm \frac{\pi}{2}\right) + 1$

(f) $x(t) = 0.2\cos\left(\frac{2\pi}{3}t + \frac{\pi}{3}\right) + 1$

$$v(t) = -\frac{0.4\pi}{3}\sin\left(\frac{2\pi}{3}t + \frac{\pi}{3}\right)$$

$$a(t) = -\frac{0.8\pi^2}{9}\cos\left(\frac{2\pi}{3}t + \frac{\pi}{3}\right)$$

$$a(1) = -\frac{0.8\pi^2}{9}\cos\left(\frac{2\pi}{3}(1) + \frac{\pi}{3}\right)$$

$$a(1) = -\frac{0.8\pi^2}{9}\cos(\pi)$$

$$a(1) = -\frac{0.8\pi^2}{9}(-1)$$

$$a(1) = 0.88 \text{ m/s}^2$$

63. (a) 3, 1, 2

(b) Because the springs are identical, the total energies are in the same rank order as the amplitudes; therefore, the order of total energies is 1, 3, 2.

(c) 3, 2, 1

(d) 1, 3, 2

65. (a) $e^{-bt/m} = 0.95$

$$-\frac{bT}{m} = \ln(0.95)$$

$$b = -\frac{m}{T}\ln(0.95)$$

$$b = -\frac{0.25}{2.0}\ln(0.95)$$

$$b = 0.0064 \text{ kg/s}$$

(b) $e^{-bt/2m} = \dfrac{0.20 \text{ m}}{0.30 \text{ m}}$

$e^{-bt/2m} = \dfrac{2}{3}$

$e^{bt/2m} = \dfrac{3}{2}$

$\dfrac{bt}{2m} = \ln\left(\dfrac{3}{2}\right)$

$t = \dfrac{2m}{b}\ln(1.5)$

$\dfrac{t}{T} = \dfrac{2m}{bT}\ln(1.5)$

$\dfrac{t}{T} = \dfrac{2(0.25)}{(0.0064)(2)}\ln(1.5)$

$\dfrac{t}{T} = 15.8$

(c) No, because the amplitude decreases according to a decreasing exponential function, not a linear function.

67. For A: $10e^{-50(0.01)} = 10e^{-0.5} = 6.1$ cm

For B: $20e^{-50(0.02)} = 20e^{-1} = 7.4$ cm

For C: $30e^{-50(0.03)} = 30e^{-1.5} = 6.7$ cm

A C B

69. Let $x(t) = A\cos(\omega t + \phi)$

$x(0) = A\cos(\phi) \Rightarrow A = \dfrac{x(0)}{\cos(\phi)}$

$x\left(\dfrac{T}{4}\right) = A\cos\left(\dfrac{\omega T}{4} + \phi\right)$

$x\left(\dfrac{T}{4}\right) = A\cos\left(\dfrac{\pi}{2} + \phi\right)$

$x\left(\dfrac{T}{4}\right) = A\left[\cos\left(\dfrac{\pi}{2}\right)\cos(\phi) - \sin\left(\dfrac{\pi}{2}\right)\sin(\phi)\right]$

$x\left(\dfrac{T}{4}\right) = -A\sin(\varphi)$

$$\left[x(0)\right]^2 + \left[x\left(\frac{T}{4}\right)\right]^2 = A^2 \cos^2(\phi) + (-A)^2 \sin^2(\phi)$$

$$\left[x(0)\right]^2 + \left[x\left(\frac{T}{4}\right)\right]^2 = A^2 \left[\cos^2(\phi) + \sin^2(\phi)\right]$$

Thus, $A = \sqrt{\left[x(0)\right]^2 + \left[x\left(\frac{T}{4}\right)\right]^2}$

Also, $\dfrac{x(T/4)}{x(0)} = \dfrac{-A\sin(\phi)}{A\cos(\phi)} = -\tan(\phi)$

Thus, $\varphi = \tan^{-1}\left[-\dfrac{x(T/4)}{x(0)}\right]$

71. $F = -\dfrac{GM_x m}{x^2}$, where $M_x = \rho V_x = \rho \cdot \dfrac{4}{3}\pi x^3$, $\rho = \dfrac{M}{V} = \dfrac{M}{\dfrac{4}{3}\pi R^3}$

$$M_x = \rho V_x = \dfrac{M}{\dfrac{4}{3}\pi R^3} \cdot \dfrac{4}{3}\pi x^3 = \left(\dfrac{M}{R^3}\right)x^3$$

$$F = -\dfrac{GM_x m}{x^2}$$

$$F = -\dfrac{Gm}{x^2} \cdot \dfrac{M}{R^3} x^3$$

$$F = -\left(\dfrac{GMm}{R^3}\right)x$$

This is SHM, with the effective spring constant being

$k = \dfrac{GMm}{R^3}$, where $M =$ mass of the earth and $R =$ radius of Earth.

$$\omega^2 = \dfrac{k}{m} = \dfrac{GM}{R^3} \rightarrow T = \dfrac{2\pi}{\omega} = 2\pi\sqrt{\dfrac{R^3}{GM}}$$

The time to reach from one pole to the other (one half cycle) is

$\dfrac{T}{2} = \pi\sqrt{\dfrac{R^3}{GM}} = 42.2$ min, and is independent of m.

73. Let ρ be the density of the liquid. When the object is floating in equilibrium, the weight of the liquid displaced by the object is equal to the weight of the object. Therefore,

$m_{\text{object}}g = V_0 \rho g$

Suppose the object is now further pushed into the liquid by a depth x. Then the additional buoyant force exerted on the object is equal to the weight of the additional liquid displaced.

$$F_B = (Ax)\rho g = (\rho A g) x$$

Notice the magnitude of the additional buoyant force is proportional to x and the direction of the force is such so as to push the object back to the equilibrium floating position. The object will therefore execute a simple harmonic motion with an effective spring constant $\rho A g$

The period of oscillation is

$$T = 2\pi \sqrt{\frac{m_{object}}{\rho A g}} = 2\pi \sqrt{\frac{\rho V_0}{\rho A g}} = 2\pi \sqrt{\frac{V_0}{A g}}$$

75. (a) Use conservation of momentum to determine the initial speed u of the block.

$$mv = -mv' + Mu \qquad (1)$$

Assuming mechanical energy is also conserved, we have

$$\frac{1}{2} m v^2 = \frac{1}{2} m (v')^2 + \frac{1}{2} M u^2 \qquad (2)$$

Solving equation (1) for v' and substituting into equation (2), we obtain

$$v' = \frac{Mu - mv}{m} \rightarrow \frac{1}{2} m v^2 = \frac{1}{2} m \left(\frac{Mu - mv}{m} \right)^2 + \frac{1}{2} M u^2$$

$$mv^2 = \frac{1}{m} \left(M^2 u^2 - 2mMuv + m^2 v^2 \right) + Mu^2$$

$$m^2 v^2 = M^2 u^2 - 2mMuv + m^2 v^2 + mMu^2$$

$$0 = Mu \left[Mu - 2mv + mu \right]$$

$$0 = u \left[(m + M)u - 2mv \right]$$

Thus, either $u = 0$ (the bullet misses the block) or

$$u = \left(\frac{2m}{m + M} \right) v$$

(b) The maximum compression of the spring (which is equal to the amplitude of the oscillation) occurs when all of the initial kinetic energy of the block is converted to elastic potential energy of the spring.

$$\frac{1}{2} M u^2 = \frac{1}{2} k A^2$$

$$A^2 = \frac{M}{k} u^2$$

$$A = u \sqrt{\frac{M}{k}}$$

$$A = \frac{2mv}{m + M} \sqrt{\frac{M}{k}}$$

Chapter 14—WAVES

1. (a) False. For a transverse wave, the particles of the medium vibrate perpendicular to the direction of the wave. For a longitudinal wave, the particles move back and forth in the direction of the wave.

    (b) False. The wave speed of a mechanical wave in a medium depends on the elastic and inertial properties of the medium.

    (c) True. The speed of a wave in a medium can vary with wavelength; this is what causes the dispersion of sunlight into rainbow colours upon refraction in a prism. However, if we ignore dispersion as we have done in this chapter, then the wave speed is independent of the wavelength of the waves.

    (d) False. Mechanical waves depend on the movement of particles of the medium through which they pass.

    (e) False. Since the speed of the wave depends on the properties of the medium, the speed will be constant in a uniform medium.

    (f) False. A traveling wave always carries energy.

    (g) False. The maximum amplitude cannot be greater than the sum of the amplitudes of individual waves.

    (h) False. To produce a standing wave, the constituent waves must be traveling in opposite directions.

3. The light wave has the longer wavelength in this case, although it would be more accurate to call it an electromagnetic wave, because at this wavelength it is in the radio wave part of the electromagnetic spectrum.

$$v_s = \lambda_s f_s$$

$$\lambda_s = \frac{v_s}{f_s} = \frac{340 \text{ m/s}}{20\,000 \text{ s}^{-1}} = 1.70 \times 10^{-2} \text{ m}$$

$$v_l = \lambda_l f_l$$

$$\lambda_l = \frac{v_l}{f_l} = \frac{3 \times 10^8 \text{ m/s}}{20\,000 \text{ s}^{-1}} = 15 \times 10^3 \text{ m}$$

5. No. The speed of individual particles in the medium is not the same as the wave speed through the medium.

7. The wave propagates outward from the point where the stone hits the water surface. As the wave front expands the energy per unit length decreases, and therefore the amplitude decreases.

9. The answer is (c) for a mechanical wave, as is shown below (see Section 14-8):

$$\Delta E = \Delta K + \Delta U$$

$$\Delta E_1 = (\mu \Delta x)\omega_1^2 A^2 \cos^2 (kx - \omega_1 t)$$

$$\Delta E_2 = (\mu \Delta x)\omega_2^2 A^2 \cos^2 (kx - \omega_2 t)$$

$$\Delta E_2 = (\mu \Delta x)(2\omega_1)^2 A^2 \cos^2 (kx - 2\omega_1 t)$$

$$\Delta E_2 = 4(\mu \Delta x)\omega_1^2 A^2 \cos^2 (kx - 2\omega_1 t)$$

$$\Delta E_2 = 4\Delta E_1$$

In words, the total energy of a mechanical wave is proportional to the square of the frequency, so if the frequency is doubled the energy increases by a factor of 4.

11. The two waves must have same frequency and amplitude to produce a standing wave. If the two waves have the same amplitude but different frequencies, interference will occur but standing waves will not be produced. If the two waves have the same frequency but different amplitudes, a pattern similar to a standing wave will be produced but without any nodes.

13. No, a difference in linear mass density could be offset by a difference in tension.

15. (a) $k = \dfrac{2\pi}{\lambda} = \dfrac{2\pi}{2.00 \text{ m}} = \pi \text{ m}^{-1}$

 (b) $\omega = 2\pi f = 2\pi(10 \text{ Hz}) = 63 \text{ rad/s}$

 (c) $v = \lambda f = 2.00\text{m}(10 \text{ s}^{-1}) = 20 \text{ m/s}$

17. (a) $\omega = kv = 2\pi f$

$$f = \frac{kv}{2\pi} = \frac{(6.00 \text{ rad/m})(150 \text{ m/s})}{2\pi} = 143 \text{ Hz}$$

 (b) $T = \dfrac{1}{f} = \dfrac{1}{143 \text{ s}^{-1}} = 6.98 \times 10^{-3} \text{ s}$

 (c) $k = \dfrac{2\pi}{\lambda}$

$$\lambda = \frac{2\pi}{k} = \frac{2\pi}{6.00 \text{ rad/m}} = 1.05 \text{ m}$$

19. $\lambda = \dfrac{v}{f}$

$$\lambda_1 = \dfrac{v}{f_1} = \dfrac{340 \text{ m/s}}{20 \text{ s}^{-1}} = 17 \text{ m}$$

$$\lambda_2 = \dfrac{v}{f_2} = \dfrac{340 \text{ m/s}}{20000 \text{ s}^{-1}} = 17 \text{ mm}$$

The range of wavelengths is from 17 mm to 17 m.

21. Wavelength range in the ocean:

$f_1 = 250 \text{ Hz}$

$f_2 = 150\,000 \text{ Hz}$

$v = 1500 \text{ m/s}$

$$\lambda_1 = \dfrac{v}{f_1} = \dfrac{1500 \text{ m/s}}{250 \text{ s}^{-1}} = 6.00 \text{ m}$$

$$\lambda_2 = \dfrac{v}{f_2} = \dfrac{1500 \text{ m/s}}{150\,000 \text{ s}^{-1}} = 10 \text{ mm}$$

Wavelength range in the air:

$f_1 = 250 \text{ Hz}$

$f_2 = 150\,000 \text{ Hz}$

$v = 340 \text{ m/s}$

$$\lambda_1 = \dfrac{v}{f_1} = \dfrac{340 \text{ m/s}}{250 \text{ s}^{-1}} = 1.4 \text{ m}$$

$$\lambda_2 = \dfrac{v}{f_2} = \dfrac{340 \text{ m/s}}{150\,000 \text{ s}^{-1}} = 2.3 \times 10^{-3} \text{ m}$$

23. $D(x,t) = \dfrac{0.3}{(x-4t)^2 + 1.2}$

$$D(1.5, 2.0) = \dfrac{0.3 \text{ m}^3}{\left((1.5 \text{ m}) - (4.0 \text{ m/s})(2.0 \text{ s})\right)^2 + 1.2 \text{ m}^2} = 6.9 \times 10^{-3} \text{ m}$$

The maximum displacement is when the quantity $(x - vt)$ is zero. The maximum displacement is 0.25 m.

The minimum displacement is zero:

$$\lim_{t \to \infty} D(x,t) = \dfrac{0.3}{(x-vt)^2 + 1.2} = 0$$

25. (a) $v = 3.0$ m/s and the pulse is travelling in the direction of the negative x-axis.

(b) $D(2,3) = \dfrac{-3.0}{6.0 + \left((2.0) + 3.0(3.0) \right)^2} = -2.36$ cm

(c) The maximum value for the displacement occurs when the value $(x + 3.0t)$ is zero, so $D = -0.5$ m.

(d) $D(x,t) = \dfrac{-3.0}{6.0 + (x + 3.0t)^2}$

$u(x,t) = \dfrac{\partial}{\partial t} D(x,t) = -\left[(-3.0)\left(6.0 + (x + 3.0t)^2\right)^{-2} \right]\left[2(3.0)(x + 3.0t) \right]$

$u(2.0,3.0) = \left[(3.0)\left(6.0 + (2.0 + 3.0(3.0))^2\right)^{-2} \right]\left[2(3.0)(2.0 + 3.0(3.0)) \right]$

$u(2.0,3.0) = 1.2$ cm/s

(e)

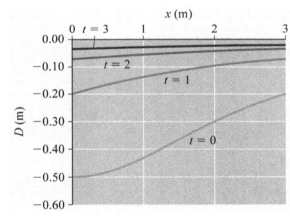

27. (a)

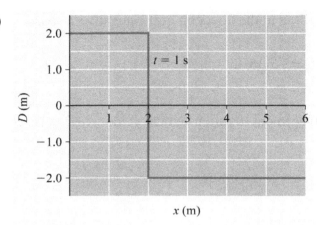

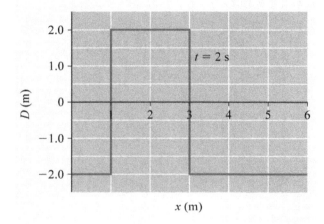

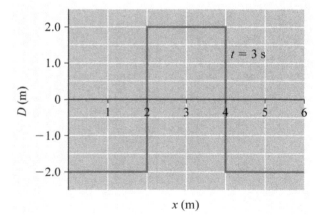

(b) Reading from the graph, the leading edge of the wave moves 1 m to the right every second, so the velocity of the pulse is 1.0 m/s to the right.

(c) For the given values of x and t, $|0.5-1.0| \le 1$, so the displacement is 2 m.

(d) $D(x,t) = \begin{cases} +2 \text{ m} & \text{if } |x+t| \le 1 \\ -2 \text{ m} & \text{if } |x+t| > 1 \end{cases}$

29. (a)

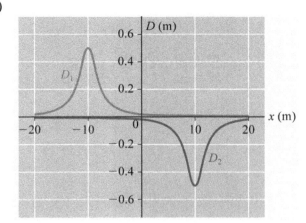

(b) The peak of each pulse occurs when the expression in parentheses is zero. At $t = 0$, the peak of the D_1 pulse is at $x = -10$ m, and the peak of the D_2 pulse is at $x = 10$ m.

(c) The condition corresponding to the cancelling of the pulses is

$$\frac{2}{(x-5t+10)^2+4}+\frac{-2}{(x+5t-10)^2+4}=0$$

$$2\left[(x+5t-10)^2+4\right]-2\left[(x-5t+10)^2+4\right]=0$$

$$(x+5t-10)^2+4-\left[(x-5t+10)^2+4\right]=0$$

$$(x+5t-10)^2+4-(x-5t+10)^2-4=0$$

$$x^2+10tx-20x+25t^2-100t+100-\left(x^2-10tx+20x+25t^2-100t+100\right)=0$$

$$20tx-40x=0$$

$$20x(t-2)=0$$

From the equation on the previous line, the pulses cancel when $t = 2$ s.

(d) From the equation at the end of the calculation in Part (c), when $x = 0$ the pulses cancel for all times.

31. First find the linear mass density of the wire.

$$\mu = \rho A$$

$$\mu = \left(7.86 \ \frac{g}{cm^3}\right)\left(\pi(0.020 \ cm)^2\right)$$

$$\mu = 9.88 \times 10^{-3} \ \frac{g}{cm}$$

Then use the formula relating velocity, tension, and linear mass density to calculate the tension in the wire.

$$v = \sqrt{\frac{T_s}{\mu}}$$

$$T_s = v^2\mu = (160 \text{ m/s})^2 \left(9.88 \times 10^{-3} \text{ g/cm}\right) = 252.855 \ \frac{\text{m}^2}{\text{s}^2} \cdot \frac{\text{g}}{\text{cm}} \cdot \frac{100 \text{ cm}}{1 \text{ m}} \cdot \frac{1 \text{ kg}}{1000 \text{ g}}$$

$$T_s = 25 \ \frac{\text{kg} \cdot \text{m}}{\text{s}^2} = 25 \text{ N}$$

33. $\mu = \dfrac{M}{L} = \dfrac{0.10 \text{ kg}}{50.0 \text{ m}} = 2.0 \times 10^{-3} \ \dfrac{\text{kg}}{\text{m}}$

$$v = \sqrt{\frac{T_s}{\mu}} = \sqrt{\frac{100.0 \text{ N}}{2.0 \times 10^{-3} \text{ kg/m}}} = 0.22 \text{ km/s}$$

35. First calculate the linear mass density of both strings in kg/m.

$$\mu_A = 2.0 \text{ g/m} = 2.0 \ \frac{\text{g} \cdot 1\text{kg}}{\text{m} \cdot 1000 \text{ g}} = 2.0 \times 10^{-3} \text{ kg/m}$$

$$\mu_B = 5.0 \text{ g/m} = 5.0 \ \frac{\text{g} \cdot 1\text{kg}}{\text{m} \cdot 1000 \text{ g}} = 5.0 \times 10^{-3} \text{ kg/m}$$

Then calculate the speed of each pulse using the same value for the tension since the strings are attached and the tension is constant.

$$v_A = \sqrt{\frac{T}{\mu_A}} = \sqrt{\frac{50.0 \text{ N}}{2.0 \times 10^{-3} \text{ kg/m}}} = 1.6 \times 10^2 \text{ m/s}$$

$$v_B = \sqrt{\frac{T}{\mu_B}} = \sqrt{\frac{50.0 \text{ N}}{5.0 \times 10^{-3} \text{ kg/m}}} = 1.0 \times 10^2 \text{ m/s}$$

Now calculate the time it takes for pulse A to travel the length of string A (10.0 m). Then calculate how far pulse B would travel in the same time. After this time, the pulses will be travelling on string B only, and will therefore travel at the same speed.

$$t_A = \frac{d_A}{v_A} = \frac{10.0 \text{ m}}{1.6 \times 10^2 \text{ m/s}} = 6.3 \times 10^{-2} \text{ s}$$

$$d_B = v_B \cdot t_A = \left(1.0 \times 10^2 \text{ m/s}\right)\left(6.3 \times 10^{-2} \text{ s}\right) = 6.3 \text{ m}$$

So the pulses are now 20.0 m − 6.3 m = 13.7 m apart, and they travel at the same speed in string B. Therefore they meet halfway between their current positions, 6.3 m + 6.85 m = 13.2 m from the far end of string B.

37. (a) $[v] = \left(\text{ms}^{-2} \cdot \text{m}\right)^{1/2} = \text{m/s}$

(b) $T = \dfrac{1}{f} = \dfrac{\lambda}{v} = \lambda\sqrt{\dfrac{2\pi}{g\lambda}} = \sqrt{\dfrac{2\pi\lambda}{g}}$

(c) From the result of part (b),

$$\frac{v}{T} = \frac{g}{2\pi}$$

And therefore

$$v = \frac{gT}{2\pi} = \frac{(9.81)(12)}{2\pi} = 19 \text{ m/s}$$

39. All four waves move in the positive x-direction, because the argument of the sinusoidal function in each case is of the form $kx - \omega t$. Because the wave number k is the reciprocal of the wavelength, the rank of the waves in order of increasing wavelength is b < a = c < d. The magnitude of the coefficient of t in the argument of the sinusoid is proportional to the frequency, so the rank of the waves in order of increasing wavelength is a = b = d < c.

41. From the graph for D_2, $T_2 = \frac{2}{3}$ s and $A = 1.0$ cm. Sine $\lambda_2 = 1.0$ m, therefore,

$$D_2(x,t) = (1.0 \text{ cm})\sin(2\pi x - 3\pi t + \phi)$$

From the graph for D_2 we notice that $D_2(1,0) = 1.0$ cm. Therefore,

$$D_2(1,0) = (1.0 \text{ cm})\sin(2\pi + \phi) = 1.0 \text{ cm}.$$

This gives $\phi = \pi/2$ rad and the equation for D_2 is

$$D_2(x,t) = (1.0 \text{ cm})\sin(2\pi x - 3\pi t + \frac{\pi}{2})$$

From the graph of D_1, $T_1 = 1.0$ s. Since $v = \frac{\lambda_1}{T_1} = \frac{\lambda_2}{T_2}$, we get

$$\lambda_1 = T_1\frac{\lambda_2}{T_2} = \frac{3}{2} \text{ m}$$

Therefore,

$$D_1(x,t) = (1.0 \text{ cm})\sin(\frac{4\pi}{3}x - 2\pi t + \phi)$$

From the graph, $D_1(1,0) \cong 0.3$ cm. Therefore,

$$D_1(1,0) = (1.0 \text{ cm})\sin(4.5 + \phi) = (0.3 \text{ cm})$$

This gives $\phi \cong -3.88$ rad or $\phi \cong -1.35$ rad. The value $\phi \cong -1.35$ rad correctly describes the shape of $D_1(1,t)$. Therefore,

$$D_1(x,t) = (1.0 \text{ cm})\sin(\frac{4\pi}{3}x - 2\pi t - 1.35)$$

43. (a) $\lambda = \dfrac{2\pi}{k} = \dfrac{2\pi}{10.0} = 0.628$ m

$f = \dfrac{\omega}{2\pi} = \dfrac{7.50}{2\pi} = 1.19$ Hz

$v = \dfrac{\omega}{k} = \dfrac{7.50}{10.0} = 0.750$ m/s

(b) $D(x,t) = (0.05 \text{ m})\sin(10.0x - 7.50t)$

45. (a) $\lambda = \dfrac{2\pi}{k} = \dfrac{2\pi}{2\pi/3.0} = 3.0$ m

$f = \dfrac{\omega}{2\pi} = \dfrac{2\pi/6.0}{2\pi} = 0.17$ Hz

$v = \dfrac{\omega}{k} = \dfrac{2\pi/6.0}{2\pi/3.0} = 0.50$ m/s

(b) The wave travels to the left.

(c)

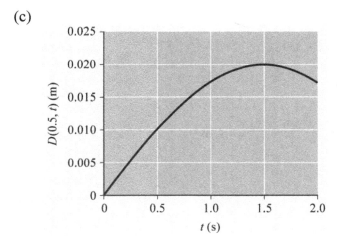

(d) $u(x,t) = \dfrac{\partial}{\partial t} D(x,t)$

$u(x,t) = (0.02 \text{ m})\cos\left(\dfrac{2\pi}{3.0}x + \dfrac{2\pi}{6.0}t - \dfrac{\pi}{3}\right) \times \dfrac{2\pi}{6.0}$

$u(0.5, 2.0) = (0.02 \text{ m})\cos\left(\dfrac{2\pi}{3.0}(0.5) + \dfrac{2\pi}{6.0}(2.0) - \dfrac{\pi}{3}\right) \times \dfrac{2\pi}{6.0}$

$u(0.5, 2.0) = (0.02 \text{ m})\cos\left(\dfrac{2\pi}{3}\right) \times \dfrac{2\pi}{6.0}$

$u(0.5, 2.0) = -0.010$ m/s

47. (a) Reading from the graph, the wavelength of the wave is 8 m.

(b) $v = f\lambda = (5.0\ \text{Hz})(8\ \text{m}) = 40\ \text{m/s}$

(c) If the wave is modelled by a cosine function, its phase constant is 0; if is modelled by a sine function, its phase constant is $\pi/2$.

(d) $D(x,t) = 1.5\sin\left(\dfrac{\pi}{4}x - 10\pi t + \dfrac{\pi}{2}\right)$

49. (a) Reading from the graph, the period is $T = 2$ s, so $f = \dfrac{1}{T} = \dfrac{1}{2} = 0.5\ \text{Hz}$.

(b) $v = f\lambda$, so $\lambda = \dfrac{v}{f} = \left(\dfrac{4.0}{0.5}\right) = 8.0\ \text{m}$

(c) The equation for the wave can this be written as

$$D(x,t) = (1.5\ \text{m})\sin\left(\dfrac{\pi}{4}x - \pi t + \phi\right)$$

To determine the phase constant ϕ notice from the graph that $D(2.5, 0) = -0.5$ m. Therefore,

$$D(2.5, 0) = (1.5\ \text{m})\sin\left(\dfrac{\pi}{4}\times 2.5 + \phi\right) = -0.5\ \text{m}$$

$$\sin(1.96 + \phi) = -0.3$$

The two possible values of ϕ are

(1) $\sin(1.96 + \phi) = -0.3$

$\quad 1.96 + \phi = \sin^{-1}(-0.3)$

$\quad \phi = \sin^{-1}(-0.3) - 1.96 = -2.3\ \text{rad}$

(2) $\sin(\pi - 1.96 - \phi) = -0.3$

$\quad \pi - 1.96 - \phi = \sin^{-1}(-0.3)$

$\quad \phi = 1.5\ \text{rad}$

The given shape for $D(2.5, t)$ plot corresponds to $\phi = 1.5$ rad.

(d) $D(x,t) = (1.5\ \text{m})\sin\left(\dfrac{\pi}{4}x - \pi t + 1.5\right)$

51. (a) From the first graph $\lambda = 8$ m and from the second graph $T = 1$ s. Thus, the equation for the wave is given by

$$D(x,t) = (1.5 \text{ m}) \sin\left(\frac{\pi}{4}x - 2\pi t + \phi\right)$$

From the first graph note that $D(0,1) = 1.5$ m. This gives $\phi = \frac{\pi}{2}$ rad. Therefore

$$D(x,t) = (1.5 \text{ m}) \sin\left(\frac{\pi}{4}x - 2\pi t + \frac{\pi}{2}\right)$$

(b) $D(x,t) = (1.5 \text{ m}) \sin\left(\frac{\pi}{4}x + 2\pi t + \frac{\pi}{2}\right)$

53. (a) From the displacement versus position graph, $\lambda = 4$ m and $A = 0.10$ m. From displacement versus time graph, $T = 6$ s. Thus, the equation of the wave is

$$D(x,t) = (0.10 \text{ m}) \sin\left(\frac{2\pi}{4}x - \frac{2\pi}{6}t + \phi\right)$$

From the displacement versus time graph, $D(1,0) = 0.025$ m. Therefore,

$$(0.10 \text{ m}) \sin\left(\frac{\pi}{2} + \phi\right) = 0.025 \text{ m}$$

Since $\sin\left(\frac{\pi}{2} + \phi\right) = \cos(\phi)$, the above equation becomes

$$\cos(\phi) = 0.25$$

Thus, the possible values of ϕ are

$\phi = \cos^{-1}(0.25) = 1.3$ rad or $\phi = 2\pi - 1.3 = 5.0$ rad

The value that reproduces the gives graphs corresponds to $\phi = 5.0$ rad. Therefore, the equation for the wave is

$$D(x,t) = (0.10 \text{ m}) \sin\left(\frac{\pi}{2}x - \frac{\pi}{3}t + 5.0\right)$$

(b) The wave travels to the right at a speed of $v = \frac{\omega}{k} = \frac{\pi/3}{\pi/2} = \frac{2}{3}$ m/s ≈ 0.66 m/s.

55. (a) The phase constant is 0.2 rad.

(b) The phase of the wave is $4.0x - 0.5t + 0.2 = 4.0(0.5) - 0.5(2.0) + 0.2 = 1.2$ rad.

(c) Yes, the phase changes linearly with time for each fixed value of x.

(d) The phase difference is

$$\left(4.0x - 0.5t + 0.2\right)_2 - \left(4.0x - 0.5t + 0.2\right)_1$$
$$= 4.0\left(x_2 - x_1\right) - 0.5\left(t_2 - t_1\right) + 0.2 - 0.2 = 4.0(0.1) = 0.4 \text{ rad}$$

(e) Yes, the phase difference is the same for all fixed times for the given difference in x-values.

(f) For a fixed time, the phase difference between two points that are one wavelength apart is

$$4.0\left(x_2 - x_1\right) = 4.0\frac{2\pi}{4.0} = 2\pi \text{ rad}$$

which is equivalent to 0 rad.

57. $P = \dfrac{1}{2}\mu v \omega^2 A^2$

$$P = \frac{1}{2}\mu\sqrt{\frac{T}{\mu}}\left(2\pi f\right)^2 A^2$$
$$P = \sqrt{T\mu}\,2\pi^2 f^2 A^2$$
$$P = \sqrt{(200.0)(0.1000)}\,2\pi^2\left(10.0\right)^2\left(0.010\right)^2$$
$$P = 0.88 \text{ W}$$

59. (a)

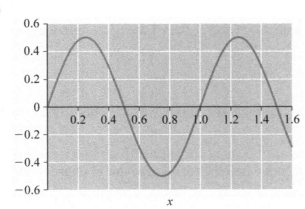

(b) $A = 0.5$, $\lambda = 1$

(c) The two component waves have the same frequency and wavelength, so they have the same speed. Consequently, the superposition of the two waves also has the same speed.

(d) The resultant wave is a travelling wave, because the two component waves travel in the same direction.

61.

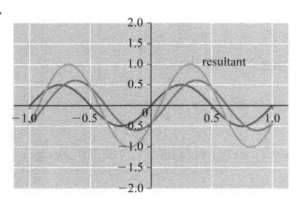

From the graph you can read the phase constant is approximately –0.4 rad.

63. Using equation (14-37),

$$2A\cos\left(\frac{\phi}{2}\right) = \frac{A}{2}$$

$$\cos\left(\frac{\phi}{2}\right) = \frac{1}{4}$$

$$\frac{\phi}{2} = \cos^{-1}\left(\frac{1}{4}\right)$$

$$\frac{\phi}{2} = 1.3181$$

$$\phi = 2.636 \text{ rad}$$

The phase difference between the resultant wave and either of the two component waves is

$$\frac{\phi}{2} = 1.318 \text{ rad}$$

65. (a) The minimum amplitude, which occurs when the two waves are out of phase by half a cycle, is $|A_1 - A_2|$.

(b) The maximum amplitude, which occurs when the two waves are in phase, is $A_1 + A_2$.

67. Using the identity

$$\sin(a) + \sin(b) = 2\cos\left(\frac{a-b}{2}\right)\sin\left(\frac{a+b}{2}\right)$$

and letting a and b represent the arguments of the constituent waves, the resultant wave can be expressed as

$$D(x,t) = 2A\cos\left(\frac{[kx - \omega t + \phi_1] - [kx - \omega t + \phi_2]}{2}\right)\sin\left(\frac{[kx - \omega t + \phi_1] + [kx - \omega t + \phi_2]}{2}\right)$$

$$D(x,t) = 2A\cos\left(\frac{\phi_1 - \phi_2}{2}\right)\sin\left(\frac{2kx - 2\omega t + \phi_1 + \phi_2}{2}\right)$$

$$D(x,t) = 2A\cos\left(\frac{\phi_1 - \phi_2}{2}\right)\sin\left(kx - \omega t + \frac{\phi_1 + \phi_2}{2}\right)$$

69. (a) For a hard reflection, the wave is inverted and travels in the opposite direction; therefore, its wave function is

$$D(x,t) = 0.2\sin(3x - 4t + \pi)$$

(b) For a soft reflection, the wave is NOT inverted and travels in the opposite direction; therefore, its wave function is

$$D(x,t) = 0.2\sin(3x - 4t)$$

71. (a) The frequency of the standing wave is $\frac{20\pi}{2\pi} = 10$ Hz and the wavelength is

$$\frac{2\pi}{0.6} = 10.5 \text{ m.}$$

(b) The distance between consecutive nodes is half of a wavelength, which is 5.2 m, and the distance between a node and an adjacent antinode is a quarter of a wavelength, which is 2.6 m.

(c) $D_1(x,t) = (0.750 \text{ cm})\sin(0.6x - 20\pi t)$

$D_2(x,t) = (0.750 \text{ cm})\sin(0.6x + 20\pi t)$

(d) $D(0.20, 3.0) = (1.50 \text{ cm})\sin(0.6(0.20))\cos(20\pi(3.0))$

$D(0.20, 3.0) = (1.50 \text{ cm})\sin(0.12)\cos(60\pi)$

$D(0.20, 3.0) = (1.50 \text{ cm})(0.1197)(1)$

$D(0.20, 3.0) = 0.18 \text{ cm}$

73. (a) $D(x,t) = (2.0 \text{ mm})\sin(\pi x)\cos(0.5\pi t)$

(b) The first three nodes on either side of $x = 0$ are located at $x = \pm 1$ m, ± 2 m, ± 3 m. The first three antinodes on either side of $x = 0$ are $x = \pm 0.5$ m, ± 1.5 m, ± 2.5 m

(c) Based on the result of part (b), the distance between two consecutive nodes is 1.0 m.

75. Using the identity

$$\sin(a) + \sin(b) = 2\cos\left(\frac{a-b}{2}\right)\sin\left(\frac{a+b}{2}\right)$$

and letting a and b represent the arguments of the constituent waves, the resultant wave can be expressed as

$$D(x,t) = 2A\cos\left(\frac{[kx - \omega t + \phi_1] - [kx + \omega t + \phi_2]}{2}\right)\sin\left(\frac{[kx - \omega t + \phi_1] + [kx + \omega t + \phi_2]}{2}\right)$$

$$D(x,t) = 2A\cos\left(-\omega t + \frac{\phi_1 - \phi_2}{2}\right)\sin\left(\frac{2kx + \phi_1 + \phi_2}{2}\right)$$

$$D(x,t) = 2A\cos\left(\omega t - \frac{\phi_1 - \phi_2}{2}\right)\sin\left(kx + \frac{\phi_1 + \phi_2}{2}\right)$$

77. $f = \dfrac{m}{2L}\sqrt{\dfrac{T}{\mu}} = \dfrac{1}{2(0.5)}\sqrt{\dfrac{T}{\mu}} = \sqrt{\dfrac{T}{\mu}} = 260 \text{ Hz}$

When the tension is increased by 4%, the new frequency is

$$f = \sqrt{\frac{1.04T}{\mu}} = \sqrt{1.04}\sqrt{\frac{T}{\mu}} = 1.0198(260) = 265 \text{ Hz}$$

79. (a) The longest wavelength standing wave possible is 4.0 m, because at least half a wavelength must fit into the length of the string.

(b) $f = \dfrac{v}{2L}$

$f = \dfrac{200.0}{2(2.0)}$

$f = 50 \text{ Hz}$

(c) No; the allowed frequencies are whole-number multiples of the fundamental frequency.

(d) The frequency of the second harmonic is 2(50) = 100 Hz, and the frequency of the fourth harmonic is 4(50) = 200 Hz.

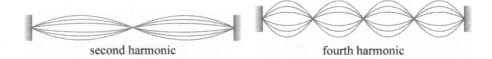

second harmonic   fourth harmonic

81. (a) The fundamental frequency is

$$f = \frac{1}{2L}\sqrt{\frac{T}{\mu}}$$

$$f = \frac{1}{2(1.000 \text{ m})}\sqrt{\frac{25.00 \text{ N}}{0.000650 \text{ kg/m}}}$$

$$f = 98.1 \text{ Hz}$$

(b) The frequency of the second harmonic is $2(98.1) = 196$ Hz, and the frequency of the third harmonic is $3(98.1) = 294$ Hz.

(c) The wave speed is

$$v = \sqrt{\frac{T}{\mu}}$$

$$v = \sqrt{\frac{25.00 \text{ N}}{0.000650 \text{ kg/m}}}$$

$$f = 196 \text{ m/s}$$

(d) The new fundamental frequency is

$$f = \frac{1}{2L}\sqrt{\frac{T}{\mu}}$$

$$f = \frac{1}{2(1.000 \text{ m})}\sqrt{\frac{35.00 \text{ N}}{0.000650 \text{ kg/m}}}$$

$$f = 116 \text{ Hz}$$

1. True. Yes, sound waves are longitudinal.

3. The speed of a wave is the product of its frequency and its wavelength, and the speed depends on properties of the medium. Thus, if the frequency doubles, then the wavelength is divided by 2, or equivalently, is multiplied by a factor of 0.5.

5. True. As for light waves, angles of incidence and reflection are equal for sound waves as well.

7. True. Yes, confining a wave between boundaries results in interference that produces a standing wave provided that the wave has an appropriate wavelength.

9. False. It is possible to have a number of different standing waves exist at the same time in an air column; they will combine to produce one sound.

11. Consider two sources of intensity I_1 and I_2, respectively. The corresponding sound intensities are β_1 and β_2. Then,

$$\beta_1 = 10\log_{10}\left(\frac{I_1}{I_0}\right) \quad \text{and} \quad \beta_2 = 10\log_{10}\left(\frac{I_2}{I_0}\right)$$

$$\beta_2 - \beta_1 = 10\log_{10}\left(\frac{I_2}{I_0}\right) - 10\log_{10}\left(\frac{I_1}{I_0}\right)$$

$$\beta_2 - \beta_1 = 10\left[\log_{10}\left(\frac{I_2}{I_0}\right) - \log_{10}\left(\frac{I_1}{I_0}\right)\right]$$

$$\beta_2 - \beta_1 = 10\left[\log_{10}\left(\frac{I_2}{I_0} \div \frac{I_1}{I_0}\right)\right]$$

$$\beta_2 - \beta_1 = 10\left[\log_{10}\left(\frac{I_2}{I_0} \times \frac{I_0}{I_1}\right)\right]$$

$$\beta_2 - \beta_1 = 10\left[\log_{10}\left(\frac{I_2}{I_1}\right)\right]$$

Now, if $I_2 = 2I_1$, then

$$\beta_2 - \beta_1 = 10\log_{10}\left(\frac{2I_1}{I_1}\right)$$

$$\beta_2 - \beta_1 = 10\log_{10}(2)$$

$$\beta_2 - \beta_1 = 10(0.301)$$

$$\beta_2 - \beta_1 = 3.01$$

Thus, the increase in sound intensity level is 3.01 dB when the sound intensity is doubled.

13. c. Sound waves can interfere both spatially and temporally.

15. $v = f\lambda$

$v = 343$ m/s for air at 20°C

$\lambda = \dfrac{v}{f} = \dfrac{343 \text{ m/s}}{1000 \text{ s}^{-1}} = 0.343$ m

17. The speed of sound in water at 20°C is 1482 m/s.

$\Delta y = v\Delta t \rightarrow \Delta t = \dfrac{\Delta y}{v}$

$\Delta t = \dfrac{30 \text{ m} + 30 \text{ m}}{1482 \text{ m/s}} = 0.040$ s

19. $s(x,t) = s_m \cos(kx - \omega t)$

$s(x,t) = s_m \cos\left(\dfrac{2\pi}{\lambda}x - 2\pi ft\right)$

Now, $f = 60$ Hz, so $2\pi f = 377$ s^{-1}.

For sound in air, $v = 343$ m/s, so

$k = \dfrac{2\pi}{\lambda} = \dfrac{2\pi f}{v} = \dfrac{2\pi(60)}{343} = 1.1 m^{-1}$

Thus, the displacement amplitude can be expressed as

$s(x,t) = s_m \cos(1.1x - 377t)$

21. The amplitude of pressure variations is $\Delta p_m = Bks_m$ where s_m is the amplitude of displacement variations, and the bulk modulus of air is $B = 1.01 \times 10^5$ Pa.

Now $k = \dfrac{2\pi}{\lambda} = \dfrac{2\pi f}{v} = \dfrac{2\pi(1000 \text{ Hz})}{343 \text{ m/s}} = 18.32$ m^{-1}

$s_m = \dfrac{\Delta p_m}{(1.01 \times 10^5 \text{ Pa})(18.32 \text{ m}^{-1})} = 1.1 \times 10^{-5}$ m

23. $\Delta p_m = Bks_m$, so when the displacement amplitude doubles, so does the pressure amplitude.

25.

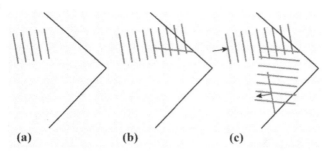

(a)	(b)	(c)

27. $f_{1x} = \dfrac{v}{2L_x} = \dfrac{343 \text{ m/s}}{2(0.10 \text{ m})} = 1.7 \text{ kHz}$

$\quad f_{1y} = \dfrac{v}{2L_y} = \dfrac{343 \text{ m/s}}{2(0.15 \text{ m})} = 1.1 \text{ kHz}$

$\quad f_{1z} = \dfrac{v}{2L_z} = \dfrac{343 \text{ m/s}}{2(0.20 \text{ m})} = 860 \text{ Hz}$

29. The beat frequency is $296 \text{ Hz} - 294 \text{ Hz} = 2 \text{ Hz}$.

31. The beat frequency is $440 \text{ Hz} - 350 \text{ Hz} = 90 \text{ Hz}$.

33. $\beta_1 = 10 \log\left(\dfrac{I_1}{I_0}\right) = 95 \text{ dB}$

$\quad \beta_2 = 10 \log\left(\dfrac{2I_1}{I_0}\right)$

$\quad \beta_2 = 10\left[\log(2) + \log\left(\dfrac{I_1}{I_0}\right)\right]$

$\quad \beta_2 = 10\log(2) + 10\log\left(\dfrac{I_1}{I_0}\right)$

$\quad \beta_2 = 10\log(2) + 95 \text{ dB}$

$\quad \beta_2 = 3.01 \text{ dB} + 95 \text{ dB}$

$\quad \beta_2 = 98 \text{ dB}$

35. $80\dfrac{\text{km}}{\text{h}} = \dfrac{80\,000 \text{ m}}{3600 \text{ s}} = 22.2 \text{ m/s}$

$\quad f_r = \dfrac{v \pm v_r}{v \mp v_s} f_s$

$\quad f_r = \dfrac{343 + 0}{343 - 22.2}(700.0 \text{ Hz})$

$\quad f_r = 748 \text{ Hz}$

37. $f_r - f_s = \left[\dfrac{v \pm v_r}{v \mp v_s} - 1\right] f_s$

$f_r - f_s = \left[\dfrac{v \pm v_r - (v \mp v_s)}{v \mp v_s}\right] f_s$

$\dfrac{f_r - f_s}{f_s} = \dfrac{\pm v_r \pm v_s}{v \mp v_s}$

$\dfrac{|f_r - f_s|}{f_s} = \dfrac{v_r + v_s}{v}$, because $v_s \ll v$

Thus $v_r + v_s = \dfrac{v|f_r - f_s|}{f_s}$

$v_r + v_s = \dfrac{343(35 \text{ Hz})}{61 \text{ kHz}}$

$v_r + v_s = 19.68 \text{ cm/s}$

Assuming v_r and v_s are comparable, then each of them is about 9.8 cm/s.

39. Light takes only very slightly more than 0 s to reach you, so this time is negligible. The distance travelled by sound in 4 s is $(343 \text{ m/s})(4 \text{ s}) = 1.4 \text{ km}$.

41. $s_1(x,t) = 10\cos(kx - \omega t)$

$s_2(x,t) = 20\cos\left(kx - \omega t \pm \dfrac{\pi}{4}\right)$

$s_1(x,t) + s_2(x,t) = 10\cos(kx - \omega t) + 20\cos\left(kx - \omega t \pm \dfrac{\pi}{4}\right)$

$s_1(x,t) + s_2(x,t) = 10\cos(kx - \omega t) + 20\cos(kx - \omega t)\cos\left(\dfrac{\pi}{4}\right)$

$\mp 20\sin(kx - \omega t)\sin\left(\dfrac{\pi}{4}\right)$

$s_1(x,t) + s_2(x,t) = (10 + 10\sqrt{2})\cos(kx - \omega t) \mp 10\sqrt{2}\sin(kx - \omega t)$ ... (1)

$s_1(x,t) + s_2(x,t) = A\cos(kx - \omega t - \theta)$, for some constants A and θ

Expanding the expression on the right side of the previous line, we obtain
$A\cos(kx - \omega t)\cos\theta + A\sin(kx - \omega t)\sin\theta$

Comparing with equation (1), we obtain
$A\cos\theta = 10 + 10\sqrt{2}$ and $A\sin\theta = \mp 10\sqrt{2}$

Squaring and adding the two equations in the previous line, we obtain

$$A^2 \cos^2 \theta + A^2 \sin^2 \theta = \left[10\left(1+\sqrt{2}\right)\right]^2 + \left[10\sqrt{2}\right]^2$$

$$A^2 \left(\cos^2 \theta + \sin^2 \theta\right) = 100\left[1+2\sqrt{2}+2+2\right]$$

$$A^2 = 100\left(5+2\sqrt{2}\right)$$

$$A = 10\sqrt{\left(5+2\sqrt{2}\right)}$$

$$A = 27.979$$

$$A = 28 \text{ nm}$$

43. $I = I_0 10^{\beta/10} = 10^{-12} \times 10^{65/10} = 10^{-12} \times 10^{6.5} = 10^{0.5} \times 10^{-6}$

$$I = 3.16 \times 10^{-6} \text{ W/m}^2$$

$$P = \left(4\pi r^2\right)I = 4\pi(23)^2\left(3.16 \times 10^{-6}\right)$$

$$P = 2.1 \times 10^{-2} \text{ W}$$

$$P = 21 \text{ mW}$$

45. The speed of the car is $130\dfrac{\text{km}}{\text{h}} = \dfrac{130\,000}{3600}\dfrac{\text{m}}{\text{s}} = 36.1 \text{ m/s}$

$$f_r = \frac{v \pm v_r}{v \mp v_s}f_s$$

The frequency received by the moving car is

$$\frac{c-36.1}{c}(10 \text{ GHz})$$

This is the same frequency as the reflected waves, which are received back at the radar unit with a frequency of

$$f_r = \left(\frac{c}{c+36.1}\right)\left(\frac{c-36.1}{c}\right)(10 \text{ GHz})$$

$$f_r = \left(\frac{c-36.1}{c+36.1}\right)(10 \text{ GHz})$$

$$f_r = \left(\frac{3\times10^8 - 36.1}{3\times10^8 + 36.1}\right)(10 \text{ GHz})$$

$$f_r = 9.999997593 \text{ GHz}$$

The beat frequency is $f_s - f_r = 24000 \text{ Hz} = 2.4 \text{ kHz}$

47. $G_4 \rightarrow 392.00$ Hz

$C_5 \rightarrow 523.25$ Hz

$E_5 \rightarrow 659.26$ Hz

$G_5 \rightarrow 783.99$ Hz

$392.00 = kf_1$

$523.25 = (k+1)f_1$

$\therefore f_1 = 523.25 - 392.00 = 131.25$ Hz

Similarly, $659.26 - 523.25 = 136.01$ Hz $= f_1$

$783.99 - 659.26 = 124.73$ Hz $= f_1$

The average value is $f_1 = \dfrac{124.73 + 136.01 + 131.25}{3} = 131$ Hz

Thus, $f_1 = \dfrac{v}{2L}$, so $L = \dfrac{v}{2f_1} = \dfrac{343}{2(131)} = 1.3$ m

49. $f_1 = \dfrac{v}{4L}$

$v = 4Lf_1$

$v = \dfrac{4L}{T_1}$

$v = \dfrac{4(270 \text{ km})}{12.5 \text{ h}}$

$v = 86.4$ km/h

$v = \dfrac{86.4}{3.6} \dfrac{\text{m}}{\text{s}}$

$v = 24$ m/s

51. $I = I_0 10^{\beta/10} \Rightarrow \dfrac{I_2}{I_1} = 10^{(\beta_2 - \beta_1)/10}$

A reduction in sound level by 60 dB corresponds to a reduction in intensity by a factor of $10^{-60/10} = 10^{-6}$. Thus, the new sound intensity is $0.2 \ \mu\text{W/m}^2$.

53. Let y represent the depth of the well. The time needed for the stone to drop is t, where

$y = \dfrac{1}{2}gt_1^2$

So $t_1 = \sqrt{\dfrac{2y}{g}}$

The additional time needed for the sound of the splash to reach the top of the well is t_2, where

$$y = vt_2$$

$$t_2 = \frac{y}{343}$$

The total time is 2.8 s, so $t_1 + t_2 = 2.8$ s

$$\sqrt{\frac{2y}{g}} + \frac{y}{343} = 2.8$$

$$\sqrt{\frac{2y}{g}} = 2.8 - \frac{y}{343}$$

$$\frac{2y}{g} = \left(2.8 - \frac{y}{343}\right)^2$$

$$\frac{2y}{g} = 2.8^2 - \frac{5.6}{343}y + \frac{y^2}{343^2}$$

$$0 = (2.8)^2 - \left(\frac{5.6}{343} + \frac{2}{g}\right)y + \frac{y^2}{343^2}$$

$$y = \frac{\left(\frac{5.6}{343} + \frac{2}{g}\right) \pm \sqrt{\left(\frac{5.6}{343} + \frac{2}{g}\right)^2 - \frac{4(2.8)^2}{343^2}}}{(2/343^2)}$$

$y = 25871$ m or 35.7 m. 25871 m is rejected because t_2 by itself would be much greater than 2.8 s. Thus, the depth of the well is 35.7 m.

55. Because the sound is loudest when you are equidistant from the sources, the sources are in phase. The condition for destructive interference is

$$\Delta d = \left(n + \frac{1}{2}\right)\lambda$$

where n is a natural number and λ is the wavelength of waves emitted by the sources. Thus,

$$\lambda = \frac{\Delta d}{n + \frac{1}{2}}$$

$$\frac{v}{f} = \frac{\Delta d}{n + \frac{1}{2}}$$

$$f = \frac{v\left(n + \frac{1}{2}\right)}{\Delta d}$$

$$f = \frac{343\left(n+\frac{1}{2}\right)}{2.8}$$

$$f = 122.5\left(n+\tfrac{1}{2}\right) \text{ s}^{-1}$$

If this is the lowest order minimum, then $n = 0$, and

$$f = \frac{122.5}{2} = 61 \text{ Hz}$$

57. (a) For the fundamental frequency, half of a wavelength fits into the length of the rod, so $\lambda_1 = 240$ cm.

   (b) $v = f_1 \lambda_1$

$$f_1 = \frac{v}{\lambda_1} = \frac{6420 \text{ m/s}}{2.4 \text{ m}} = 2675 \text{ Hz} = 2.68 \text{ kHz}$$

59. Consider two waves:

$$s_1(x,t) = s_m \cos\left(k_1 x - \omega_1 t + \phi_1\right)$$

$$s_2(x,t) = s_m \cos\left(k_2 x - \omega_2 t + \phi_2\right)$$

To find the resulting wave, we apply the Principle of Superposition and add the two waves:

$$s_{Total}(x,t) = s_1(x,t) + s_2(x,t)$$
$$= s_m \cos\left(k_1 x - \omega_1 t + \phi_1\right) + s_m \cos\left(k_2 x - \omega_2 t + \phi_2\right)$$
$$= s_m \left[\cos\left(k_1 x - \omega_1 t + \phi_1\right) + \cos\left(k_2 x - \omega_2 t + \phi_2\right)\right]$$
$$= 2 s_m \cos\frac{1}{2}\left(k_1 x - \omega_1 t + \phi_1 + k_2 x - \omega_2 t + \phi_2\right) \times \cos\frac{1}{2}\left(k_1 x - \omega_1 t + \phi_1 - k_2 x - \omega_2 t + \phi_2\right)$$
$$= 2 s_m \cos\left[\left(\frac{k_1 + k_2}{2}\right)x - \left(\frac{\omega_1 + \omega_2}{2}\right)t + \frac{\phi_1 + \phi_2}{2}\right] \times \cos\left[\left(\frac{k_1 - k_2}{2}\right)x - \left(\frac{\omega_1 - \omega_2}{2}\right)t + \frac{\phi_1 - \phi_2}{2}\right]$$

To simplify this expression, we define the following quantities:

$$\bar{k} = \frac{k_1 + k_2}{2} \quad \text{Mean wave number}$$

$$\bar{\omega} = \frac{\omega_1 + \omega_2}{2} \quad \text{Mean angular frequency}$$

$$\Delta k = \frac{k_1 - k_2}{2} \quad \text{Wave number difference}$$

$$\Delta\omega = \frac{\omega_1 - \omega_2}{2} \quad \text{Angular frequency difference}$$

$$\bar{\phi} = \frac{\phi_1 + \phi_2}{2} \text{ Average phase constant}$$

$$\Delta\phi = \frac{\phi_1 - \phi_2}{2} \text{ Phase constant difference}$$

We then get

$$s_{Total}(x,t) = 2s_m \cos(\Delta kx - \Delta\omega t + \Delta\phi)\cos(\bar{k}x - \bar{\omega}t + \bar{\phi})$$

Since we are concerned with temporal interference, we will consider the expression at $x = 0$ for simplicity.

$$s_{Total}(x,t) = 2s_m \cos(\Delta\phi - \Delta\omega t)\cos(\phi - \bar{\omega}t)$$

This is a wave that had a frequency of $\bar{\omega} = \frac{\omega_1 + \omega_2}{2}$ that is amplitude modulated at a frequency of $\Delta\omega = \frac{\omega_1 - \omega_2}{2}$.

61. $I_1 + I_2 + I_3 = I_0 10^{\beta_1/10} + I_0 10^{\beta_2/10} + I_0 10^{\beta_3/10}$

$I_{total} = I_0\left[10^{85/10} + 10^{110/10} + 10^{110/10}\right]$

$I_{total} = I_0\left[10^{8.5} + 10^{11} + 10^{11}\right]$

$I_{total} = I_0\left[2.00316 \times 10^{11}\right]$

This is the total intensity at a distance of 10 m. The total power produced is thus

$$P_{Total} = I_0 4\pi r^2 = I_0\left[2.00316 \times 10^{11}\right]400\pi$$

We now find the distance, R_{85}, at which this power produced a sound intensity level of 85 dB. The intensity corresponding to 85 dB is $I_0 10^{85/10}$. Thus, we equate

$$\frac{P_{Total}}{4\pi R_{85}^2} = I_0 10^{8.5}$$

We can solve this for

$$R_{85} = \sqrt{\frac{P_{Total}}{4\pi I_0 10^{8.5}}} = \sqrt{\frac{I_0 2.000316 \times 10^{11} 400\pi}{4\pi I_0 10^{8.5}}}$$

$$= \sqrt{\frac{2.000316 \times 10^{11} 100}{10^{8.5}}} = 252\,\text{m}$$

Chapter 16—TEMPERATURE AND THE ZEROTH LAW OF THERMODYNAMICS

1. Generally speaking, two different thermometers will not agree at all intermediate temperatures.

3. The most likely value is not necessarily the average value; it depends on the shape of the probability distribution.

(a) **Symmetric Probability Distribution with Equal Mean and Most Probable Values**

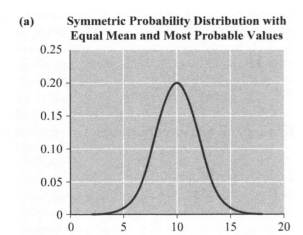

(b) **Skewed Probability Distribution with Unequal Mean and Most Probable Values**

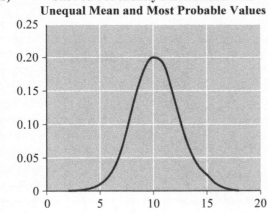

5. The triple point of water exists at exactly 0.01°C or 273.16 K. If liquid water did not expand when cooled, the melting point of ordinary ice would not decrease as a function of pressure, and the triple point would be the limit below which water would not exist in liquid state.

7. No, because the gases do not necessarily have equal concentrations (N/V).

9. No; for an ideal graph the corresponding graph would be a straight line that would pass through the origin if extended.

11. As the Sun rises, the temperature of the land increases more rapidly than that of the water, and the air just above the land becomes warmer than the air above the water. As a result, the air pressure on land is lower than the pressure above water. This pressure difference forces the cooler air above the water to move toward the land, which we feel as a wind.

13. For an ideal gas the relationship between pressure, volume and temperature is given by $PV = NkT$. Therefore, if the temperature increases while the volume is reduced by half, the pressure will more than double.

15. The average volume occupied by each gas particle is

$$\frac{22.4 \text{ L}}{6.02 \times 10^{23}} = \frac{22.4 \times 10^3 \text{ cm}^3}{6.02 \times 10^{23}} = 3.72 \times 10^{-20} \text{ cm}^3$$

The average spacing between centres of particles is

$$\left(3.72 \times 10^{-20} \text{ cm}^3\right)^{1/3} = 3.34 \times 10^{-7} \text{ cm} = 3.34 \text{ nm}$$

17. No; when a system is in thermodynamic equilibrium, all macroscopic properties, including temperature, remain constant.

19. (a) $100 \text{ psi} = 110 \text{ psi} \times \dfrac{101325 \text{ Pa}}{14.69595 \text{ psi}} = 758423 \text{ Pa} = 758 \text{ kPa}$

   (b) $V = 2\pi r \times A = 2\pi(35.0 \text{ cm})(4.00 \text{ cm}^2) = 880 \text{ cm}^3$

   (c) $\dfrac{3PV}{N} = mv^2 = 1.24 \times 10^{-20} \text{ J}$

   $$N = \frac{3PV}{1.24 \times 10^{-20} \text{ J}} = \frac{3(758423 \text{ Pa})(880 \times 10^{-4} \text{ m}^3)}{1.24 \times 10^{-20} \text{ J}} = 1.61 \times 10^{23} \text{ particles}$$

   $$N = \frac{1.61 \times 10^{23} \text{ particles}}{6.02 \times 10^{23} \text{ particles/mol}}$$

   $$N = 0.268 \text{ mol}$$

   (d) mass of nitrogen = 0.268 mol × 28.014 g/mol = 7.51 g
   mass of helium = 0.268 mol × 4.0026 g/mol = 1.07 g

21. $\alpha = \dfrac{\Delta L}{L} \cdot \dfrac{1}{\Delta T} \Rightarrow \Delta L = \alpha L \Delta T$

   $$\Delta L = \left(\frac{17 \times 10^{-6}}{\text{°C}}\right)(2.00 \text{ m})(40.0°C - 20.0°C)$$

   $$\Delta L = 680 \times 10^{-6} \text{ m}$$

   $$\Delta L = 0.68 \text{ mm}$$

23. $-40^\circ\text{F} = \left(-40^\circ\text{F} - 32^\circ\text{F}\right) \times \dfrac{5\ ^\circ\text{C}}{9\ ^\circ\text{F}} = -40^\circ\text{C}$

25. (a) $58^\circ\text{C} = 58^\circ\text{C} \times \dfrac{9\ ^\circ\text{F}}{5\ ^\circ\text{C}} + 32^\circ\text{F} = 136^\circ\text{F}$

(b) $58^\circ\text{C} + 273.15 = 331\ \text{K}$

27. (a) $PV = NkT \rightarrow N = \dfrac{PV}{kT}$

$N = \dfrac{\left(5000\ \text{Pa}\right)\left(792\ \text{m}^3\right)}{\left(1.38\times10^{-23}\ \text{J}\cdot\text{K}^{-1}\right)\left(220\ \text{K}\right)}$

$N = 1.30\times10^{27}$ atoms

The mass of helium is

$m = \dfrac{1.30\times10^{27}\ \text{atoms}}{6.02\times10^{23}\ \text{atoms/mol}} \times 4.00\ \text{g/mol} = 8.67\ \text{kg}$

(b) $PV = NkT$

$V = \dfrac{NkT}{P} = \dfrac{\left(1.30\times10^{27}\right)\left(1.38\times10^{-23}\right)\left(293.16\right)}{101000}$

$V = 52.2\ \text{m}^3$

(c) The number of moles of helium lost is

$N = \dfrac{PV}{kT} = \dfrac{\left(101000\right)\left(65.0 - 52.2\right)}{\left(1.38\times10^{-23}\right)\left(293.16\right)} \times \dfrac{1}{6.02\times10^{23}}\ \text{mol}$

$N = 530\ \text{mol}$

29. (a) $PV = NkT \rightarrow V = \dfrac{NkT}{P}$

$V = \dfrac{\left(6.02\times10^{23}\right)\left(1.38\times10^{-23}\right)\left(273.16\right)}{101325}$

$V = 0.0224\ \text{m}^3$

$V = 22.4\ \text{L}$

(b) $\left(v^2\right)_{\text{avg}} = \dfrac{3PV}{mN} = \dfrac{3\left(101325\right)\left(0.0224\right)}{0.028}$

$v_{\text{rms}} = 493\ \text{m/s}$

(c) $\left(v^2\right)_{avg} = \dfrac{3PV}{mN} = \dfrac{3(101325)(0.0224)}{0.032}$

$v_{rms} = 461$ m/s

(d) $\left(v_{rms}\right)_O : \left(v_{rms}\right)_N = 1:1.069$

31. For a temperature of $-81.4°F$,

$$\frac{1}{2}m\left(v_{avg}\right)^2 = \frac{3}{2}kT$$

$$\left(v_{avg}\right)^2 = \frac{3kT}{m} = \frac{\left(1.38\times10^{-23}\right)(210)}{28\times10^{-3}\left(6.02\times10^{23}\right)^{-1}} = 186\,921$$

$v_{rms} = 432$ m/s

For a temperature of $58°C$,

$$\frac{1}{2}m\left(v_{avg}\right)^2 = \frac{3}{2}kT$$

$$\left(v_{avg}\right)^2 = \frac{3kT}{m} = \frac{3\times\left(1.38\times10^{-23}\right)(331)}{28\times10^{-3}\left(6.02\times10^{23}\right)^{-1}} = 294\,623$$

$v_{rms} = 543$ m/s $= 0.54$ km/s

33. The average molecular speed is

$$v_{avg} = \int_0^\infty vf(v)\,dv$$

$$v_{avg} = \int_0^\infty v \cdot 4\pi\left(\frac{m}{2\pi kT}\right)^{3/2} v^2 e^{-\left[mv^2/2kT\right]}\,dv$$

$$v_{avg} = 4\pi\left(\frac{m}{2\pi kT}\right)^{3/2} \int_0^\infty v^3 e^{-\left[mv^2/2kT\right]}\,dv$$

Let $x = \dfrac{m}{2kT}v^2$, so $dx = \dfrac{m}{kT}v\,dv$

Thus,

$$v_{avg} = 4\pi\left(\frac{m}{2\pi kT}\right)^{3/2}\left(\frac{kT}{m}\right)\left(\frac{2kT}{m}\right)\int_0^\infty xe^{-x}\,dx$$

$$v_{avg} = 8\pi\frac{m^{3/2}k^2T^2}{(2\pi)^{3/2}k^{3/2}T^{3/2}m^2}\left[-xe^{-x}\Big|_0^\infty + e^{-x}\,dx\right]$$

$$v_{avg} = \frac{8\pi}{2\pi}\cdot\frac{k^{1/2}T^{1/2}}{(2\pi)^{1/2}m^{1/2}}\,[1]$$

$$v_{avg} = 4\sqrt{\frac{kT}{2\pi m}}$$

$$v_{avg} = \sqrt{\frac{8kT}{\pi m}}$$

35. Reading from the phase diagrams, water, oxygen, and carbon dioxide are in the solid phase, and nitrogen is in the liquid phase.

37. Answers will vary. For a 75-kg person, the mass of water in the body is $(0.6)(75) = 45$ kg.

39. $\frac{1}{2}m\left(v^2\right)_{avg} = \frac{3}{2}kT$

$$v_{rms} = \sqrt{\frac{3kT}{m}}$$

$$v_{rms} = \sqrt{\frac{3\left(1.38\times10^{-23}\right)\left(1\times10^{-9}\right)}{\left(0.13290546\,\frac{kg}{mol}\right)\left(6.02\times10^{23}\ mol^{-1}\right)^{-1}}}$$

$v_{rms} = 4.33\times10^{-4}$ m/s

$v_{rms} = 0.4$ mm/s

41. $PV = NkT$

$$T = \frac{PV}{Nk} = \frac{\left(150000\ \text{Pa}\right)\left(30\times10^{-3}\ \text{m}^3\right)}{\left(3\times6.02\times10^{23}\right)\left(1.38\times10^{-23}\right)}$$

$T = 180$ K

43. The distance travelled in time Δt by a particle in an ideal gas travelling at speed v is $v\Delta t$, so the average number of collisions with the walls of a cubical enclosure

of length L in time Δt is $\dfrac{v\Delta t}{L}$.

Thus, the number of collisions per unit time per unit area is

$$\frac{\left(\dfrac{v\Delta t}{L}\right)}{\Delta t L^2} = \frac{v}{L^3} = \frac{v}{V}$$

The total number of collisions for all particles, per unit time per unit area, is

$$\frac{N\left(v_{avg}\right)}{V}$$

From equation (16-13),

$$P = \tfrac{1}{3}\frac{mN\left(v^2\right)_{avg}}{V}$$

So $\left(v^2\right)_{avg} = \dfrac{3PV}{mN}$

and therefore $v_{avg} \approx \sqrt{\dfrac{3PV}{mN}}$

The result is not exact, because $\left(v_{avg}\right)^2 \neq \left(v^2\right)_{avg}$.

Thus, the total number of collisions per unit time per unit area is

$$\frac{N}{V}\left(v_{avg}\right) = \frac{N}{V}\sqrt{\frac{3PV}{mN}} = \frac{\left(6.02\times10^{23}\right)}{\left(0.0224\ \text{m}^3\right)}\sqrt{\frac{3\left(101325\ \text{Pa}\right)\left(0.0224\ \text{m}^3\right)}{0.028\ \text{kg}}}$$

$$\frac{N}{V}\left(v_{avg}\right) = 1.3\times10^{28}\ \text{m}^{-2}\text{s}^{-1}$$

A more precise calculation introduces a factor of $\tfrac{1}{4}$, and yields the more accurate result of $3.3\times10^{27}\ \text{m}^{-2}\text{s}^{-1}$.

45. (a) $PV = NkT \rightarrow P = \left(\dfrac{N}{V}\right)kT$

Now, for hydrogen atoms,

$$\frac{N}{V} = \frac{\rho N_A}{m} = \frac{150000\,\tfrac{\text{kg}}{\text{m}^3}\times\dfrac{1000\ \text{g}}{\text{kg}}\times6.02\times10^{23}\ \text{particles/mole}}{1.00794\,\dfrac{\text{g}}{\text{mole}}}$$

$$\frac{N}{V} = 8.96\times10^{31}$$

$$P = \left(\frac{N}{V}\right)kT = \left(8.96\times10^{31}\right)\left(1.38\times10^{-23}\right)\left(13.6\times10^{6}\right)$$

$$P = 1.68\times10^{16}\ \text{Pa}$$

If instead we assume that in such extreme conditions hydrogen still exists in the form of molecules, then the pressure is half as much, $P = 8.4\times10^{15}$ Pa.

(b) $P = \frac{1}{3} m \frac{N}{V} \left(v^2 \right)_{avg}$

$\left(v^2 \right)_{avg} = \frac{3PN}{mN} \left(\frac{N}{V} \right)^{-1} = 3.359 \times 10^{11}$

$v_{rms} = 5.8 \times 10^5$ m/s

If we consider the hydrogen to exist in molecules, then m is twice as large, and $v_{rms} = 4.1 \times 10^5$ m/s.

47.

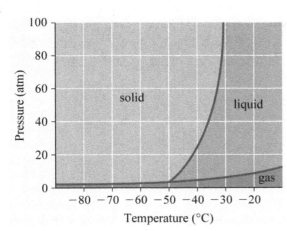

49. Using software, we obtain the line of best fit $R = 0.19T + 117.24$. For a temperature of 400 K, the predicted resistance is

$R = 0.19 \times 400 + 117.24$

$R = 193 \ \Omega$

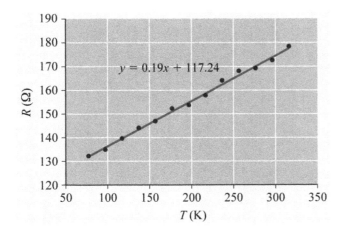

51. We found in question 43 that the collision rate was given approximately by $\dfrac{N}{V}\sqrt{\dfrac{3PV}{mN}}$. We will assume we maintain the volume and temperature of our container and remove particles. Thus we need to write our expression for the collision rate in terms of constants (V, T) and N, the number of particles. We will do this by assuming the gas behaves as an ideal gas and then we can replace the PV in the numerator of the square root with NkT to find the collision rate goes as $\dfrac{N}{V}\sqrt{\dfrac{3NkT}{mN}} = \dfrac{N}{V}\sqrt{\dfrac{3kT}{m}}$. In question 43 we found a collision rate of roughly 10^{28} collisions/second/m^2 at STP. If we assume our sensor has an area of roughly 10^{-4} m^2, we would see about 10^{24} collisions/second at STP. If the sensor has a response time of 1 ms, we would need a collision rate on the order of 1 ms, so we would have to reduce the concentration of particles by a factor of 10^{28}. At STP, N/V is roughly 10^{25} particles/m^3. To see these fluctuations, we would need a concentration on the order of 10^{-3} particles/m^3.

Chapter 17—HEAT, WORK, AND THE FIRST LAW OF THERMODYNAMICS

1. A constant volume process should be used, so that none of the supplied thermal energy goes into doing work on the sample of gas.

3. A: solid only

   B: solid and liquid

   C: liquid only

   D: liquid and gas

   E: gas only

5. During the winter, the interior surfaces of our houses tend to be cooler than in the summer. As a consequence, they radiate less heat into the room and therefore we absorb less heat and feel cooler.

7. The force done by the system on its surroundings is positive.

9. The energy conducted through a thermal insulator is proportional to the temperature difference across the insulator (see equation 17-39), so if the temperature difference doubles, the energy conducted also doubles.

11. The food is in thermal equilibrium with the water bath. The net flow of energy into the food from the bath is zero and its temperature therefore does not increase beyond the temperature of the water bath. Since the degree of "doneness" only depends upon the temperature that the food is brought to, it can be held at a constant temperature for an extended period of time without further cooking.

13. Because the amount of thermal energy that flows into one object is the same as the amount of thermal energy that flows into the other object, we can write

    $$m_1 c_1 |\Delta T| = m_2 c_2 |\Delta T|$$

    and therefore

    $$m_1 c_1 = m_2 c_2$$

    Thus, for the final temperature to be the average of the initial temperatures of each object, the product of the mass and specific heat capacity of each object should be the same. In other words, the heat capacities of the two objects are the same.

15. The total energy that must flow is

    $$Q = mc\Delta T$$

    $$Q = (1.0)(4186)(95)$$

    $$Q = 397670 \text{ J}$$

The time needed is

$$\Delta t = \frac{397\,670 \text{ J}}{500 \text{ J/s}}$$

$$\Delta t = 795 \text{ s}$$

$$\Delta t = 8.0 \times 10^2 \text{ s}$$

17. $m_{Al} c_{Al} \Delta T_{Al} = -m_{water} c_{water} \Delta T_{water}$

$m_{Al} c_{Al} (T_f - 100) = -m_{water} c_{water} (T_f - 10)$

$m_{Al} c_{Al} T_f - 100 m_{Al} c_{Al} = -m_{water} c_{water} T_f + 10 m_{water} c_{water}$

$m_{Al} c_{Al} T_f + m_{water} c_{water} T_f = 100 m_{Al} c_{Al} + 10 m_{water} c_{water}$

$T_f (m_{Al} c_{Al} + m_{water} c_{water}) = 100 m_{Al} c_{Al} + 10 m_{water} c_{water}$

$$T_f = \frac{100 m_{Al} c_{Al} + 10 m_{water} c_{water}}{m_{Al} c_{Al} + m_{water} c_{water}}$$

$$T_f = \frac{100(0.1)(897) + 10(0.5)(4186)}{(0.1)(897) + (0.5)(4186)}$$

$$T_f = 14$$

The final temperature of the aluminum and water is 14°C.

19. $Q = mc\Delta T$

$$c = \frac{Q}{m\Delta T}$$

$$c = \frac{65 \text{ J}}{(0.1 \text{ kg})(5 \text{ K})}$$

$$c = 130 \text{ J/(kg} \cdot \text{K)}$$

21. The amount of thermal energy needed to warm 20 g of ice to a temperature of 0°C is

$Q_1 = mc\Delta T$

$Q_1 = (0.020)(2100)(4)$

$Q_1 = 168 \text{ J}$

The amount of thermal energy needed to melt 20 g of ice at a temperature of 0°C is

$Q_2 = mL$

$Q_2 = (0.020)(3.33 \times 10^5)$

$Q_2 = 6660 \text{ J}$

The amount of thermal energy that would need to be removed from the 100 g of water to cool it to 0°C is

$$Q_3 = mc\Delta T$$
$$Q_3 = (0.1)(4186)(30)$$
$$Q_3 = 12558 \text{ J}$$

Thus, there is plenty of thermal energy available to melt the ice completely. The remaining energy,

$$Q_3 - (Q_1 + Q_2) = 12558 - (168 + 6660) = 5730 \text{ J}$$

will be used to warm the entire 120 g of liquid water from 0°C to a final temperature:

$$5730 \text{ J} = 0.12 \text{ kg} \times 4180 \frac{\text{J}}{\text{kg} \cdot \text{K}}(T_f - 0°\text{C})$$

$$\therefore T_f = \frac{5730 \text{ J}}{0.12 \text{ kg} \times 4180 \frac{\text{J}}{\text{kg} \cdot \text{K}}} = 11.4°\text{C}$$

The final temperature is $11\,°\text{C}$.

23. Because the process is isothermal, T is constant, and

$$PV = nRT$$
$$P = \frac{nRT}{V}$$

The work done in the process is

$$W = \int PdV$$
$$W = \int \frac{nRT}{V} dV$$
$$W = nRT \int \frac{1}{V} dV$$
$$W = nRT \left[\ln(V)\right]_{V_1}^{V_2}$$
$$W = nRT \left[\ln(V_2) - \ln(V_1)\right]$$
$$W = nRT \ln\left(\frac{V_2}{V_1}\right)$$
$$W = (3)(8.314)(373.16)\ln(1.3)$$
$$W = 2400 \text{ J}$$

25. By the first law of thermodynamics (see equation 17-16), in an isothermal process the work done by the system is equal to the thermal energy that flows into the system; thus, 100 J of energy flows into the system.

27. Using equation (17-36),

$$P = \sigma \varepsilon A T^4$$

$$\frac{P}{A} = \sigma \varepsilon T^4$$

$$\frac{P}{A} = \left(5.67 \times 10^{-8}\right)(0.5)(650 + 273.16)^4$$

$$\frac{P}{A} = 21 \text{ kW/m}^2$$

29. $$P = \frac{\kappa A \Delta T}{x}$$

$$P = \frac{(401)\left(4 \times 10^{-4}\right)(400 - 100)}{0.005}$$

$$P = 9624 \text{ W}$$

$$P = 9.6 \text{ kW}$$

31. $$Q = C \Delta T$$

$$\frac{Q}{\Delta t} = C \frac{\Delta T}{\Delta t}$$

$$400 = C(0.05)$$

$$C = \frac{400 \text{ J/s}}{0.05 \text{ K/s}}$$

$$C = 8.0 \text{ kJ/K}$$

$$c = \frac{C}{m}$$

$$c = \frac{8000 \text{ J/K}}{0.1 \text{ kg}}$$

$$c = 80 \text{ kJ/(kg} \cdot \text{K)}$$

33. The amount of energy needed to evaporate 4.0 kg of water is

$$Q = mL$$

$$Q = (4.0 \text{ kg})\left(2.26 \times 10^6 \text{ J/kg}\right)$$

$$Q = 9.0 \times 10^6 \text{ J}$$

If all of this water is evaporated in 2 h, then the power delivered to the water is

$$P = \frac{Q}{\Delta t} = \frac{9.0 \times 10^6 \text{ J}}{(2.0 \text{ h})(3600 \text{ s/h})} = 1.3 \text{ kW}$$

If all of the heat to evaporate the water is generated by the athlete, then she must actually generate more power, because she will have to power her movements and her other body processes.

35. $W = \int P dV$

$W = \int \dfrac{nRT}{V} dV$

$W = nRT \int \dfrac{1}{V} dV$

$W = nRT \left[\ln(V) \right]_{V_1}^{V_2}$

$W = nRT \left[\ln(V_2) - \ln(V_1) \right]$

$W = nRT \ln \left(\dfrac{V_2}{V_1} \right)$

$W = (1.5)(8.314)(273.16) \ln \left(\dfrac{1}{3} \right)$

$W = -3740 \text{ J}$

37. Assume for simplicity that the shape of the shrew is a cylinder of length 0.04 m and radius 0.01 m, and that the shrew's emissivity is 0.5. Then the power radiated by the shrew is

$P = \sigma \varepsilon A T^4$

$P = \sigma \varepsilon \left(2\pi R^2 + 2\pi R h \right) T^4$

$P = \left(5.67 \times 10^{-8} \right)(0.5) \left(2\pi \left[0.01^2 \right] + 2\pi [0.01][0.04] \right)(273 + 36)^4$

$P = \left(5.67 \times 10^{-8} \right)(\pi)(0.0005)(309)^4$

$P = 0.81 \text{ W}$

39. Because the internal pressure is constant, the work done by the gas on the piston is

$W = P \Delta V$

$W = P(V_2 - V_1)$

$W = (101000 \text{ Pa})(15.0 - 10.0) \text{ m}^3$

$W = 505 \text{ kJ}$

The change in internal energy of the gas is

$\Delta U = Q - W$

$\Delta U = 1000 \text{ kJ} - 505 \text{ kJ}$

$\Delta U = 495 \text{ kJ}$

41. The work done in the constant-pressure process is

$$W = P\Delta V$$
$$W = (1909 \text{ kPa})(7-11) \times 10^{-3} \text{ m}^{-3}$$
$$W = -7.6 \text{ kJ}$$

No work is done in the constant volume process, so this is the net work done.

The initial temperature is

$$P_1 V_1 = nRT_1$$
$$T_1 = \frac{P_1 V_1}{nR}$$
$$T_1 = \frac{(1909 \text{ kPa})(11 \times 10^{-3} \text{ m}^{-3})}{n(8.314)}$$
$$T_1 = \frac{2526}{n}$$

The final temperature is

$$P_2 V_2 = nRT_2$$
$$T_2 = \frac{P_2 V_2}{nR}$$
$$T_2 = \frac{(3000 \text{ kPa})(7 \times 10^{-3} \text{ m}^{-3})}{n(8.314)}$$
$$T_2 = \frac{2526}{n}$$

The net change in internal energy of the system is 0 because the initial and final temperatures are the same. Thus, there is a net flow of thermal energy out of the system:

$$Q = W + \Delta U$$
$$Q = W + 0$$
$$Q = -7.6 \text{ kJ}$$

43. Because the process from A to B is isothermal, we can calculate the volume of the gas at B as follows:

$$P_1 V_1 = P_2 V_B$$
$$V_B = \frac{P_1 V_1}{P_2}$$
$$V_B = \frac{(1.00 \times 10^5 \text{ Pa})(0.50 \text{ m}^3)}{1.50 \times 10^5 \text{ Pa}}$$
$$V_B = 0.33 \text{ m}^3$$

Similarly, the process from C to D is isothermal, we can calculate the volume of the gas at C as follows:

$$P_1 V_2 = P_2 V_C$$

$$V_C = \frac{P_1 V_2}{P_2}$$

$$V_C = \frac{\left(1.00 \times 10^5 \text{ Pa}\right)\left(2.00 \text{ m}^3\right)}{1.50 \times 10^5 \text{ Pa}}$$

$$V_C = 1.33 \text{ m}^3$$

Now we have the information we need to calculate the work done in each phase of the process:

$$W_{AB} = nRT \ln\left(\frac{V_B}{V_1}\right)$$

$$W_{AB} = P_1 V_1 \ln\left(\frac{V_B}{V_1}\right)$$

$$W_{AB} = \left(1.00 \times 10^5 \text{ Pa}\right)\left(0.50 \text{ m}^3\right) \ln\left(\frac{1/3}{1/2}\right)$$

$$W_{AB} = \left(1.00 \times 10^5 \text{ Pa}\right)\left(0.50 \text{ m}^3\right) \ln\left(\frac{2}{3}\right)$$

$$W_{AB} = -20.3 \text{ kJ}$$

$$W_{BC} = P_2\left(V_C - V_B\right)$$

$$W_{BC} = \left(1.50 \times 10^5 \text{ Pa}\right)\left(1.33 \text{ m}^3 - 0.33 \text{ m}^3\right)$$

$$W_{BC} = 150 \text{ kJ}$$

$$W_{CD} = nRT \ln\left(\frac{V_2}{V_C}\right)$$

$$W_{CD} = P_1 V_2 \ln\left(\frac{V_2}{V_C}\right)$$

$$W_{CD} = \left(1.00 \times 10^5 \text{ Pa}\right)\left(2.0 \text{ m}^3\right) \ln\left(\frac{2.0}{4/3}\right)$$

$$W_{CD} = \left(1.00 \times 10^5 \text{ Pa}\right)\left(2.0 \text{ m}^3\right) \ln\left(1.5\right)$$

$$W_{CD} = 81.1 \text{ kJ}$$

$$W_{DA} = P_1\left(V_1 - V_2\right)$$

$$W_{DA} = \left(1.00 \times 10^5 \text{ Pa}\right)\left(0.50 \text{ m}^3 - 2.00 \text{ m}^3\right)$$

$$W_{DA} = -150 \text{ kJ}$$

Thus, the total work done in the entire process is

$$W_{AB} + W_{BC} + W_{CD} + W_{DA} = -20.3 \text{ kJ} + (150 \text{ kJ}) + 81.1 \text{ kJ} + (-150 \text{ kJ})$$
$$W_{AB} + W_{BC} + W_{CD} + W_{DA} = 61 \text{ kJ}$$

45. The initial volume can be calculated as

$$PV_1 = nRT_1$$
$$V_1 = \frac{nRT_1}{P}$$
$$V_1 = \frac{(1)(8.314)(273.16 + 50)}{100\,000}$$
$$V_1 = 0.02687 \text{ m}^3$$

The final volume is three times the initial volume:

$$V_2 = 3 \times 0.02687 \text{ m}^3$$
$$V_2 = 0.0806 \text{ m}^3$$

The final temperature is

$$T_2 = \frac{PV_2}{nR}$$
$$T_2 = \frac{(100\,000 \text{ Pa})(0.0806 \text{ m}^3)}{(1)(8.314)}$$
$$T_2 = 970 \text{ K}$$

The work done is

$$W = P\Delta V$$
$$W = (100\,000 \text{ Pa})(0.0806 \text{ m}^3 - 0.02687 \text{ m}^3)$$
$$W = 5.37 \text{ kJ}$$

The change in internal energy is

$$\Delta U = \frac{3}{2}nR\Delta T$$
$$\Delta U = \frac{3}{2}nR(T_2 - T_1)$$
$$\Delta U = \frac{3}{2}(1)(8.314)(969.48 \text{ K} - 323.16 \text{ K})$$
$$\Delta U = 8.06 \text{ kJ}$$

Thus, the amount of thermal energy that flows is

$Q = \Delta U + W$

$Q = 8.06 \text{ kJ} + 5.37 \text{ kJ}$

$Q = 13.4 \text{ kJ}$

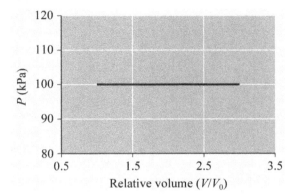

47. $Q = mc\Delta T$

$$\frac{Q}{\Delta t} = mc \left(\frac{\Delta T}{\Delta t} \right)$$

The quantity on the left side of the previous equation is known (it's the power delivered by the heater), and the quantity in parentheses is the slope of the plotted data.

A plot of the data is shown below.

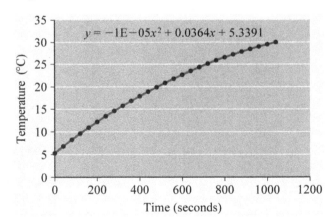

We can see that it is not simply a straight line. This is because the container absorbs heat from the room when it is below room temperature and loses heat to the room when it is above room temperature. Thus, we need the rate of change of the temperature at room temperature where the container is only absorbing heat from the heater. A simple second order fit to the data yields

$T = -1 \times 10^{-5} t^2 + 0.0363t + 5.3391$

$$\therefore \frac{dT}{dt} \bigg|_{t=480 \text{ s}} = -2 \times 10^{-5} 480 + 0.0363 = 0.0267 \text{ K/s}$$

Thus,

$$m = \dfrac{\dfrac{Q}{\Delta t}}{c\dfrac{\Delta T}{\Delta t}} = \dfrac{10\ \text{J/s}}{\left(4186\ \dfrac{\text{J}}{\text{K}\times\text{kg}}\right)\left(0.0267\ \dfrac{\text{K}}{\text{s}}\right)} = 0.09\ \text{kg}$$

Chapter 18—HEAT ENGINES AND THE SECOND LAW OF THERMODYNAMICS

1. This is not a good idea, because more thermal energy will be deposited into the room (from the heating coils on the back of the refrigerator) than will be removed by cooling, as the refrigerator is not perfectly efficient.

3. The waste heat from a heat engine is delivered to a low temperature reservoir. In order to get that heat to the high temperature input reservoir, some work would need to be done by a heat pump which, at best, would offset any additional work extracted from the heat engine.

5. No. If the temperature difference between the waste energy source and the heat reservoir of the heat engine is too great, the heat pump will use more energy than it transfers to the heat reservoir.

7. In order to clean your room, you need to convert energy that is stored in your body to make your muscles move, and in the process heat is released. This conversion of stored energy into heat causes an increase in the entropy of the universe that exceeds the decrease associated with the tidying.

9. Another example is a pendulum swinging in a vacuum on a frictionless pivot. The time-reversed motion is also perfectly realistic.

11. (a) $5.8 \dfrac{L}{100 \text{ km}} \times 100 \dfrac{\text{km}}{\text{h}} \times \dfrac{1 \text{ h}}{3600 \text{ s}} = 1.6 \text{ mL/s}$

    (b) $33 \dfrac{\text{MJ}}{\text{L}} \times 1.6 \dfrac{\text{mL}}{\text{s}} = 53 \text{ kJ/s}$

13. (a) $P = \dfrac{A\Delta T}{R}$

    $P = \dfrac{(2)(20)}{3.2}$

    $P = 12.5 \text{ W}$

    (b) The refrigerator must remove thermal energy from the interior of the refrigerator at the same rate that it flows in through the walls, so as to maintain a constant temperature.

    (c) The amount of energy released per second by the propane combustion is

    $50 \dfrac{\text{MJ}}{\text{kg}} \times 0.50 \dfrac{\text{kg}}{\text{day}} \times \dfrac{1 \text{ day}}{24 \text{ h}} \times \dfrac{1 \text{ h}}{3600 \text{ s}} = 290 \text{ W}$

(d) The amount of thermal energy delivered to the surroundings external to the refrigerator is the energy delivered by the propane burner and the energy delivered by the back of the refrigerator, which is

290 W + 12.5 W = 300 W

15. (a) $\eta = 1 - \dfrac{T_C}{T_H}$

$\eta = 1 - \dfrac{(500 + 273)}{(1000 + 273)}$

$\eta = 39.28\%$

(b) $\eta = \dfrac{|W|}{|Q_H|}$

$|W| = \eta |Q_H|$

$|W| = 0.3928 \times 100.0 \text{ kJ}$

$|W| = 39.28 \text{ kJ}$

17. $CP_H = \dfrac{|Q_H|}{|W|}$

$|W| = \dfrac{|Q_H|}{CP_H}$

$|W| = \dfrac{63 \text{ MJ}}{16.3}$

$|W| = 3.9 \text{ MJ}$

19. (a) The constant pressure molal specific heat of a monatomic ideal gas is $C_p = \dfrac{5}{2}R$. The total amount of thermal energy that flows is

$$Q = n\dfrac{5}{2}R\Delta T$$

and therefore the thermal energy that flows per mole is

$$\dfrac{Q}{n} = \dfrac{5}{2}R10 = (25 \text{ K})R$$

(b) The change in entropy of the reservoir, per mole of gas cooled is

$$\dfrac{\Delta S_{reservoir}}{n} = \dfrac{Q}{T} = \dfrac{25R}{200} = 0.125 \ R$$

(c) The change in entropy of the gas, per mole, is

$$\frac{\Delta S_{gas}}{n} = \int_{T_i}^{T_f} \frac{dQ}{T} = \int_{T_i}^{T_f} \frac{\frac{5}{2}RdT}{T} = \frac{5}{2}R\ln\frac{T_f}{T_i} = -0.122R$$

(d) The overall change in entropy of the system, per mole of gas cooled, is

$$\frac{\Delta S}{n} = \frac{\Delta S_{gas}}{n} + \frac{\Delta S_{reservoir}}{n} = -0.122R + 0.125R = 0.003R$$

(e) The net change in entropy of the system is positive, so the process is not reversible.

21. (a) Using a value of 43 J/(m·K) for the thermal conductivity of steel, the rate at which thermal energy flows through the bottom of the pot is

$$P = \frac{\kappa A \Delta T}{x}$$

$$P = \frac{(43)(0.030)(500.0 - 10.0)}{0.010}$$

$$P = 63 \text{ kW}$$

(b) The rate of change of entropy of the burner (assuming its temperature is constant) is

$$\frac{\Delta S}{\Delta t} = \frac{Q/\Delta t}{T}$$

$$\frac{\Delta S}{\Delta t} = \frac{P}{T}$$

$$\frac{\Delta S}{\Delta t} = \frac{-63 \text{ kW}}{(500 + 273)K}$$

$$\frac{\Delta S}{\Delta t} = -82 \text{ J/(K·s)}$$

(c) The rate of change of entropy of the water is calculated as follows:

$$dS = \frac{1}{T}dQ$$

$$\frac{dS}{dt} = \frac{1}{T} \cdot \frac{dQ}{dt}$$

$$\frac{dS}{dt} = \frac{1}{(273 + 10)\text{ K}} \times 63 \text{ kW}$$

$$\frac{dS}{dt} = 220 \text{ J/(K·s)}$$

Note that the rate of change of the entropy of the water depends on its temperature; the calculated value is the initial rate of change of entropy.

(d) The net rate of change of entropy (initially) is $220 \text{ J/(K·s)} - 82 \text{ J/(K·s)} = 140 \text{ J/(K·s)}$.

23. The volume of such a cube is

$$V = (20 \text{ nm})^3$$
$$V = (2 \times 10^{-8} \text{ m})^3$$
$$V = 8 \times 10^{-24} \text{ m}^3$$

The density of silicon is 2329 kg/m³, so the approximate mass of silicon atoms in the cube is

$$m = \rho V = (2329)(8 \times 10^{-24}) = 1.86 \times 10^{-20} \text{ kg} = 1.86 \times 10^{-17} \text{ g}$$

Because the atomic mass of silicon is 28.085 g/mol, the number of moles of silicon in the cube is

$$n = \frac{1.86 \times 10^{-17} \text{ g}}{28.085 \text{ g/mol}} = 6.6 \times 10^{-19} \text{ mol}$$

Thus, the approximate number of silicon atoms in the cube is

$$6.6 \times 10^{-19} \text{ mol} \times 6.02 \times 10^{23} \text{ atoms/mol} = 4 \times 10^5 \text{ atoms}$$

Thus, there are about 400 000 silicon atoms in a cube with edge length 20 nm.

25. (a) $\Delta S = mc \ln\left(\frac{T_f}{T_i}\right)$

$$\Delta S = (200.0 \times 10^{-3})(385)\ln\left(\frac{99}{100}\right)$$
$$\Delta S = -0.77 \text{ J/K}$$

(b) $\Delta S = mc \ln\left(\frac{T_f}{T_i}\right)$

$$\Delta S = (200.0 \times 10^{-3})(385)\ln\left(\frac{9}{10}\right)$$
$$\Delta S = -8.1 \text{ J/K}$$

(c) $\Delta S = mc \ln\left(\frac{T_f}{T_i}\right)$

$$\Delta S = (200.0 \times 10^{-3})(385)\ln\left(\frac{1}{2}\right)$$
$$\Delta S = -53 \text{ J/K}$$

(d) $\Delta S = mc \ln\left(\dfrac{T_f}{T_i}\right)$

$\Delta S = \left(200.0 \times 10^{-3}\right)(385)\ln\left(\dfrac{0.01}{0.02}\right)$

$\Delta S = -53$ J/K

27. No. Heat engines and heat pumps are never 100 percent efficient. The amount of work that the heat pump could do to restore a temperature difference would always be less than the energy input into the engine, which would make the temperature difference even smaller.

29. The data supplied indicate that the engine would have an efficiency of 55.5%, which is greater than the maximum possible efficiency of 50% for the given temperatures. Since the data are erroneous, the application should be rejected.

31. For simplicity, we will assume that if there are N particles they are evenly spaced out along the length of the system L. We will also assume that the spacing of the particles does not get less than ΔL, the width of the detector. There will then be N intervals of time, each of duration $\Delta L/v$, where a particle is in the detector and the detector reads Y_0. The average detector reading will then be

$$\overline{Y} = \frac{1}{T}\int_0^T Y(t)\,dt = NY_0\frac{\Delta L}{L}$$

Similarly, we can calculate the square of the standard deviation as

$$\sigma_Y^2 = \frac{1}{T}\int_0^T \left(Y - \overline{Y}\right)^2 dt = \frac{1}{T}\int_0^T \left(Y^2 - 2Y\overline{Y} + \overline{Y}^2\right)dt$$

$$= \frac{1}{T}\left(Y_0^2 \frac{N\Delta L}{v} - 2NY_0 \frac{\Delta L}{L} Y_0 \frac{N\Delta L}{v} + N^2 Y_0^2 \left(\frac{\Delta L}{L}\right)^2 \frac{L}{v}\right)$$

$$= \frac{N\Delta L}{L} Y_0^2 \left(1 - 2N\frac{\Delta L}{L} + N\frac{\Delta L}{L}\right) = Y_0^2 N \frac{\Delta L}{L}\left(1 - N\frac{\Delta L}{L}\right)$$

To see how this behaves as the number of particles increases, it is most interesting to examine the ratio

$$\frac{\sigma_Y}{\overline{Y}} = \frac{\sqrt{Y_0^2 N \dfrac{\Delta L}{L}\left(1 - N\dfrac{\Delta L}{L}\right)}}{N\dfrac{\Delta L}{L}} = \frac{\sqrt{\left(1 - N\dfrac{\Delta L}{L}\right)}}{\sqrt{N\dfrac{\Delta L}{L}}}$$

This ratio goes to zero when $N \dfrac{\Delta L}{L} = 1$. At this point, there is one particle in the detector at all times. Clearly, the simple model we have used here breaks down when the number of particles $N > \dfrac{L}{\Delta L}$.

33. The increase in the entropy of the water is the sum of two terms, the increase in entropy as its temperature is increased to the boiling point, and the increase in entropy as it vaporizes.

$$\Delta S = mc \ln\left(\frac{T_f}{T_i}\right) + \frac{Q}{T_f}$$

$$\Delta S = mc \ln\left(\frac{T_f}{T_i}\right) + \frac{mL}{T_f}$$

$$\Delta S = (1\text{ kg})(4186\text{ J/(kg} \cdot \text{K)}) \ln\left(\frac{373}{293}\right) + \frac{(1\text{ kg})(22.6 \times 10^5\text{ J/kg})}{373\text{ K}}$$

$$\Delta S = 7\text{ kJ/K}$$

35. The change in entropy of the wood is

$$\Delta S = \frac{Q}{T}$$

$$\Delta S = \frac{-20 \times 10^9\text{ J}}{(500 + 273)\text{ K}}$$

$$\Delta S = -2.6 \times 10^7\text{ J/K}$$

The change in entropy of the room is

$$\Delta S = \frac{Q}{T}$$

$$\Delta S = \frac{20 \times 10^9\text{ J}}{(20 + 273)\text{ K}}$$

$$\Delta S = 6.8 \times 10^7\text{ J/K}$$

The net change in entropy of the entire system is
$6.8 \times 10^7\text{ J/K} - 2.6 \times 10^7\text{ J/K} = 4.2 \times 10^7\text{ J/K}$.

37. Because the volume is constant, the work done in the process is zero. Thus, the amount of thermal energy that flows is equal to the change in internal energy of the ideal gas.

Thus, the change in entropy is

$$\Delta S = \int_{T_i}^{T_f} \frac{dQ}{T}$$

$$\Delta S = \int_{T_i}^{T_f} \frac{dU}{T}$$

$$\Delta S = \frac{3}{2} nR \int_{T_i}^{T_f} \frac{dT}{T}$$

$$\Delta S = \frac{3}{2} nR \left[\ln T \right]_{T_i}^{T_f}$$

$$\Delta S = \frac{3}{2} nR \left[\ln \left(T_f \right) - \ln \left(T_i \right) \right]$$

$$\Delta S = \frac{3}{2} nR \ln \left(\frac{T_f}{T_i} \right)$$

$$\Delta S = (1.5)(10)(8.314) \ln \left(\frac{400 \text{ K}}{300 \text{ K}} \right)$$

$$\Delta S = 35.9 \text{ J/K}$$

39. Assuming that the coefficient of performance depends on the difference in the temperatures as follows

$$CP_H = \frac{A}{\Delta T}$$

where A is a constant. Using the given data, the value of A is

$$A = CP_H \Delta T$$
$$A = (3.5)(20 - 10)$$
$$A = 35$$

When the air temperature outside drops to −10°C, the coefficient of performance is

$$CP_H = \frac{A}{\Delta T}$$

$$CP_H = \frac{35}{20 - (-10)}$$

$$CP_H = \frac{35}{30}$$
$$CP_H = 1.17$$

41. (a) The rate at which thermal energy crosses from water to air is

$$\frac{P}{A} = \frac{\kappa \Delta T}{z}$$

where z is the thickness of the ice layer in metres, κ is the thermal conductivity of the ice and ΔT is the difference in temperature between the lake water below the ice and the air.

(b) The rate at which the entropy of the air changes, per square metre of ice per second, is

$$\frac{1}{A}\frac{dS_{air}}{dt} = \frac{\frac{1}{A}\frac{dQ}{dt}}{T_{air}} = \frac{\frac{P}{A}}{T_{air}} = \frac{\kappa\Delta T}{zT_{air}}$$

(c) The rate at which the entropy of the water changes, per square metre of ice per second, is

$$\frac{1}{A}\frac{dS_{water}}{dt} = \frac{-\frac{1}{A}\frac{dQ}{dt}}{T_{water}} = \frac{-\frac{P}{A}}{T_{water}} = \frac{-\kappa\Delta T}{zT_{water}}$$

(d) As thermal energy leaves the water, the water can freeze. The amount of thermal energy that must be removed to freeze a volume of water V is

$$Q = \rho L V$$

where ρ is the density of the water, L is the latent heat. We can write the volume of the water in terms of the area A and the thickness z $V = Az$. We can insert this in the previous equation and solve for z to find

$$z = \frac{Q}{\rho L A}$$

To find the rate of change of the ice thickness, we differentiate this with respect to time:

$$\frac{dz}{dt} = \frac{\frac{dQ}{dt}}{\rho L A} = \frac{P}{A}\frac{1}{\rho L}$$

(e) We substitute the expression for power per unit area from part (a) into the rate of change of thickness equation we found in part (d) to get

$$\frac{dz}{dt} = \frac{\kappa\Delta T}{z}\frac{1}{\rho L} = \frac{\kappa\Delta T}{\rho L}\frac{1}{z}$$

If we set $z = 0$ when $t = 0$, we can solve this (use Maple) to find

$$z(t) = \sqrt{\frac{2\kappa\Delta T}{\rho L}t}$$

and we see that the ice grows as the square root of time. The ice acts as an insulator and as it gets thicker it reduces the growth rate.

43. We assume the ball starts at a temperature T and after a long time in contact with the ground it ends up at a temperature T. All of the kinetic energy the ball had when it hit the ground is dissipated as thermal energy in the ground. Thus, the change in entropy of the whole system results from a flow of energy into the ground at a temperature T:

$$\Delta S = \frac{mgh}{T}$$

The initial state is more ordered because the energy is all stored in the ball in terms of potential energy. Afterwards, the energy is dissipated as heat into the ground.

45. No, the second law of thermodynamics is not violated, because only minute portions of the universe are currently highly ordered; overall, the amount of order has decreased.

47. In a cycle of the Carnot engine, thermal energy $|Q_H|$ is transferred from the hot reservoir, thermal energy $|Q_C|$ is transferred to the cold reservoir, and work $|W|$ is done, where

$$|Q_H| = |W| + |Q_C|$$

This work done is then used to drive a heat engine that operates between the same two reservoirs, transferring thermal energy back from the cold reservoir to the hot reservoir.

If the heat pump is more efficient than the Carnot engine, then the amount of thermal energy transferred back to the hot reservoir is greater than $|Q_H|$. This means that with each cycle of the combined processes, thermal energy is transferred from the cold reservoir to the hot reservoir with no work being done. This violates the Clausius statement of the second law of thermodynamics, and so we can conclude that the efficiency of the heat pump cannot be greater than the efficiency of the Carnot engine.

If the efficiency of the heat pump is equal to the efficiency of the Carnot engine, then the combined process can be cycled indefinitely, returning the hot and cold reservoirs to the same state at the end of each cycle, with no expenditure of work. But this contradicts the assumption that the heat engine is irreversible, so this is not possible.

This leaves the third option as the only possibility: the efficiency of the irreversible engine must be less than the efficiency of the Carnot engine.

49. Answers will vary. For a sea with a temperature around 22°C at the surface and 4°C at some depth below the surface, the maximum theoretical efficiency would be about 6%. Extracting energy would lower the water temperature, and could affect local sea life. Large-scale operations could alter ocean currents, and affect coastal erosion, silt deposits, and possibly even Earth's climate. The core of Earth has a temperature of about 5430°C, so the theoretical efficiency with a surface temperature of about 22°C is around 95%.

Large-scale operations could affect aquifers and local wildlife habitat.

1. c. According to Table 19-1, negative charge will be transferred from glass to polyethylene, so polyethylene will end up with a negative charge and glass will end up with a positive charge when the tape is removed.

3. The answer is (b) plastic wrap and glass, since these materials appear the furthest away from each other in the triboelectric series.

5. c. $I = \dfrac{\Delta q}{\Delta t} = \dfrac{50 \times 10^{-6} \text{ C}}{0.05 \text{ s}} = 1.0 \times 10^{-3} \dfrac{\text{C}}{\text{s}} = 1 \text{ mA}$

7. b. You can argue this by quoting Newton's third law of motion, or as follows:

$$\left| \vec{F}_{1 \to 2} \right| = \frac{1}{4\pi\varepsilon_0} \frac{q \cdot 4q}{r^2} \hat{r}_{1 \to 2} = F$$

$$\left| \vec{F}_{2 \to 1} \right| = \frac{1}{4\pi\varepsilon_0} \frac{4q \cdot q}{r^2} \hat{r}_{2 \to 1} = F = \left| \vec{F}_{1 \to 2} \right|$$

9. a. For the first situation with two $+Q$ charges, the magnitude of the force is

$$\left| \vec{F}_{+Q \to +Q} \right| = \left| \frac{1}{4\pi\varepsilon_0} \frac{(+Q)(+Q)}{L^2} \hat{r} \right| = \frac{Q^2}{4\pi\varepsilon_0 L^2}$$

For the second situation involving the two charges $+2Q$ and $-2Q$,
the magnitude of the force is

$$\left| \vec{F}_{+2Q \to -2Q} \right| = \left| \frac{1}{4\pi\varepsilon_0} \frac{(+2Q)(-2Q)}{(2L)^2} (\hat{r}) \right| = \frac{Q^2}{4\pi\varepsilon_0 L^2}$$

Therefore, the magnitudes of the forces are the same.

11. c. The electric field at the point $x = 0$, $y = -1$, $z = 0$ is

$$\vec{E} = \frac{1}{4\pi\varepsilon_0} \frac{Q}{r^2} \hat{r} = \frac{1}{4\pi\varepsilon_0} \frac{(+Q)}{(1)^2} (-\hat{j}) = \frac{Q}{4\pi\varepsilon_0} (-\hat{j})$$

Therefore, the electric field points in the $(-\hat{j})$ direction.

13. b. The electric field due to the positive charge is

$$\vec{E}_1 = \frac{1}{4\pi\varepsilon_0} \frac{+Q}{r^2} (\hat{i})$$

The electric field due to the negative charge is

$$\vec{E}_2 = \frac{1}{4\pi\varepsilon_0} \frac{-Q}{r^2} (\hat{i}) = \vec{E}_1$$

The sum of the electric fields midway between the two charges is

$$\vec{E}_1 + \vec{E}_2 = \frac{1}{4\pi\varepsilon_0}\frac{+Q}{r^2}(\hat{i}) + \frac{1}{4\pi\varepsilon_0}\frac{-Q}{r^2}(\hat{i}) = \frac{1}{2\pi\varepsilon_0}\frac{Q}{r^2}(\hat{i})$$

Therefore, the direction of the electric field midway between the two charges is $\hat{i}$.

15. b. The positive ends of the dielectric molecules are attracted toward the negative external charges (and the negative ends of the dielectric molecules are attracted to the positive external charges).

17. $I = \dfrac{\Delta q}{\Delta t} = \dfrac{(250\times10^9 e)}{1\text{ s}} = \dfrac{(250\times10^9)(1.60\times10^{-19}\text{ C})}{1\text{ s}} = 40\times10^{-9}$ C/s $= 40$ nA

19. (a) $\left|\vec{F}_{e^- \to p^+}\right| = \left|\dfrac{1}{4\pi\varepsilon_0}\dfrac{(+e)(-e)}{(53.0\times10^{-12}\text{ m})^2}\right|$

$\left|\vec{F}_{e^- \to p^+}\right| = \dfrac{1}{4\pi}\left|\dfrac{(1.60\times10^{-19}\text{ C})(-1.60\times10^{-19}\text{ C})}{(8.85 \times 10^{-12}\text{ m}^{-3}\text{kg}^{-1}\text{s}^4\text{A}^2)(2.8\times10^{-21}\text{ m}^2)}\right|$

$\left|\vec{F}_{e^- \to p^+}\right| = 8.22\times10^{-8}\ \dfrac{\text{C}^2}{\text{m}^{-3}\text{kg}^{-1}\text{s}^4\text{A}^2\text{m}^2}$

$\left|\vec{F}_{e^- \to p^+}\right| = 8.22\times10^{-8}\ \dfrac{\text{C}^2}{\text{m}^{-1}\text{kg}^{-1}\text{s}^4\left(\frac{\text{C}}{\text{s}}\right)^2}$

$\left|\vec{F}_{e^- \to p^+}\right| = 8.22\times10^{-8}\ \dfrac{\text{kg}\cdot\text{m}}{\text{s}^2}$

$\left|\vec{F}_{e^- \to p^+}\right| = 8.22\times10^{-8}$ N

The force direction is attractive, along the line joining the particles.

(b) $m\dfrac{v^2}{r} = 8.22\times10^{-8}$ N

$v = \sqrt{\dfrac{r\times8.22\times10^{-8}\text{ N}}{m}}$

$v = \sqrt{\dfrac{(53.0\text{ pm})\times8.22\times10^{-8}\text{ N}}{9.11\times10^{-31}\text{ kg}}}$

$v = 2.19\times10^6$ m/s

21. The positive charge must be placed to the left of the 1.00 μC charge in order for the net force on the latter charge to be zero. Let x represent the position of the new charge. Then,

$$\frac{1}{4\pi\varepsilon_0}\frac{2.00\ \mu\text{C}}{(3.00\ \text{cm})^2} = \frac{1}{4\pi\varepsilon_0}\frac{4.00\ \mu\text{C}}{x^2}$$

$$x^2 = 2(3.00\ \text{cm})^2$$

$$x = \sqrt{2}(3.00\ \text{cm})$$

$$x = 4.24\ \text{cm}$$

Thus, the new charge must be 4.24 cm left of the 1.00 μC charge.

23. $F = qE$

$$F = 2(1.60\times10^{-19}\ \text{C})(5.00\ \text{N}/\mu\text{C})$$

$$F = 1.60\times10^{-12}\ \text{N}$$

25. $p = qd$

$$1.75\times3.34\times10^{-30}\ \text{C}\cdot\text{m} = ed$$

$$d = \frac{1.75\times3.34\times10^{-30}\ \text{C}\cdot\text{m}}{1.60\times10^{-19}\ \text{C}}$$

$$d = 3.65\times10^{-11}\ \text{m}$$

$$d = 0.0365\ \text{nm}$$

27. The linear charge density of the charge on the rod is

$$dq = \mu dx$$

$$dq = \frac{75.0\ \text{nC}}{25.0\ \text{mm}}dx$$

$$dq = (3.00\ \mu\text{C/m})\,dx$$

The magnitude of the electric force exerted by the rod on the point charge is

$$F = \int_{0.100}^{0.125}\frac{1}{4\pi\varepsilon_0}\frac{Qdq}{x^2}$$

$$F = \int_{0.100}^{0.125}\frac{1}{4\pi\varepsilon_0}\frac{Q\big((3.00\ \mu\text{C/m})dx\big)}{x^2}$$

$$F = \frac{Q(3.00\ \mu\text{C/m})}{4\pi\varepsilon_0}\int_{0.100}^{0.125}\frac{dx}{x^2}$$

$$F = \frac{(1.00\ \mu\text{C})(3.00\ \mu\text{C/m})}{4\pi\varepsilon_0}\left[-\frac{1}{x}\right]_{0.100}^{0.125}$$

$$F = \frac{(1.00 \ \mu\text{C})(3.00 \ \mu\text{C/m})}{4\pi(8.85 \times 10^{-12})} \left[-\frac{1}{0.125} + \frac{1}{0.100} \right]$$

$$F = 0.0540 \text{ N}$$

29. $I = \frac{\Delta Q}{\Delta t} \quad \Rightarrow \quad \Delta t = \frac{\Delta Q}{I}$

$$\Delta t = \frac{350 \text{ C}}{120 \text{ kA}}$$

$$\Delta t = 2.92 \text{ ms}$$

31. $\dfrac{Gm_1 m_2}{r^2} = \dfrac{1}{4\pi\varepsilon_0} \dfrac{Q^2}{r^2}$

$$Q = \sqrt{4\pi\varepsilon_0 Gm_1 m_2}$$

$$Q = \sqrt{4\pi(8.85 \times 10^{-12})(6.67 \times 10^{-11})(5.98 \times 10^{24} \text{ kg})(7.35 \times 10^{22})}$$

$$Q = 5.71 \times 10^{13} \text{ C}$$

33. By symmetry, the force on each charge is directed away from the centre of the triangle. The magnitude of the force on each charge is

$$F = \frac{1}{4\pi\varepsilon_0} \frac{Q^2}{a^2} \cos(30°) + \frac{1}{4\pi\varepsilon_0} \frac{Q^2}{a^2} \cos(30°)$$

$$F = \frac{2\cos(30°)}{4\pi\varepsilon_0} \frac{Q^2}{a^2}$$

$$F = \frac{2(0.8660)}{4\pi(8.85 \times 10^{-12})} \frac{(0.250 \ \mu\text{C})^2}{(0.0325 \text{ m})^2}$$

$$F = 0.922 \text{ N}$$

35. The two positive charges are equal and are equidistant from the point P, so the field from each of them balances to zero at P. Similarly, the two negative charges are equal and are also equidistant from the point P, so the field from each of them also balances to zero at P. Thus, the net electric field at P due to the four charges is zero.

37. $F = eE = ma$

$$E = \frac{ma}{e}$$

$$E = \frac{(1.67 \times 10^{-27})(4.0 \times 10^{8})}{1.60 \times 10^{-19}}$$

$$E = 4.2 \text{ N/C}$$

The direction of the electric field is in the positive x-direction.

39. Sketching a diagram and considering the directions and magnitudes of the fields will lead you to conclude that the only possible place at which the electric field is zero is on the x-axis to the right of the electron. Let x represent the position at which the electric field is zero. Then,

$$\frac{1}{4\pi\varepsilon_0}\frac{3e}{x^2} = \frac{1}{4\pi\varepsilon_0}\frac{e}{(x-0.20 \ \mu m)^2}$$

$$3\left(x-0.20\times10^{-6}\right)^2 = x^2$$

$$\sqrt{3}\left(x-0.20\times10^{-6}\right) = x$$

$$\left(\sqrt{3}-1\right)x = \sqrt{3}\left(0.20\times10^{-6}\right)$$

$$x = \frac{\sqrt{3}\left(0.20\times10^{-6}\right)}{\left(\sqrt{3}-1\right)}$$

$$x = 0.473 \ \mu m$$

This is equivalently 0.273 μm to right of the electron.

41. The deflection of the drop in the y-direction is

$$\Delta y = \frac{1}{2}a\left(\Delta t\right)^2$$

and therefore

$$a = \frac{2\Delta y}{\left(\Delta t\right)^2}$$

The drop is deflected while it is in the deflection region, so the time interval is

$$\Delta t = \frac{\Delta x}{v}$$

and thus, the acceleration of the drop in the deflection region is

$$a = \frac{2\Delta y}{\left(\Delta x/v\right)^2}$$

$$a = \frac{2v^2\Delta y}{\left(\Delta x\right)^2}$$

By Newton's second law,

$$a = \frac{F}{m} = \frac{qE}{m}$$

and therefore the charge on the droplet is

$$\frac{qE}{m} = \frac{2v^2 \Delta y}{(\Delta x)^2}$$

$$q = \frac{2v^2 \Delta y m}{E(\Delta x)^2}$$

$$q = \frac{2v^2 \Delta y \rho V}{E(\Delta x)^2}$$

$$q = \frac{2(25.0 \text{ m/s})^2 (0.225 \times 10^{-3} \text{ m})(1.00 \text{ kg/L})(2.00 \times 10^{-12} \text{ L})}{(95.0 \text{ kN/C})(0.0125 \text{ m})^2}$$

$$q = 3.79 \times 10^{-14} \text{ C}$$

43. $E = [30 \text{ kN/C}]\hat{i}$

The torque on the dipole is

$$\vec{\tau} = (r_1 F_1 + r_2 F_2)(-\hat{k})$$

$$\vec{\tau} = (r_1 eE + r_2 eE)(-\hat{k})$$

$$\vec{\tau} = 2reE(-\hat{k})$$

$$\vec{\tau} = 2(25 \text{ nm})(1.6 \times 10^{-19} \text{ C})(30 \text{ kN/C})(-\hat{k})$$

$$\vec{\tau} = (2.4 \times 10^{-22} \text{ N} \cdot \text{m})(-\hat{k})$$

45. By symmetry, the component of the electric field in the direction perpendicular to the z-axis is zero. Let Q represent the entire charge on the ring; then the amount of charge on an element of the ring of length ds is

$$dQ = \left(\frac{Q}{2\pi R}\right) ds$$

where R is the radius of the ring. The magnitude of the electric field at the indicated field point is

$$\vec{E} = \hat{k} \int \left(\frac{1}{4\pi\varepsilon_0}\right) \left(\frac{Q}{2\pi R}\right) \frac{\cos\theta}{r^2} ds$$

$$\vec{E} = \hat{k} \left(\frac{1}{4\pi\varepsilon_0}\right) \left(\frac{Q}{2\pi R}\right) \int \frac{1}{r^2} \cdot \frac{0.10}{r} ds$$

$$\vec{E} = \hat{k} \left(\frac{1}{4\pi\varepsilon_0}\right) \left(\frac{Q}{2\pi R}\right) \left(\frac{0.10}{r^3}\right) \int ds$$

$$\vec{E} = \hat{k}\left(\frac{1}{4\pi\varepsilon_0}\right)\left(\frac{Q}{2\pi R}\right)\left(\frac{0.10}{r^3}\right)(2\pi R)$$

$$\vec{E} = \hat{k}\left(\frac{1}{4\pi\varepsilon_0}\right)\frac{0.10Q}{\left[R^2 + 0.10^2\right]^{3/2}}$$

$$\vec{E} = \hat{k}\left(\frac{1}{4\pi\left(8.85\times10^{-12}\right)}\right)\cdot\left(\frac{(0.10)(25.0\ \mu C)}{\left[(0.04)^2 + (0.10)^2\right]^{3/2}}\right)$$

$$\vec{E} = \left(1.80\times10^7\ \text{N/C}\right)\hat{k}$$

47. The electric field becomes weaker toward the right, so the point charge must be located to the left of $x = 2$ m. Therefore, the point charge must have a negative charge, because the electric field points toward the point charge.

$$E = \frac{Q}{4\pi\varepsilon_0}\cdot\frac{1}{(x-a)^2}$$

$$Q = 4\pi\varepsilon_0 E_a (x-a)^2$$

$$Q = 4\pi\varepsilon_0 E_b (x-b)^2$$

Dividing the two previous equations yields

$$1 = \frac{E_a(x-a)^2}{E_b(x-b)^2} \quad \text{so } E_a(x-a)^2 = E_b(x-b)^2$$

$$-180(x-4)^2 = -80(x-9)^2$$

$$9(x^2 - 8x + 16) = 4(x^2 - 18x + 81)$$

$$9x^2 - 72x + 144 = 4x^2 - 72x + 324$$

$$5x^2 = 180$$

$$x^2 = 36$$

$$x = \pm 6$$

$$x = -6.0\ \text{m}$$

$$Q = 4\pi\varepsilon_0 E_a (x-a)^2$$

$$Q = 4\pi\left(8.85\times10^{-12}\right)(-180\ \text{kN/C})(-6-4)^2$$

$$Q = -2.0\ \text{mC}$$

1. a

3. c. The potential energy is negative, and since total energy is constant the kinetic energy will be greatest (therefore highest speed) when the particles are closest together. This is at point C.

5. a. $\dfrac{Q}{4\pi\varepsilon_0\,|r|} - \dfrac{4Q}{4\pi\varepsilon_0\,|3-r|} = 0$

$$\frac{Q}{4\pi\varepsilon_0\,|r|} = \frac{4Q}{4\pi\varepsilon_0\,|3-r|}$$

$$\frac{1}{|r|} = \frac{4}{|3-r|}$$

$$|3-r| = 4|r|$$

$$r = -1,\ r = \frac{3}{5}$$

7. b. Regions of higher electric potential would repel and slow down an injected positive charge; on the other hand, regions of lower electric potential attract and speed up an injected positive charge.

9. d. The electric field is the negative of the gradient of the potential function, and so is zero in the y-direction and z-direction, and has x-component equal to $-6x$.

11. d. The flux is positive because electric field lines are directed outwards from the charge, and the result is independent of the placement of the charge because the flux integral is independent of the placement of the charge.

13. a. The net charge inside the surface is zero, so the net flux through the surface is also zero, by Gauss's law.

15. a. See the discussion on page 541.

17. $U_E = \dfrac{Qq}{4\pi\varepsilon_0 r}$

$$U_E = \frac{\left(1.602\times10^{-19}\right)\left(-1.602\times10^{-19}\right)}{4\pi\left(8.854\times10^{-12}\right)\left(53.0\times10^{-12}\ \text{m}\right)}$$

$$U_E = -4.35\times10^{-18}\ \text{J}$$

$$U_E = -27.2\ \text{eV}$$

19. $V = 4\left(\dfrac{Q}{4\pi\varepsilon_0 r}\right) = \dfrac{4 \times 2.00 \times 10^{-9}\ \text{C}}{4\pi\left(8.85 \times 10^{-12}\right)\left(0.05\sqrt{2}\right)} = 1.02\ \text{kV}$

21.

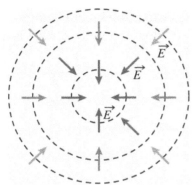

Electric field is stronger near the centre since larger change in potential for same change in distance.

23. (a) $V = -100.0 + 25.0y$

$E_x = -\dfrac{\partial V}{\partial x} = 0$

$E_y = -\dfrac{\partial V}{\partial y} = -25.0\ \text{N/C}$

$E_z = -\dfrac{\partial V}{\partial z} = 0$

(b) $\vec{E} = -25.0\,\hat{j}\ \text{N/C}$

25. (a) The equipotential surfaces are planes perpendicular to the x-axis.

(b) $-\Delta V = E\Delta x$

$\Delta x = -\dfrac{\Delta V}{E} = -\dfrac{(75.0 - 25.0)\ \text{V}}{-20\ \text{V/m}} = 2.5\ \text{m}$

Thus, $V = 75.0\text{V}$ at a position of $x = 2.5$ m.

27. (a) $\dfrac{2.45\ \text{eV}}{\text{molecule}} = 2.45 \times 1.6 \times 10^{-19}\ \dfrac{\text{J}}{\text{molecule}} = 3.92 \times 10^{-19}\ \text{J/molecule}$

(b) $\dfrac{3.92 \times 10^{-19}\ \text{J/molecule}}{4.184\ \text{J/cal}} = 9.38 \times 10^{-20}\ \text{cal/molecule}$

29. $\mu = \dfrac{165\ \mu\text{C}}{2\pi\left(0.0400\ \text{m}\right)} = 657\ \mu\text{C/m}$

31. $\rho = \dfrac{Q}{V} = \dfrac{3.00\ \mu C}{\frac{4}{3}\pi r^3} = \dfrac{3.00\ \mu C}{\frac{4}{3}\pi\left(0.0275\ \text{m}\right)^3} = 0.0344\ \text{C/m}^3$

33. The flux through the surface of the sphere is equal to the flux through the semicircular diameter of the sphere, which is

$$\Phi_E = \pi R^2 E$$

where E is the magnitude of the electric field and R is the radius of the sphere.

35. For $r < r_0$, a Gaussian sphere of radius r centred at the centre of the charge distribution encloses an amount of charge equal to

$$Q_r = \left(\tfrac{4}{3}\pi r^3\right)\rho$$

Thus, by Gauss's law,

$$E \cdot 4\pi r^2 = \dfrac{Q_r}{\varepsilon_0}$$

$$E = \dfrac{Q_r}{4\pi\varepsilon_0 r^2}$$

$$E = \dfrac{\left(\tfrac{4}{3}\pi r^3\right)\rho}{4\pi\varepsilon_0 r^2}$$

$$E = \left(\dfrac{\rho}{3\varepsilon_0}\right)r$$

The direction of the electric field is radially outward.

For $r \ge r_0$, a Gaussian surface of radius r centred at the centre of the charge distribution encloses all of the charge, so by Gauss's law

$$E \cdot 4\pi r^2 = \dfrac{\left(\tfrac{4}{3}\pi r_0^3\right)}{\varepsilon_0}\rho$$

$$E = \dfrac{\left(\tfrac{4}{3}\pi r_0^3\right)}{4\pi r^2 \varepsilon_0}\rho$$

$$E = \left(\dfrac{\rho r_0^3}{3\varepsilon_0}\right)\dfrac{1}{r^2}$$

The direction of the electric field is radially outward.

37. Because the inner cylinder is conducting, all of its excess charge lies on its surface. Thus, by Gauss's law, the electric field inside the inner conducting cylinder is zero.

In the gap between the two cylinders, draw a Gaussian surface that is a cylinder of radius r and length L coaxial with the other cylinders. By symmetry, the electric field is constant on the curved part of the cylinder, and the flux of the electric field through the ends is zero. Thus, by Gauss's law,

$$E \cdot A = Q / \varepsilon_0$$

$$E \cdot 2\pi r L = \frac{(15 \; \mu C) L}{\varepsilon_0}$$

$$E = \left(\frac{15 \; \mu C}{2\pi\varepsilon_0} \right) \frac{1}{r}$$

$$E = \frac{2.70 \times 10^5}{r} \frac{N}{C}$$

Similarly, outside the outer cylinder, draw a Gaussian surface that is a cylinder of radius r and length L coaxial with the other cylinders. By symmetry, the electric field is constant on the curved part of the cylinder, and the flux of the electric field through the ends is zero. Thus, by Gauss's law,

$$E \cdot A = Q / \varepsilon_0$$

$$E \cdot 2\pi r L = \frac{(5 \; \mu C) L}{\varepsilon_0}$$

$$E = \left(\frac{5 \; \mu C}{2\pi\varepsilon_0} \right) \frac{1}{r}$$

$$E = \frac{8.99 \times 10^4}{r} \frac{N}{C}$$

In both cases, the direction of the electric field is radially outward.

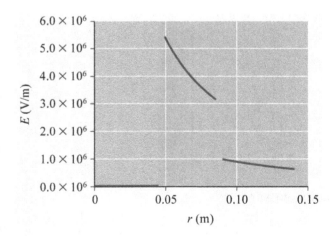

39. $U = \dfrac{q_1 q_2}{4\pi\varepsilon_0 r} = \dfrac{\left(1.602 \times 10^{-19}\right)^2}{4\pi\left(8.85 \times 10^{-12}\right)\left(3.00 \times 10^{-15} \; m\right)} = 7.69 \times 10^{-14} \; J$

41. The charged particles are identical, so by conservation of momentum they will have the same kinetic energy. The initial potential energy of the system is

$$U = \frac{q_1 q_2}{4\pi\varepsilon_0 r}$$

$$U = \frac{(25.0 \ \mu C)^2}{4\pi (8.85 \times 10^{-12})(35.0 \times 10^{-3})}$$

$$U = 160.568 \text{ J}$$

Thus, the kinetic energy of each particle is

$$K = \tfrac{1}{2} mv^2 = \tfrac{1}{2}(160.568 \text{ J})$$

$$v = \sqrt{\frac{160.568}{5.00 \times 10^{-3}}} = 179 \text{ m/s}$$

43. $\tfrac{1}{2} mv^2 = q\Delta V$

$$\Delta V = \frac{\tfrac{1}{2} mv^2}{e}$$

$$\Delta V = \frac{\tfrac{1}{2}(9.11 \times 10^{-31} \text{ kg})(3.0 \times 10^{5} \text{ m/s})^2}{1.60 \times 10^{-19} \text{ C}}$$

$$\Delta V = 0.256 \text{ V}$$

45. $V = \dfrac{Q_A}{4\pi\varepsilon_0 r_A} + \dfrac{Q_B}{4\pi\varepsilon_0 r_B} + \dfrac{Q_C}{4\pi\varepsilon_0 r_C} + \dfrac{Q_D}{4\pi\varepsilon_0 r_D}$

$$V = \frac{1}{4\pi\varepsilon_0}\left[\frac{-12 \ \mu C}{\sqrt{(0.30)^2 + (0.15)^2} \text{ m}} + \frac{10 \ \mu C}{0.15 \text{ m}} + \frac{12 \ \mu C}{0.15 \text{ m}} + \frac{14 \ \mu C}{\sqrt{(0.30)^2 + (0.15)^2} \text{ m}} \right]$$

$$V = \frac{1}{4\pi (8.85 \times 10^{-12})}\left[\frac{20 \times 10^{-6} \text{ C}}{0.15 \text{ m}} \right]$$

$$V = 1.20 \text{ MV}$$

47. When the alpha particle reaches its closest approach to the gold nucleus, it stops momentarily, and its kinetic energy is zero. At this distance,

$$\frac{1}{2} mv^2 = \frac{q_1 q_2}{4\pi\varepsilon_0 r}$$

$$r = \frac{q_1 q_2}{4\pi\varepsilon_0 \left(\tfrac{1}{2} mv^2 \right)}$$

$$r = \frac{(2e)(79e)}{2\pi\varepsilon_0 \left(2\times1.673\times10^{-27} + 2\times1.675\times10^{-27}\right)\left(1.50\times10^7\right)^2}$$

$$r = 4.84\times10^{-14} \text{ m}$$

49. $E = \dfrac{Q}{4\pi\varepsilon_0 r^2}$, where r is the radius of Earth.

$$-150 \text{ N/C} = \frac{Q}{4\pi\varepsilon_0 r^2} \rightarrow Q = -150\left(4\pi\varepsilon_0\right)\left(6.37\times10^6\right)^2 = -6.77\times10^5 \text{ C}$$

As a surface charge density this is $1.33\times10^{-9}\,\text{C/m}^2$

51. (a) $q\Delta V = \frac{1}{2}mv^2$

$$\Delta V = \frac{mv^2}{2q}$$

$$\Delta V = \frac{\left(9.11\times10^{-31} \text{ kg}\right)\left(2.25\times10^7 \text{ m/s}\right)^2}{2\left(1.60\times10^{-19}\text{C}\right)}$$

$$\Delta V = 1.44 \text{ kV}$$

(b) $\frac{1}{2}mv^2 = q\Delta V$

$$\frac{1}{2}mv^2 = \left(1.6\times10^{-19} \text{ C}\right)\left(1.44 \text{ kV}\right)$$

$$\frac{1}{2}mv^2 = 1.44 \text{ keV}$$

53. The spherical cavity is irrelevant to the calculation of field and potential in this problem. Because the sphere is conducting, the electric field inside the sphere is zero. Outside the sphere, the electric field is the same as that of a point charge at the centre of the sphere, according to Gauss's law. Therefore, the electric field outside the sphere has magnitude

$$E = \frac{Q}{4\pi\varepsilon_0 r^2}$$

$$E = \frac{240 \text{ nC}}{4\pi\left(8.85\times10^{-12}\right)r^2}$$

$$E = \frac{2160}{r^2}$$

and is directed radially outward.

The potential outside the sphere is therefore

$$V = \frac{2160}{r}$$

and so the potential at the surface of the sphere is

$$V = \frac{2160}{0.12}$$
$$V = 18.0 \text{ kV}$$

Thus, the potential at all points within the sphere, including in the cavity, is the same constant V = 18.0 kV.

55. For a cylindrically symmetrical charge distribution, one can use a coaxial cylindrical Gaussian surface of radius r and length L to determine the electric field as follows:

$$E \cdot 2\pi r L = \frac{Q}{\varepsilon_0}$$

$$E \cdot 2\pi r L = \frac{\mu L}{\varepsilon_0}$$

$$E = \frac{\mu}{2\pi \varepsilon_0 r}$$

where μ is the linear charge density. Thus, the magnitude of the electric field between the inner wire and the first conductor is (the direction is radially outward)

$$E = \frac{15.0 \ \mu\text{C/m}}{2\pi \left(8.85 \times 10^{-12}\right) r}$$

$$E = \frac{270}{r} \text{ kN/C}$$

Similarly, the magnitude of the electric field between the two conductors is (again, directed radially outward)

$$E = \frac{(15.0 - 7.00) \ \mu\text{C/m}}{2\pi \left(8.85 \times 10^{-12}\right) r}$$

$$E = \frac{144}{r} \text{ kN/C}$$

Similarly, the magnitude of the electric field outside the largest conductor is

$$E = \frac{(15.0 - 7.00 - 8.00) \ \mu\text{C/m}}{2\pi \left(8.85 \times 10^{-12}\right) r}$$

$$E = 0$$

57. In each case, the x-component of the electric field is the negative of the slope of the given graph. The approximate values are obtained graphically.

(a) −11 V/m

(b) 0

(c) −30 V/m

(d) +16 V/m

59. For a uniform sphere, the potential and electric fields near the sphere are

$$V = \frac{Q}{4\pi\varepsilon_0 r} \quad \text{and} \quad E = \frac{Q}{4\pi\varepsilon_0 r^2}$$

Thus,

$$E = \frac{V}{r}$$

$$E = \frac{800 \text{ V}}{1.5\times10^{-3} \text{ m}}$$

$$E = 533 \text{ kV/m}$$

61. (a) $V = \dfrac{Q}{4\pi\varepsilon_0 r}$

$$V = \frac{50\times10^{-9} \text{ C}}{4\pi\left(8.85\times10^{-12}\right)\left(250\times10^{-6} \text{ m}\right)}$$

$$V = 1.80 \text{ MV}$$

(b) When the drops coalesce, the volume of the single drop is three times the volume of each small drop, so the radius of the new drop is

$$R = 3^{\frac{1}{3}}r = 3^{\frac{1}{3}}\left(250\times10^{-6} \text{ m}\right) = 361\times10^{-6} \text{ m}$$

The new potential is

$$V = \frac{3Q}{4\pi\varepsilon_0 R}$$

$$V = \frac{3\left(50\times10^{-9} \text{ C}\right)}{4\pi\left(8.85\times10^{-12}\right)\left(361\times10^{-6} \text{ m}\right)}$$

$$V = 3.74 \text{ MV}$$

(c) For the small drops,

$$E = \frac{Q}{4\pi\varepsilon_0 r^2} = \frac{V}{r} = \frac{1.80 \text{ MV}}{250\times10^{-6} \text{ m}} = 7.19\times10^{9} \text{ V/m}$$

For the large drop,

$$E = \frac{3Q}{4\pi\varepsilon_0 r^2} = \frac{3.74 \text{ MV}}{361\times10^{-6} \text{ m}} = 1.04\times10^{10} \text{ V/m}$$

63. Using Gauss's law, and a Gaussian surface that is a cylinder of radius r and length L coaxial with the wire, we can determine the electric field a perpendicular distance r from the wire, as follows:

$$E \cdot 2\pi rL = \frac{Q}{\varepsilon_0}$$

$$E \cdot 2\pi rL = \frac{\mu L}{\varepsilon_0}$$

$$E = \frac{\mu}{2\pi\varepsilon_0} \cdot \frac{1}{r}$$

The direction of the electric field is in the radial direction.

The potential a distance s from the wire is

$$V = -\int_a^s E \, dr$$

$$V = -\int_a^s \frac{\mu}{2\pi\varepsilon_0} \cdot \frac{1}{r} \, dr$$

$$V = -\frac{\mu}{2\pi\varepsilon_0} \int_a^s \frac{1}{r} \, dr$$

$$V = -\frac{\mu}{2\pi\varepsilon_0} \Big[\ln(r) \Big]_a^s$$

$$V = -\frac{\mu}{2\pi\varepsilon_0} \Big[\ln(s) - \ln(a) \Big]$$

The potential at $s = a/2$ is

$$V = -\frac{\mu}{2\pi\varepsilon_0} \Big[\ln(a/2) - \ln(a) \Big]$$

$$V = -\frac{\mu}{2\pi\varepsilon_0} \left[\ln\left(\frac{a}{2a}\right) \right]$$

$$V = -\frac{\mu}{2\pi\varepsilon_0} \left[\ln\left(\frac{1}{2}\right) \right]$$

$$V = \frac{\mu \ln(2)}{2\pi\varepsilon_0}$$

Similarly, the potential at $s = 2a$ is

$$V = -\frac{\mu}{2\pi\varepsilon_0}\left[\ln(2a) - \ln(a)\right]$$

$$V = -\frac{\mu}{2\pi\varepsilon_0}\left[\ln\left(\frac{2a}{a}\right)\right]$$

$$V = -\frac{\mu}{2\pi\varepsilon_0}\left[\ln(2)\right]$$

$$V = -\frac{\mu\ln(2)}{2\pi\varepsilon_0}$$

65. Dividing the disks into thin concentric rings at radius r, we can make use of the result of question 64. If the total charge on the disk is Q, then the charge on each ring is

$$dQ = Q \cdot \frac{2\pi r dr}{\pi a^2}$$

$$dQ = Q \cdot \frac{2r dr}{a^2}$$

The potential at a distance z above the centre of the disk is

$$V = \int \frac{dQ}{4\pi\varepsilon_0 R}$$

$$V = \int_0^a \frac{2Q r dr}{4\pi\varepsilon_0 a^2 R}$$

$$V = \frac{Q}{2\pi\varepsilon_0 a^2}\int_0^a \frac{r dr}{R}$$

$$V = \frac{Q}{2\pi\varepsilon_0 a^2}\int_0^a \frac{r dr}{\sqrt{z^2 + r^2}}$$

$$V = \frac{Q}{2\pi\varepsilon_0 a^2}\left[\sqrt{z^2 + r^2}\right]_0^a$$

$$V = \frac{\pi a^2 \sigma}{2\pi\varepsilon_0 a^2}\left[\sqrt{z^2 + a^2} - \sqrt{z^2}\right]$$

$$V = \frac{\sigma}{2\varepsilon_0}\left[\sqrt{z^2 + a^2} - z\right]$$

1. b. By definition, the capacitance C is given by

$$C = \frac{Q}{V} = \frac{160 \times 10^{-6}\,\text{C}}{40\,\text{V}} = 4.0\,\mu\text{F}$$

3. d. The electric field intensity E can be calculated as

$$E = \frac{V}{d} = \frac{10\,\text{V}}{10^{-3}\,\text{m}} = 10^4\,\text{V/m}$$

5. b. The separation between the cylindrical plates is small compared to their radii for A (but not B); therefore, the parallel plate approximation is approximately valid for A.

7. c. Let us say the capacitor C is added to the existing capacitor C_0. The new effective capacitances for a series connection C_{series} and a parallel connection $C_{parallel}$ are given by

$$C_{series} = \frac{1}{\dfrac{1}{C_0} + \dfrac{1}{C}}$$

$$C_{parallel} = C_0 + C$$

It follows that the only way that the new capacitance is slightly smaller than C_0 is in series connection with a large added capacitance C which makes the quantity $1/C$ small.

9. b. The following equation can be written:

$$C_{dielectric} = \frac{\varepsilon_0 \varepsilon_r A}{d} = \frac{2\varepsilon_0 A}{d} = 2C$$

The voltage across the capacitor X does not change because the capacitor is still connected to the battery.

11. d. The following equation can be written:

$$E_0 = \frac{CV^2}{2}$$

Hence,

$$V = 2V_0 \Rightarrow E = 4E_0$$

13. b. Improvement in rectifying the power supply output can be achieved by increasing the RC constant. Therefore, the addition of an extra capacitor in parallel will increase the effective capacitance C and therefore increase the effective RC constant of the rectifying circuit.

15. b. The following relationships can be written:

$$f_{cutoff} = \frac{1}{2\pi RC} \Rightarrow \frac{f_{cutoff}}{2} = \frac{1}{2\pi(2R)C}$$

17. $C = \dfrac{Q}{V} = \dfrac{6\times10^{-3}\,\text{C}}{12.0\,\text{V}} = 0.5\times10^{-3}\,\text{F} = 0.500\,\text{mF}$

19. (a) $V = Ed = 2400\dfrac{\text{V}}{\text{m}}\times2\times10^{-3}\,\text{m} = 4.80\,\text{V}$

    (b) $C = \dfrac{Q}{V} = \dfrac{56\times10^{-6}\,\text{C}}{4.8\,\text{V}} = 11.7\times10^{-6}\,\text{F} = 11.7\,\mu\text{F}$

21. The following equation can be written:

$$C = \frac{\varepsilon_{air}A}{d} \approx \frac{\varepsilon_0 A}{d}$$

It follows that

$$d = \frac{\varepsilon_0 A}{C} = \frac{\varepsilon_0 \pi r^2}{C} = \frac{8.85\times10^{-12}\,\text{C}^2\text{N}^{-1}\text{m}^{-2}\times\pi\times7.5^2\times10^{-4}\,\text{m}^2}{115\times10^{-12}\,\text{F}} = 1.36\times10^{-3}\,\text{m} = 1.36\,\text{mm}$$

23. The following two equations can be written for the equivalent capacitances of n identical capacitors C connected in parallel (C_p) and in series (C_s):

$$C_p = C + C + C + \cdots + C = nC \tag{1}$$

$$C_s = \left(\frac{1}{C} + \frac{1}{C} + \frac{1}{C} + \cdots + \frac{1}{C}\right)^{-1} = \left(\frac{n}{C}\right)^{-1} = \frac{C}{n} \tag{2}$$

The following answers can be derived based on equations (1) and (2):

(a) $n = \dfrac{C_p}{C} = \dfrac{2\times10^{-9}\,\text{F}}{5\times10^{-10}\,\text{F}} = 4$ (parallel)

(b) $n = \dfrac{C}{C_s} = \dfrac{5\times10^{-10}\,\text{F}}{1\times10^{-10}\,\text{F}} = 5$ (series)

(c) $n_1 = \dfrac{C_p}{C} = \dfrac{10^{-9}\,\text{F}}{5\times10^{-10}\,\text{F}} = 2$ (parallel)

   $n_2 = \dfrac{C}{C_s} = \dfrac{5\times10^{-10}\,\text{F}}{2.5\times10^{-10}\,\text{F}} = 2$ (series)

   The circuit should have three parallel branches, where the first two branches each contain a single 500 pF capacitor, and the third branch contains two 500 pF connected in series.

25. The following equation can be written:

$$C = \frac{\varepsilon A}{d} = \frac{\kappa \varepsilon_0 A}{d}$$

It follows that

$$\kappa = \frac{Cd}{\varepsilon_0 A} = \frac{169 \times 10^{-12}\,\text{F} \times 2 \times 10^{-3}\,\text{m}}{8.85 \times 10^{-12}\,\text{C}^2\text{N}^{-1}\text{m}^{-2} \times 6.5 \times 10^{-2}\,\text{m}^2 \times 8.5 \times 10^{-2}\,\text{m}^2} = 6.91$$

Based on the values from Table 21-1, the dielectric material might be mica.

27. (a) $C_s = \dfrac{C}{2} = \dfrac{5 \times 10^{-6}\,\text{F}}{2} = 2.5\,\mu\text{F}$

$$U = \frac{1}{2}C_s V^2 = 0.5 \times 2.5 \times 10^{-6} \times 12^2\,\text{J} = 180\,\mu\text{J}$$

(b) $C_p = 2C = 2 \times 5 \times 10^{-6}\,\text{F} = 10\,\mu\text{F}$

$$U = \frac{1}{2}C_p V^2 = 0.5 \times 10 \times 10^{-6} \times 12^2\,\text{J} = 720\,\mu\text{J}$$

29. $U = \dfrac{1}{2}CV^2 \Rightarrow V = \sqrt{\dfrac{2U}{C}} = \sqrt{\dfrac{2 \times 125 \times 10^{-3}\,\text{J}}{50 \times 10^{-6}\,\text{F}}} = 70.7\,\text{V}$

31. The permittivity of free space ε_0 is equal to

$$\varepsilon_0 = 8.85 \times 10^{-12}\,\text{C}^2\text{N}^{-1}\text{m}^{-2} = 8.85 \times 10^{-12}\,\frac{\text{C} \cdot \text{C}}{\text{N} \cdot \text{m} \cdot \text{m}} = 8.85 \times 10^{-12}\,\frac{\text{C} \cdot \text{C}}{\text{J} \cdot \text{m}}$$

Further,

$$\varepsilon_0 = 8.85 \times 10^{-12}\,\frac{\text{C}}{\dfrac{\text{J}}{\text{C}} \cdot \text{m}} = 8.85 \times 10^{-12}\,\frac{\text{C}}{\text{V} \cdot \text{m}} = 8.85 \times 10^{-12}\,\frac{\text{C/V}}{\text{m}} = 8.85 \times 10^{-12}\,\frac{\text{F}}{\text{m}}$$

Alternatively, one can express each quantity in terms of the SI base units (see Chapter 1) to prove the result or use the expression for the parallel plate capacitor to solve for the units of permittivity.

33. (a) $Q = CV = 90 \times 10^{-15}\,\text{F} \times 3.3\,\text{C} = 300\,\text{fC}$

(b) $N_e = \dfrac{Q}{e} = \dfrac{297 \times 10^{-15}\,C}{1.602 \times 10^{-19}\,C} = 1.9 \times 10^6$ electrons

35. Draw a Gaussian cylinder of radius r and length L. The electric field between the two cylinders is obtained. If the inner cylinder carries a positive charge Q and the outer cylinder a negative charge $-Q$, the direction of the electric field will be radially outward (positive).

$$\oint E \cdot dA = \frac{q_{enc}}{\varepsilon_0}$$

$$E = \frac{Q}{2\pi r L \varepsilon_0}$$

We can integrate to find the potential difference between the two surfaces.

$$V_b - V_a = -\int_a^b E \circ dr = -\frac{Q}{2\pi L \varepsilon_0} \int_{0.010}^{0.025} \frac{dr}{r} = -\frac{Q}{2\pi L \varepsilon_0} \ln(2.50)$$

If the inner cylinder is the positively charged one, then the outer cylinder will be at a lower electric potential. However, in finding capacitance it is just the magnitude of the potential difference which is important (since it could be charged either way).

We use $Q = CV$ to find the capacitance. We are looking for the capacitance per length so divide each side by L.

$$\frac{C}{L} = \frac{Q}{LV} = \frac{Q}{L\dfrac{Q}{2\pi L \varepsilon_0} \ln(2.50)} = \frac{2\pi \varepsilon_0}{\ln(2.50)} = 6.07 \times 10^{-11}\,\text{F/m}$$

This can be written as 60.7 pF/m.

37. The capacitor with the two dielectric insertions is equivalent to two capacitors connected in parallel, each space between the plates filled with one of the dielectric materials, and with plate areas equal to half of the initial plate area. Hence, one can write that the equivalent capacitance C_e is given by

$$C_e = C_1 + C_2 = \frac{\kappa_1 \varepsilon_0 A / 2}{d} + \frac{\kappa_2 \varepsilon_0 A / 2}{d} = \frac{\varepsilon_0 A}{2d}\left(\kappa_1 + \kappa_2\right)$$

39. (a) $C_e = \dfrac{C}{5} = \dfrac{2\mu F}{5} = 0.400\mu F$

$Q_0 = C_e V = 0.4 \times 10^{-6}\,\text{F} \times 9\,\text{V} = 3.60\mu\text{C}$

(b) $V_0 = \dfrac{V}{5} = 1.80\,\text{V}$

(c) $U_0 = \dfrac{Q_0^2}{2C} = 3.24\mu\text{J}$

41. (a) Two equations can be written:

$$F = ma \qquad (1)$$

$$F = eE \qquad (2)$$

From equations (1) and (2), it follows that

$$E = \frac{m}{e}a = \frac{7 \times 10^{15}\,\text{m} \cdot \text{s}^{-2} \times 9.109 \times 10^{-31}\,\text{kg}}{1.602 \times 10^{-19}\,\text{C}} = 39.8\,\text{kV/m}$$

(b) $V = Ed = 40 \times 10^3\,\text{V/m} \times 4 \times 10^{-3}\,\text{m} = 159\,\text{V}$

(c) Electrons move in the direction of the plate with higher electric potential; that is, toward the positively charged plate.

(d) The electrons were initially at rest; the following kinematics equation can be written:

$$v = at \Rightarrow t = \frac{v}{a} = \frac{2.50 \times 10^7\,\text{m/s}}{7 \times 10^{15}\,\text{m/s}^2} = 3.57\,\text{ns}$$

(e) The following kinematics equation can be used:

$$v^2 = 2as \Rightarrow s = \frac{v^2}{2a} = \frac{2.50^2 \times 10^{14}\,\text{m/s}}{2 \times 7 \times 10^{15}\,\text{m/s}^2} = 44.6\,\text{mm}$$

Hence, the distance travelled by electron s is larger than the separation between the plates of the capacitor d (so it is not possible to reach this speed).

43. All capacitors in Figure 21-22 have the same charge Q on their plates. The following relationships can be written:

$$C_1 = 2\,\mu\text{F} = \frac{Q}{8\,\text{V}} \Rightarrow Q = 16\,\mu\text{C}$$

$$C_2 = \frac{16\,\mu\text{C}}{(12-8)\,\text{V}} = 4\,\mu\text{F}$$

$$C_3 = \frac{16\,\mu\text{C}}{(14-12)\,\text{V}} = 8\,\mu\text{F}$$

$$C_4 = \frac{16\,\mu\text{C}}{(16-14)\,\text{V}} = 8\,\mu\text{F}$$

$$C_5 = \frac{16\,\mu\text{C}}{(24-16)\,\text{V}} = 2\,\mu\text{F}$$

45. (a) $C = \dfrac{Q}{V} = \dfrac{150\,\text{nC}}{60\,\text{V}} = 2.50\,\text{nF}$

(b) $C = \dfrac{\kappa \varepsilon_0 A}{d} \Rightarrow d = \dfrac{\kappa \varepsilon_0 A}{C} = \dfrac{4.0 \times 8.85 \times 10^{-12}\,\text{F/m} \times 20 \times 10^{-4}\,\text{m}^2}{2.5 \times 10^{-9}\,\text{F}} = 28\,\mu\text{m}$

(c) $U = \dfrac{CV^2}{2} = \dfrac{2.5 \times 10^{-9}\,\text{F} \times 36 \times 10^2\,V^2}{2} = 4.5\,\mu\text{J}$

(d) $E = \dfrac{V}{d} \cong \dfrac{60\,\text{V}}{28.32 \times 10^{-6}\,\text{m}} = 2.1\,\text{MV/m}$

47. The energy density u in a region of space where the magnitude of the electric field is E is given by

$$u = \frac{1}{2}\varepsilon_0 E^2 \qquad\qquad (1)$$

Using Gauss's law in the region just outside the sphere with charge Q gives

$$E4\pi r^2 = \frac{Q}{\varepsilon_0} \Rightarrow E = \frac{Q}{4\pi \varepsilon_0 r^2} \qquad\qquad (2)$$

Combining equations (1) and (2), it follows that

$$u = \frac{1}{2}\varepsilon_0 \frac{Q^2}{16\pi^2 \varepsilon_0^2 r^4} = \frac{Q^2}{32\pi^2 \varepsilon_0 r^4} = \frac{\left(3.50 \times 10^{-9}\right)^2 C^2}{32\pi^2 \times 8.85 \times 10^{-12}\,\text{F/m} \times \left(3.50 \times 10^{-2}\right)^4 \text{m}^4} = 2.92\,\text{mJ} \cdot \text{m}^{-3}$$

49. The following diagram and notations can be used for this problem.

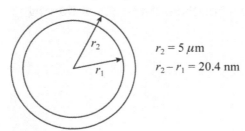

$r_2 = 5\,\mu\text{m}$
$r_2 - r_1 = 20.4\,\text{nm}$

(a) $C_{cell} = \dfrac{4\pi \varepsilon}{\dfrac{1}{r_1} - \dfrac{1}{r_2}} = 4\pi \kappa \varepsilon_0 \dfrac{r_1 r_2}{r_2 - r_1} = 4\pi \kappa \varepsilon_0 \dfrac{(r_2 - t) r_2}{t} = 4\pi \kappa \varepsilon_0 r_2 \left(\dfrac{r_2}{t} - 1\right)$

Numerically,

$$C_{cell} = 4\pi \times 4 \times 8.85 \times 10^{-12}\,\text{F/m} \times 5 \times 10^{-6}\,\text{m} \times \left(\dfrac{5 \times 10^{-6}\,\text{m}}{20.4 \times 10^{-9}\,\text{m}} - 1\right) = 0.543\,\text{pF}$$

(b) $Q = CV = 40.7\,\text{fC}$

(c) $N_e = \dfrac{Q}{e} = 2.54 \times 10^5$ elementary charges

51. The energy stored in the cylindrical capacitor U is given by

$$U = u \times \text{volume} = u\pi r^2 L = \frac{1}{2}CV^2$$

It follows that

$$C = \frac{2\pi u r^2 L}{V^2} = \frac{2\pi \times 2 \times 10^6 \, \text{J} \cdot \text{m}^{-3} \times 25 \times 10^{-4} \, \text{m}^2 \times 10 \times 10^{-2} \, \text{m}}{750^2 \, V^2} = 5.59 \, \text{mF}$$

Chapter 22—ELECTRIC CURRENT AND FUNDAMENTALS OF DC CIRCUITS

1. (a) False. Historically, the direction of the electric current is in the direction of motion of positive charges if the charge carriers were positive.

    (b) False. The resistance depends also on the wire dimensions ($R = \rho \dfrac{l}{A}$).

    (c) False. Free electron density n_e is a characteristic of the metal independent of temperature.

    (d) True. $\rho = \rho_0 (1 + \alpha \Delta T)$, $\sigma = 1/\rho$.

    (e) True. The following equation is obtained by combining equations (22-7) and (22-11):
    $$v_d = \frac{E}{\rho e n_e}$$

3. b. Assuming that the resistance of one light bulb is R, the equivalent resistances for circuits II (the series circuit) and III (the parallel circuit) are $R_{eq\_series} = R_{eq\_II} = 10R$ and

    $R_{eq\_parallel} = R_{eq\_III} = \dfrac{R}{10}$, respectively. Given that $P = \dfrac{V^2}{R_{eq}}$, it follows that the power output

    of circuit III (the parallel circuit) is larger than the power output of circuit I (the original circuit) which is larger than the power output of circuit II (the series circuit):

    $$\frac{P_{parallel}}{P} = \frac{P}{P_{series}} = 10 \Rightarrow P_{parallel} = 100 P_{series}; P_{parallel} = 10P; P_{series} = \frac{P}{10}$$

5. e. $R_{eq} = \dfrac{1}{\dfrac{1}{R/N} + \dfrac{1}{R/N} + \cdots + \dfrac{1}{R/N}} = \dfrac{R}{N^2}$

7. d. $E = I^2 Rt = VIt = 12\,\text{V} \times 100\,\text{A} \times 3600\,\text{s} = 432 \times 10^4\,\text{J} = 4320\,\text{kJ}$

9. b. The equivalent resistance for a circuit powered by a 3 V battery in which the current is $I = 4.5\,\text{A}$ is given by

    $$R_{eq} = \frac{V}{I} = \frac{3\,\text{V}}{4.5\,\text{A}} = \frac{2}{3}\,\Omega$$

    This equivalent resistance is obtained by connecting all six of the 4 Ω resistors in parallel:

    $$R_e = \frac{4\,\Omega}{6} = \frac{2}{3}\,\Omega$$

11. c

13. c. Consider the circuit as consisting of three components connected in series with each other: light bulb M, two light bulbs N and O connected in series, and three light bulbs P, O, and R connected in parallel (P and O are connected in series with each other and in parallel with light bulb R). These three circuit components have the resistances of $R, \dfrac{R}{2}$, and $\dfrac{2R}{3}$ respectively. Therefore, the equivalent resistance of the entire circuit can be calculated as

$$R_{eq} = R + \frac{R}{2} + \frac{2R}{3} = \frac{13R}{6} = \frac{130\ \Omega}{6} = 21.7\ \Omega$$

15. e. $\quad V_{AB} = IR_{AB} = I\dfrac{R}{2} = \dfrac{120}{130}\ \text{A} \times \dfrac{10}{2}\ \Omega = 4.62\ \text{V}$

17. d. The following series of equivalent circuit diagrams show the step-by-step calculations of the equivalent resistance R_{eq}.

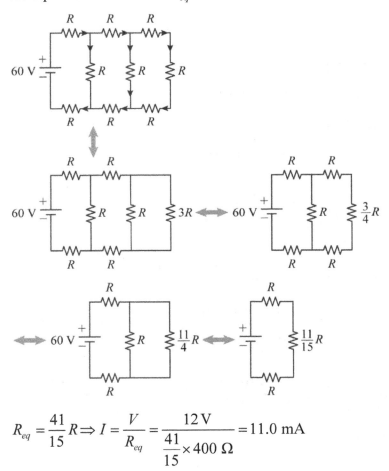

$$R_{eq} = \frac{41}{15}R \Rightarrow I = \frac{V}{R_{eq}} = \frac{12\ \text{V}}{\dfrac{41}{15} \times 400\ \Omega} = 11.0\ \text{mA}$$

19. b. The time it takes to charge or discharge a capacitor depends only on its capacitance and the resistance of the circuit. It does not depend on the voltage of the battery; thus, the answer remains the same.

21. a. The current is greatest early in the experiment, thus the wires are more likely to burn at that time. See equation (22-46): $I_{max} = \dfrac{\varepsilon}{R}$.

23. (a) The rate of flow of electrons is measured in electrons per second and is called electron current:

Diameter = 2
I = 2A

$$I_e = \frac{I}{e} = \frac{2\ C/s}{1.6 \times 10^{-19}\ C/\text{electron}} = \frac{2\ A}{1.6 \times 10^{-19}\ C/\text{electron}} = 1.3 \times 10^{19}\ \text{electrons/s}$$

(b) $$J = \frac{I}{A} = \frac{I}{\dfrac{\pi d^2}{4}} = \frac{2\ A}{\dfrac{\pi \times 4 \times 10^{-6}\ m^2}{4}} = 6.4 \times 10^5\ A/m^2$$

(c) $Q = It = 2\ A \times 2\ s = 4\ C$

25. (a) Using data from Table 22-2:

$$\rho(15°C) = \rho(20°C)(1 + \alpha(T - T_0)) =$$
$$= (1.72 \times 10^{-8}\ \Omega \cdot m)(1 + 4.3 \times 10^{-3} K^{-1}\ (288\ K - 293K))$$
$$= 1.683 \times 10^{-8}\ \Omega \cdot m = 1.7 \times 10^{-8}\ \Omega \cdot m$$

$$\Rightarrow \sigma(15°C) = \rho(15°C)^{-1} = 5.9 \times 10^7\ \Omega^{-1} \cdot m^{-1} = 5.9 \times 10^7\ (\Omega \cdot m)^{-1}$$

(b) $$J = \frac{I}{A} = n_e v_d e \Rightarrow v_d = \frac{I}{A e n_e} = \frac{I}{\dfrac{\pi d^2}{4} e n_e} = \frac{4I}{\pi d^2 e n_e}$$

$$v_d = \frac{4 \times 2\ A}{\pi \times 4 \times 10^{-6}\ m^2 \times 1.6 \times 10^{-19}\ C \times 8.4 \times 10^{28}\ m^{-3}} = 4.7 \times 10^{-5}\ m/s$$

(c) $$\sigma = \frac{e^2 n_e \tau}{m} \Rightarrow \tau = \frac{\sigma m}{e^2 n_e} = \frac{5.93 \times 10^7\ \Omega^{-1} \cdot m^{-1} \times 9.11 \times 10^{-31}\ kg}{(1.60 \times 10^{-19}\ C)^2 \times 8.4 \times 10^{28}\ m^{-3}} = 2.51 \times 10^{-14}\ s = 25\ fs$$

(d) $$v_d = \frac{eE}{m}\tau \Rightarrow E = \frac{v_d m}{e\tau} = \frac{4.7 \times 10^{-5}\ m/s \times 9.11 \times 10^{-31}\ kg}{1.6 \times 10^{-19}\ C \times 2.51 \times 10^{-14}\ s} = 10.7 \times 10^{-3}\ \frac{V}{m} = 11\ \frac{mV}{m}$$

(e) (a) The answer will not change.

(b) The drift velocity v_d will double its value.

(c) The answer will not change.

(d) The electric field inside the wire E will double its value.

(f) No change in any of the quantities.

(g) (a) The resistivity ρ will increase and the conductivity σ will decrease.

(b) The answer will not change.

(c) The average time between collisions τ will decrease.

(d) The electric field inside the wire E will increase.

27. Let us denote the density of copper as d_{Cu} and the resistivity of copper as ρ_{Cu}. The cross sectional area and the length of copper are denoted as A and l respectively. Then, the following two equations can be written:

$$m = d_{Cu}V = d_{Cu}Al \Rightarrow Al = \frac{m}{d_{Cu}} \tag{1}$$

$$R = \rho_{Cu}\frac{l}{A} \Rightarrow \frac{l}{A} = \frac{R}{\rho_{Cu}} \Rightarrow Al = A^2\frac{l}{A} = A^2\frac{R}{\rho_{Cu}} = \frac{A^2R}{\rho_{Cu}} \tag{2}$$

From Table 22-2, $\rho_{cu} = 1.72 \times 10^{-8}\,\Omega \cdot m$. Also, $d_{Cu} = 8.96\,g/cm^3$.

Plugging the wire length l obtained from equation (1) into equation (2) gives

$$\frac{A^2R}{\rho_{Cu}} = \frac{m}{d_{Cu}} \Rightarrow A = \sqrt{\frac{m\rho_{Cu}}{Rd_{Cu}}} = \sqrt{\frac{1.78\times10^{-3}kg\times1.72\times10^{-8}\,\Omega\cdot m}{34\,\Omega\times8.96\times10^{3}\,kg\cdot m^{-3}}} = 0.01mm^2 = 10^{-8}\,m^2$$

(This means that the diameter of the wire is about 0.1 mm).

$$l = \frac{R}{\rho_{Cu}}A = \frac{34\,\Omega}{1.72\times10^{-8}\,\Omega\cdot m}\times10^{-8}m^2 = 20\,m$$

29. $I = \dfrac{P}{V} = \dfrac{10^3\,W}{120\,V} = 8.3\,A$

$$I_{max} = \frac{NP}{V} \Rightarrow N = \frac{I_{max}V}{P} = 1.8 < 2$$

Hence, only one toaster can be operated safely.

31. (a) $R = \rho\dfrac{l}{A} = \rho\dfrac{l}{\pi\dfrac{d^2}{4}} = \dfrac{4\rho l}{\pi d^2} = \dfrac{4\times0.45\times10^{-6}\,\Omega\cdot m\times56\times10^{-2}\,m}{\pi\times0.4^2\times10^{-6}\,m^2} = 2.0\ \Omega$

(b) $I = \dfrac{V}{R} = \dfrac{3\,V}{2.0\,\Omega} = 1.5\,A$

33. For the parallel connection of n resistors the equivalent resistance R_n is given by

$$R_{eq} = \left(\frac{1}{R_1} + \frac{1}{R_2} + \cdots + \frac{1}{R_n} \right)^{-1}$$

For convenience we can assume that the first resistor has the least resistance: $R_1 = R_{min}$.
Then,

$$R_{eq} = \left(\frac{1}{R_{min}} + \frac{1}{R_2} + \cdots + \frac{1}{R_n} \right)^{-1} = \frac{1}{\dfrac{1}{R_{min}} + \displaystyle\sum_{i=2}^{n} \frac{1}{R_i}} < \frac{1}{\dfrac{1}{R_{min}}} = R_{min}$$

35. (a) The energy saved per year is

$$E = Pt_{saved} = 1000 \text{ W} \times 2 \times 60 \text{ s/day} \times 365 \text{ days/year} = 44 \times 10^6 \text{ J} = 12 \text{ kW} \cdot \text{h}$$

   (b) The annual monetary savings is

$$\text{savings/year} = 12.17 \text{ kW} \cdot \text{h} \times 6.5 \text{ cents/kW} \cdot \text{h} = 79 \text{ cents}$$

   (c) Effective ways to reduce your electricity bill are replace incandescent light bulbs with compact fluorescent lamp (CFL) bulbs, unplug the fridge and other appliances which run on stand-by mode (TVs, computers, toasters with electronic display, etc.) when leaving for long periods of time, turn off the light when leaving the room, check the heating insulation of your house (particularly important for electric heating, but keep in mind that other forms of heating use electricity as well).

37. (a) Connect the four 1.5 V batteries in parallel.

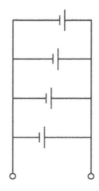

   (b) Two 1.5 V batteries are connected in series which are then connected in parallel with the other two 1.5 V batteries connected in series.

(c) Option 1: Two 1.5 V batteries are connected in parallel and the combination is then connected in series with the remaining two 1.5 V batteries.

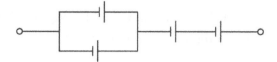

Option 2: Another possibility is use only three 1.5 V batteries connected in series.

(d) All four 1.5 V batteries are connected in series.

39. (a) Based on the circuit diagram from Figure 22-30 one can calculate the equivalent R_{eq} as follows:

$$R_{eq} = R + \left(\frac{1}{2R} + \frac{1}{2R}\right)^{-1} + \left(\frac{1}{2R} + \frac{1}{3R}\right)^{-1} + \left(\frac{1}{2R} + \frac{1}{2R}\right)^{-1} = R + R + \frac{6}{5}R + R = \frac{21}{5}R = 21\ \Omega$$

$$P = \frac{\varepsilon^2}{R_{eq}} = \frac{12^2\ \text{V}}{21\ \Omega} = 6.9\ \text{W}$$

(b) Let us label the resistors A, B, C, D, E, F, G, H, and I, as shown in the diagram below:

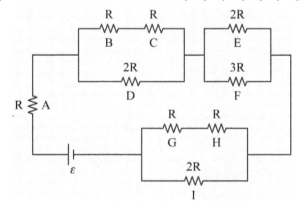

$$I_A = \frac{\varepsilon}{R_{eq}} = \frac{12\ \text{V}}{21\ \Omega} = 0.57\ \text{A}$$

Since the resistors B and C are connected in parallel to the resistor D, and the resistances of the two branches are equal, the current flowing through the battery (and resistor A) will split in half in these branches. Therefore, $I_B = I_C = I_D = \frac{I_A}{2}$. The same applied to resistors G, H and I. Thus, we conclude that the current trough the battery I_A is divided in two through the resistors B, C, D, H, G, and I:

$$I_B = I_C = I_D = I_G = I_H = I_I = \frac{I_A}{2} = \frac{12}{42} \text{A} = 0.29 \text{A}$$

The currents through resistors E and F, I_E and I_F can be calculated using Kirchhoff's laws in the loop with resistors E and F, yielding the next two equations:

$$I_E + I_F = I_A$$

$$2R I_E = 3R I_F \Rightarrow \frac{I_E}{I_F} = \frac{3}{2}$$

Solving the two equations above gives

$$I_F = \frac{2I_A}{5} = \frac{24}{105} \text{A} = 0.23 \text{A}$$

$$I_E = I_A - I_F = \left(\frac{12}{21} - \frac{24}{105}\right) \text{A} = \frac{36}{105} \text{A} = 0.34 \text{A}$$

Another way of thinking about it without directly using Kirchhoff's laws is as follows:

Since resistors E and F are connected in parallel, the voltages across them must be equal. Therefore,

$$V_E = V_F \Rightarrow I_E R_E = I_F R_F \Rightarrow \frac{R_F}{R_E} = \frac{I_E}{I_F} \Rightarrow \frac{I_E}{I_F} = \frac{3}{2} \Rightarrow I_E = \frac{3}{2} I_F, \text{ in addition, } I_E + I_F = I_A$$

Therefore, $I_F + \dfrac{3}{2} I_F = I_A$

$$\frac{5}{2} I_F = I_A \Rightarrow I_F = \frac{2}{5} I_A \text{ and } I_E = \frac{3}{5} I_A$$

$$I_E = 0.34 \text{ A and } I_F = 0.23 \text{ A and } I_A = 0.57 \text{ A}$$

(c) $V_A = I_A R = \dfrac{12}{21} \text{A} \times 5\Omega = 2.86 \text{V}$

$$V_B = V_C = V_H = V_G = \frac{I_A}{2} R = \frac{12}{42} \text{A} \times 5\Omega = 1.43 \text{V}$$

$$V_D = V_I = \frac{I_A}{2} 2R = V_A = \frac{60}{21} \text{V} = 2.86 \text{ V}$$

$$V_E = V_F = 2R I_E = 2 \times 5 \ \Omega \times \frac{36}{105} \text{A} = 3.43 \text{ V}$$

Notice, to check the answer you can add the voltages across the resistors and see if they add up to the potential difference across the battery:

$$V_A + V_D + V_F + V_I = 2.86 \text{ V} + 2.86 \text{ V} + 3.43 \text{ V} + 2.86 \text{ V} = 12.0 \text{ V} \equiv \varepsilon$$

(d)　$P_A = I_A^2 R = \left(\dfrac{12}{21}\right)^2 \times 5\,\text{W} = 1.63\,\text{W}$

$P_B = P_C = P_H = P_G = \left(\dfrac{I_A}{2}\right)^2 R = \left(\dfrac{12}{42}\right)^2 \times 5\,\text{W} = 0.41\,\text{W}$

$P_D = P_I = \left(\dfrac{I_A}{2}\right)^2 2R = \left(\dfrac{12}{42}\right)^2 \times 10\,\text{W} = 0.82\,\text{W}$

$P_E = I_E^2\, 2R = \left(\dfrac{36}{105}\right)^2 \times 10\,\text{W} = 1.18\,\text{W}$

$P_F = I_F^2\, 3R = \left(\dfrac{24}{105}\right)^2 \times 15\,\text{W} = 0.78\,\text{W}$

(e)　$P_{total} = P_A + P_B + P_C + P_D + P_E + P_F + P_G + P_H + P_I = 6.9\,\text{W}$, which is equal to the power value from part (a) using the equivalent resistance R_{eq}, as expected.

41. The following equivalent circuit diagrams can be used to calculate the equivalent resistance R_{eq}.

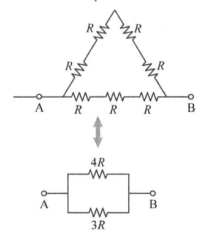

$R_{eq} = \left(\dfrac{1}{4R} + \dfrac{1}{3R}\right)^{-1} = \dfrac{12}{7}R$

43. The figure below is based on the circuit diagram from Figure 22-33; upper left-hand side loops 1 and 2 are indicated along with currents I, I_1, I_2, I_3, and I_4, and left-hand side nodes P and Q.

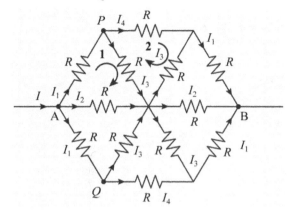

Kirchhoff's laws in loops 1 and 2 and node P indicated in the circuit diagram above yield the following equations:

$$I_1 R + I_3 R = I_2 R \Rightarrow I_2 = I_1 + I_3 \qquad (1)$$

$$I_1 = I_3 + I_4 \qquad (2)$$

$$I_4 R = I_3 R + I_3 R = 2I_3 R \Rightarrow I_4 = 2I_3 \qquad (3)$$

Using equations (1), (2), and (3), the following relationships between the four currents can be derived:

$$I_1 = 3I_3$$

$$I_2 = 4I_3$$

$$I_4 = 2I_3$$

Hence, the total current I is given by

$$I = 2I_1 + I_2 = 6I_3 + 4I_3 = 10I_3$$

Based on these relationships, the equivalent circuit diagram is shown below.

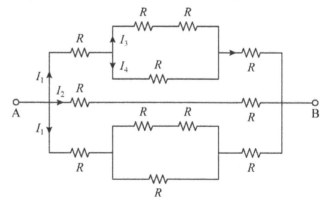

The equivalent resistance R_{eq} is then given by

$$R_{eq} = \left[2\frac{1}{2R + \left(\frac{1}{R} + \frac{1}{2R}\right)^{-1}} + \frac{1}{2R} \right]^{-1} = \left(\frac{2}{2R + \frac{2R}{3}} + \frac{1}{2R} \right)^{-1} = \left(\frac{3}{4R} + \frac{1}{2R} \right)^{-1}$$

$$= \frac{4}{5}R = \frac{4}{5} \times 3\,\Omega = 2.4\ \Omega$$

It is often helpful to make sense of the complex problem. You can do it by choosing any path from point A to point B and calculating potential difference along it. For example, if the pass is from point A to the centre of the diagram and to point B, the potential difference along it can be calculated as: $V_{AB} = I_2 R + I_2 R = 2(4I_3 R) = 8I_3 R$. If you choose an alternative path – going from point A to point Q and then to the centre and to point B, you will find the potential difference across it to be

$$V_{AB} = I_1 R + I_3 R + + I_2 R = 3I_3 R + I_3 R + 4I_3 R = 8I_3 R$$

You can check that potential difference along any path connecting points A and B will also be equal to $V_{AB} = 8I_3 R$.

45. (a) $q(t) = q_{max}\left(1 - e^{-t/\tau}\right) \Rightarrow e^{-t/\tau} = 1 - \dfrac{q(t)}{q_{max}}$

$$\Rightarrow t = -\tau \ln\left[1 - \frac{q(t)}{q_{max}}\right] = -\tau \times \ln(1 - 0.999) = -\tau \times \ln(10^{-3}) = \tau \times 3\ln 10 = 6.9\tau$$

(b) $q_{max} = CV = 300 \times 10^{-6}\ \text{F} \times 12\,\text{V} = 3.6\,\text{mC}$

$$q(\tau) = q_{max}\left(1 - e^{-1}\right) = 3.6\,\text{mC} \times \left(1 - e^{-1}\right) = 2.3\ \text{mC}$$

$$q(2\tau) = q_{max}\left(1 - e^{-2}\right) = 3.6\,\text{mC} \times \left(1 - e^{-2}\right) = 3.1\ \text{mC}$$

$$q(3\tau) = q_{max}\left(1 - e^{-3}\right) = 3.6\,\text{mC} \times \left(1 - e^{-3}\right) = 3.4\ \text{mC}$$

These values are 64, 86, and 94% of q_{max}, respectively.

(c) $I(\tau) = I_0 e^{-1} = \dfrac{V}{R} e^{-1} = \dfrac{12\,\text{V}}{5 \times 10^3} \times e^{-1} = 2.4\,\text{mA} \times e^{-1} = 0.88\,\text{mA}$

$$I(2\tau) = 2.4\,\text{mA} \times e^{-2} = 0.32\,\text{mA}$$

$$I(3\tau) = 2.4\,\text{mA} \times e^{-3} = 0.12\,\text{mA}$$

47. (a) The energy stored in the dry-cell battery is approximately equal to

$$E_{battery} = IVt = 0.67\,\text{A} \times 1.5\,\text{V} \times 5\frac{\text{h}}{\text{s}} \times 3600\ \text{s} = 18090\ \text{J} = 18\ \text{kJ}$$

$$1\,\text{kW}\cdot\text{h} = 10^3\,\text{W} \times 3600\ \text{s} = 3.6\ \text{MJ}$$

If the energy from the dry cell were taken from the electrical outlet it would cost:

$$\text{cost} = \frac{18090\,\text{J}}{3.6 \times 10^6\ \dfrac{\text{J}}{\text{kW}\cdot\text{h}}} \times 10\frac{\text{cents}}{\text{kW}\cdot\text{h}} = 0.05025\,\text{cents}$$

Given the cost of the D battery of approximately $1, the energy stored in the battery is almost 2000 times more expensive than the energy purchased from the electrical energy grid suppliers.

(b) Using the example given in Figure 22-10, the average Canadian household electrical energy monthly consumption is approximately $\dfrac{935}{2}\text{kW}\cdot\text{h} = 1683\,\text{MJ}$. The price of this energy using D batteries would be:

$$\text{cost}_{D-battery} = \frac{1683 \times 10^6\,\text{J}}{18090\,\dfrac{\text{J}}{\text{battery}}} \times \frac{\$1}{\text{battery}} = \$93\,000$$

Analyzing these numbers make us want to use rechargeable batteries.

49. Notations and directions for currents are indicated in the circuit diagram shown below.

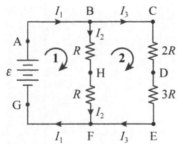

Applying Kirchhoff's laws in the loops 1 and 2, and node B, which are indicated in the figure above, one obtains the following equations:

Loop 1: $2I_2R = \varepsilon$ (1)

Loop 2: $2I_3R + 3I_3R = 2I_2R$ (2)

Node B: $I_1 = I_2 + I_3$ (3)

Using equation (1),

$$I_2 = \frac{\varepsilon}{2R} = \frac{10\,\text{V}}{10\,\Omega} = 1\,\text{A} \tag{4}$$

Dividing both sides of the equation (2) by R, one finds:

$$I_3 = \frac{2}{5}I_2 \overset{eq.\ (4)}{\Rightarrow} I_3 = 0.4\,\text{A} \tag{5}$$

Using equation (3),

$$I_1 = 1\,\text{A} + 0.4\,\text{A} = 1.4\,\text{A} \tag{6}$$

Using Ohm's law, the electric potential differences between different points of the circuit can be calculated as follows:

$$V_{AB} = V_{FG} = V_{EF} = 0\,\text{V}$$

$$V_{AG} = \varepsilon = 10\,\text{V}$$

$$V_{BH} = V_{HF} = I_2 R = 5\,\text{V}$$

$$V_{CD} = 2I_3 R = 4\,\text{V}$$

$$V_{DE} = 3I_3 R = 6\,\text{V}$$

$$V_{HD} = V_{HF} + V_{FE} + V_{ED} = (5 + 0 - 6)\,\text{V} = -1\,\text{V}$$

$$V_{BE} = V_{BH} + V_{HF} + V_{FE} = (5 + 5 + 0)\,\text{V} = 10\,\text{V}$$

$$V_{CF} = V_{CD} + V_{DE} + V_{EF} = (4 + 6 + 0)\,\text{V} = 10\,\text{V}$$

It makes sense that the potential differences between points B and E and C and F are equal: $V_{BE} = V_{CF}$, as points B and C and points F and E have the same potential.

51. Current directions and notations in the circuit diagram (b) are indicated in the figure below.

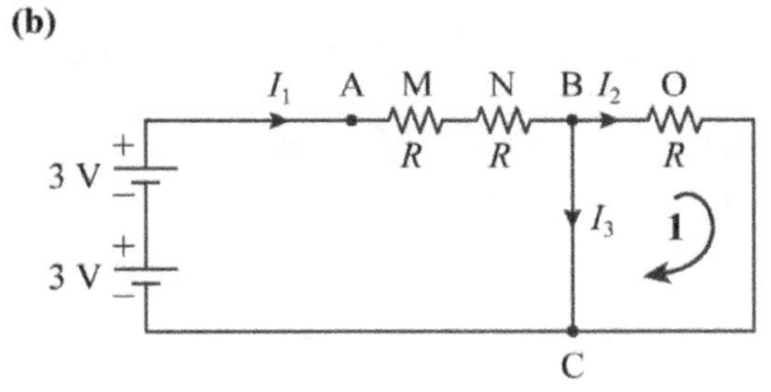

(b)

(a) The current through the batteries in circuit (a) $I_{batteries\ (a)}$ is equal to

$$I_{batteries\ (a)} = \frac{3V + 3V}{3R} = \frac{2V}{R} \tag{1}$$

In circuit (b), one of the parallel branches (BC) has zero resistance; thus, all the current will flow through it (the branch that has light bulb O is shorted). This will make the equivalent resistance of the circuit become 2R.

Thus:

$$I_{batteries\ (b)} = \frac{3V + 3V}{2R} = \frac{3V}{R}$$

This will make the light bulbs in circuit (b) to be brighter than in circuit (a).

Alternatively, you can resolve this circuit using Kirchhoff's laws.

Applying Kirchhoff's laws in the right-hand side loop of circuit (b) and node B, one obtains the following results:

Loop: $\quad\quad I_2 R = 0\,V \Rightarrow I_2 = 0\,A \quad\quad$ (2)

Node B: $\quad\quad I_1 = I_2 + I_3 \overset{eq.\ (1)}{\Rightarrow} I_1 = I_3 \quad\quad$ (3)

Based on equation (3), the current through the batteries in circuit (b) $I_{batteries\ (b)}$ is equal to

$$I_{batteries\ (b)} = I_1 = \frac{3V + 3V}{2R} = \frac{3V}{R} \quad\quad (4)$$

Using equations (1) and (4) the ratio of the currents through the batteries in the two circuits is equal to

$$\frac{I_{batteries\ (a)}}{I_{batteries\ (b)}} = \frac{2}{3}$$

(b) The ratio of the electrical potential differences between points A and B corresponding to the two circuits is given by

$$\frac{V_{AB\ (a)}}{V_{AB\ (b)}} = \frac{I_{batteries\ (a)}\,2R}{I_{batteries\ (b)}\,2R} = \frac{2}{3}$$

The ratio is less than 1 due to the higher current in circuit (b) (short-circuit of bulb O).

(c) The following equations can be written for the power output (proportional to light brightness) of the bulbs in circuits (a) and (b), respectively:

Circuit (a): $P_M = P_N = P_O = \dfrac{4V^2}{R}$

Circuit (b): $P_M = P_N = \dfrac{9V^2}{R}$, $P_O = 0\,W$

The brightness of bulbs M and N in circuit (b) is $9/4 = 2.25$ times greater than the brightness of bulbs M, N, and O in circuit (a).

53. The equilibrium temperature will be reached when the power dissipated by the resistor is equal to the rate of the heat loss to the surroundings. The increase in the temperature of the resistor is curbed by two processes: the increase in resistivity and the heat loss. Since the heat loss is the same for the two wires at the same temperature, it follows that the ratio of the power dissipated by the two wires is approximately proportional to the ratio of the equilibrium temperatures reached in the two wires. The following relationships can be derived for the two cases: parallel connection (a) and series connection (b).

(a) $\dfrac{T_{Al}}{T_W} \cong \dfrac{P_{Al}}{P_W} = \dfrac{\varepsilon^2 / R_{Al}(T_{Al})}{\varepsilon^2 / R_W(T_W)} = \dfrac{R_W(T_W)}{R_{Al}(T_{Al})} = \dfrac{R_W(T_0)\left[1 + \alpha_W(T_W - T_0)\right]}{R_{Al}(T_0)\left[1 + \dfrac{\alpha_W}{2}(T_{Al} - T_0)\right]} > 1 \Rightarrow T_{Al} > T_W$

(b) $\dfrac{T_{Al}}{T_W} \cong \dfrac{P_{Al}}{P_W} = \dfrac{I^2 R_{Al}(T_{Al})}{I^2 R_W(T_W)} = \dfrac{R_{Al}(T_{Al})}{R_W(T_W)} = \dfrac{R_W(T_0)\left[1 + \alpha_W(T_W - T_0)\right]}{R_{Al}(T_0)\left[1 + \dfrac{\alpha_W}{2}(T_{Al} - T_0)\right]} < 1 \Rightarrow T_W > T_{Al}$

55. The following useful notations are indicated in the circuit diagram below.

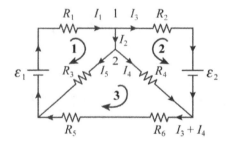

(a) Applying Kirchhoff's laws in loops 1, 2, and 3, and nodes 1 and 2, the following equations are obtained:

Loop 1: $\quad \varepsilon_1 = I_1 R_1 + I_5 R_3$ $\hspace{4cm}$ (1)

Loop 2: $\quad \varepsilon_2 = I_4 R_4 - I_3 R_2$ $\hspace{4cm}$ (2)

Loop 3: $\quad I_4 R_4 + (I_3 + I_4)(R_5 + R_6) = I_5 R_3$ $\hspace{2.3cm}$ (3)

Node 1: $\quad I_1 = I_2 + I_3$ $\hspace{4.7cm}$ (4)

Node 2: $\quad I_2 = I_4 + I_5$ $\hspace{4.7cm}$ (5)

Using equations (4) and (5) it follows that

$I_1 = I_3 + I_4 + I_5 \overset{eq.(1)}{\Rightarrow} \varepsilon_1 = (I_3 + I_4 + I_5)R_1 + I_5 R_3 \Rightarrow I_5 = \dfrac{\varepsilon_1 - (I_3 + I_4)R_1}{R_1 + R_3}$ $\hspace{1cm}$ (6)

Plugging the current I_5 value from equation (6) into equation (3) one obtains:

$$I_4 R_4 + (I_3 + I_4)(R_5 + R_6) = \frac{R_3}{R_1 + R_3}\left[\varepsilon_1 - (I_3 + I_4)R_1\right] \tag{7}$$

Equations (3) and (7) have two unknowns: currents I_3 and I_4, which can be solved numerically to give

$$I_3 = -\frac{39}{60}\,\text{A} = -0.65\,\text{A} \tag{8}$$

$$I_4 = \frac{1}{2}(I_3 + 2) = \frac{81}{120} = 0.675\,\text{A} \tag{9}$$

Using equation (6), one can find current I_5:

$$I_5 = \frac{15 - (0.675 - 0.65) \times 10}{20 + 5}\,\text{A} = \frac{14.75}{25}\,\text{A} = 0.59\,\text{A} \tag{10}$$

Using equations (5) and (4), respectively, one can obtain the following values of I_1 and I_2:

$$I_2 = I_4 + I_5 = 1.265\,\text{A} \tag{11}$$

$$I_1 = I_2 + I_5 = 0.615\,\text{A} \tag{12}$$

We didn't use equation (2) in our solution. However, we can use it to check if our answers are correct:

$$\varepsilon_2 = I_4 R_4 - I_3 R_2$$
$$\varepsilon_2 = 0.675\ \text{A} \times 20\ \Omega - (-0.65\ \text{A}) \times 10\ \Omega$$
$$\varepsilon_2 = 20\ \text{V}$$

(b) Voltages across resistors can be calculated using Ohm's law:

$$V_{R_1} = I_1 R_1 = 0.615\ \text{A} \times 10\ \Omega = 6.15\ \text{V}$$

$$V_{R_2} = I_3 R_2 = -0.65\ \text{A} \times 10\ \Omega = -6.5\ \text{V}$$

$$V_{R_3} = I_5 R_3 = 0.59\ \text{A} \times 15\ \Omega = 8.85\ \text{V}$$

$$V_{R_4} = I_4 R_4 = 0.675\ \text{A} \times 20\ \Omega = 13.5\ \text{V}$$

$$V_{R_5} = (I_3 + I_4) R_5 = (-0.65 + 0.675)\ \text{A} \times 20\ \Omega = 0.5\ \text{V}$$

$$V_{R_6} = (I_3 + I_4) R_6 = (-0.65 + 0.675)\ \text{A} \times 10\ \Omega = 0.25\ \text{V}$$

Once again, we can check if the values we have obtained make sense. For example, the sum of voltages of all the elements in each loop must equal zero if we consider the voltages across the battery as positive and the voltages across the resistors as negative (second Kirchhoff's circuit law—a consequence of energy conservation—equation (22-32)). Let us consider loop 1:

$$\varepsilon_1 - V_{R_1} - V_{R_3} = 15\,\text{V} - 6.15\,\text{V} - 8.85\,\text{V} = 0$$

This tells us that we solved the equations correctly.

(c) If the batteries are real, then small resistors will be connected in series with resistors R_1 and R_2. Hence, currents I_1 and I_3 will decrease.

57. The following useful notations are indicated in the circuit diagram below.

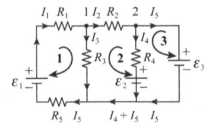

The following equations can be obtained by applying Kirchhoff's laws in loops 1, 2, and 3, and nodes 1 and 2:

Loop 1: $\varepsilon_1 = I_1 R_1 + I_3 R_3 + I_1 R_5$ (1)

Loop 2: $\varepsilon_2 = -I_4 R_4 - I_2 R_2 + I_3 R_3$ (2)

Loop 3: $\varepsilon_3 - \varepsilon_2 = I_4 R_4$ (3)

Node 1: $I_1 = I_2 + I_3$ (4)

Node 2: $I_2 = I_4 + I_5$ (5)

From equation (3) it follows:

$$I_4 = \frac{\varepsilon_3 - \varepsilon_2}{R_4} = \frac{(12-10)\,\text{V}}{20\,\Omega} = 0.1\,\text{A} \qquad (6)$$

Combining equations (1), (4), and (5), one obtains the following equation:

$$\varepsilon_1 = (I_3 + I_4 + I_5)(R_1 + R_5) + I_3 R_3 \qquad (7)$$

Combining equations (2) and (5), one obtains the following equation:

$$\varepsilon_2 = -I_4 R_4 - (I_4 + I_5) R_2 + I_3 R_3 \qquad (8)$$

Equations (7) and (8) can be solved numerically to find the following values of electrical currents I_3 and I_5:

$$I_3 = -\frac{27}{95}\,\text{A} = 0.28\,\text{A} \qquad (9)$$

$$I_5 = 2.5 + 3I_3 = \frac{31.3}{19} \text{ A} = 1.65 \text{ A} \tag{10}$$

Using equations (5) and (4), respectively, currents I_2 and I_1 can be calculated:

$$I_2 = I_4 + I_5 = \left(0.1 + \frac{31.2}{19}\right) \text{ A} = \frac{33.2}{19} \text{ A} = 1.75 \text{ A} \tag{11}$$

$$I_1 = I_2 + I_3 = \left(\frac{33.2}{19} - \frac{27}{95}\right) \text{ A} = \frac{139}{19} \text{ A} = 7.32 \text{ A} \tag{12}$$

The voltages across all the resistors are

$$V_{R_1} = I_1 R_1 = \frac{139}{95} \text{ A} \times 10 \text{ } \Omega = \frac{278}{19} \text{ V} = 14.6 \text{ V}$$

$$V_{R_2} = I_2 R_2 = \frac{33.2}{19} \text{ A} \times 5 \text{ } \Omega = \frac{166}{19} \text{ V} = 8.74 \text{ V}$$

$$V_{R_3} = I_3 R_3 = \left(-\frac{27}{95}\right) \text{ A} \times 15 \text{ } \Omega = -\frac{81}{19} \text{ V} = -4.26 \text{ V}$$

$$V_{R_4} = I_4 R_4 = 0.1 \text{ A} \times 20 \text{ } \Omega = 2.0 \text{ V}$$

$$V_{R_5} = I_1 R_5 = \frac{139}{95} \text{ A} \times 10 \text{ } \Omega = 14.6 \text{ V}$$

59. The following useful notations are indicated in the circuit diagram below.

(a)

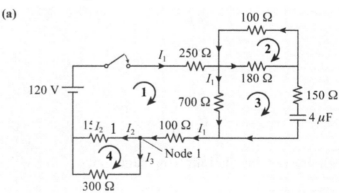

(a) After the switch was closed for a long time the capacitor is fully charged and the current through it is zero. Hence, the electrical currents in the capacitor branch of loop 3 and in loop 2 are also zero.

The following numerical equations are obtained by applying Kirchhoff's laws in loops 1, 3, and 4, and at node 1:

Loop 1: $\quad 120 = (250 + 100 + 700)I_1 + 150I_2 \Rightarrow 120 = 1050I_1 + 150I_2 \tag{1}$

Loop 3: $\quad 700I_1 = V_C \tag{2}$

Loop 4: $150I_2 = 300I_3 \Rightarrow I_2 = 2I_3$ (3)

Node 1: $I_1 = I_2 + I_3$ (4)

From equations (3) and (4),

$$I_1 = \frac{3}{2}I_2$$ (5)

Using equations (1) and (5):

$$120 = 1050 \times 1.5 \times I_2 + 150 \times I_2 \Rightarrow I_2 = \frac{120}{1725}\,\text{A} = \frac{8}{115}\,\text{A} = 0.07\,\text{A}$$ (6)

Using equation (5):

$$I_1 = \frac{12}{115}\,\text{A} = 0.10\,\text{A}$$ (7)

Using equation (3),

$$I_3 = \frac{1}{2}I_2 = \frac{4}{115}\,\text{A} = 0.04\,\text{A}$$ (8)

The voltages across all resistors and capacitor are

$$V_C = V_{700\,\Omega} = 700\,\Omega \times \frac{12}{115}\,\text{A} = 73.0\,\text{V}$$

$$V_{100\,\Omega} = 100\,\Omega \times \frac{12}{115}\,\text{A} = 10.4\,\text{V}$$

$$V_{150\,\Omega} = 150\,\Omega \times \frac{8}{115}\,\text{A} = 10.4\,\text{V}$$

$$V_{250\,\Omega} = 250\,\Omega \times \frac{12}{115}\,\text{A} = 26.1\,\text{V}$$

$$V_{300\,\Omega} = 300\,\Omega \times \frac{4}{115}\,\text{A} = 10.4\,\text{V}$$

We can check that these answers make sense as the sum of voltages of all the elements connected in series must equal the terminal voltage of the battery:

$$V_{250\,\Omega} + V_{700\,\Omega} + V_{100\,\Omega} + V_{150\,\Omega} = 73.0\,\text{V} + 26.1\,\text{V} + 10.4\,\text{V} + 10.4 = 119.9\,\text{V} \approx 120.0\,\text{V}$$

(b) When the switch is reopened the capacitor discharges through the resistors from circuit loops 2 and 3. Therefore, the time constant of the circuit $\tau_{circuit}$ is given by

$$\tau_{circuit} = R_{eq}C$$

The equivalent resistance R_{eq} in the above equation is equal to

$$R_{eq} = 700\ \Omega + 150\ \Omega + \left(\frac{1}{180\ \Omega} + \frac{1}{100\ \Omega}\right)^{-1} = 914\ \Omega$$

Hence, the time constant of the circuit after the switch is reopened is equal to

$$\tau_{circuit} = 914\ \Omega \times 4 \times 10^{-6}\ F = 3.66\ ms$$

61. (a) The effective resistance of the circuit R_{eff} is equal to

$$R_{eff} = \frac{V_{out}}{I_{out}} = \frac{10\ V}{250 \times 10^{-3}\ A} = 40.0\ \Omega$$

(b) The minimal voltage the capacitor V_{min} is allowed to discharge during the time period $T = \frac{1}{60}$ s is equal to

$$V_{min} = 10\ V - 0.100\ V = 9.900\ V$$

The discharging of the capacitor during the period of time T is described by the following exponential time decay:

$$V_{min} = V(T) = V_{out} e^{-\frac{T}{RC_{min}}} \Rightarrow C_{min} = \frac{T}{R \ln\left(\dfrac{V_{out}}{V_{min}}\right)} = \frac{(1/60)\ s}{40\ \Omega \times \ln\left(\dfrac{10}{9.9}\right)} = 41.5\ mF$$

(c) The energy stored by the capacitor when fully charged U is equal to

$$U = \frac{1}{2} C_{min} V_{out}^2 = 2.07\ J$$

(d) The energy drawn from the capacitor when its voltage dropped by 0.100 V is

$$\Delta U = \frac{1}{2} C_{min} \left(V_{out}^2 - V_{min}^2\right) = 0.5 \times \left(\frac{1/60\,s}{40\ \Omega \times \ln\left(\dfrac{10}{9.9}\right)}\right) F \times \left(10^2 - 9.9^2\right) V^2 =$$

$$= 0.5 \times (41.5\ mF) \times 1.99\ V^2 = 41.3\ mJ$$

1. Yes, there could be a nonzero magnetic field in a region where a charged particle moves in a straight line if the magnetic field lines are parallel to the particle's velocity.

3. The direction of the magnetic force $\vec{F_B}$ for each diagram from Figure 23-47 is indicated in the diagrams from the figure shown below.

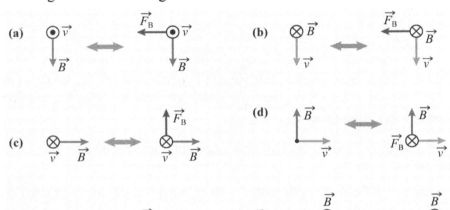

5. Because point a has a higher potential than point b ($V_a > V_b$), it follows that there is a surplus of positive charges on the back face of the rectangular piece of semiconductor. Using the historical convention that electric current direction is established by the bulk motion of positive charge carriers, it follows that current flows from left to right in the rectangular piece of semiconductor from the circuit diagram of Figure 23-48. Using the right-hand rule, the direction of the magnetic force $\vec{F_B}$ is from the front to the back of the rectangular semiconductor as shown in the figure below. Therefore, if the charge carriers are positive, an excess of positive charges will accumulate on the back face of the semiconductor, consistent with the measured potential difference: $V_a > V_b$. If the charge carriers are negative (i.e., electrons) the direction of the current will be reversed, the magnetic force direction $\vec{F_B}$ will be the same; however, the potential inequality will change: $V_a < V_b$.

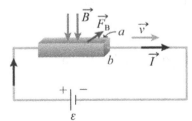

7. b. The diagram shown below is based on Figure 23-50 and indicates the magnetic field lines for the coil, thus leading to the polarity of the permanent magnet which results in an attractive force.

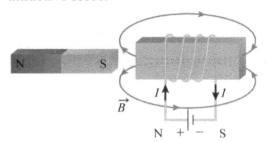

9. f. The magnetic field is oriented from right to left (from the north pole of the right magnet to the south pole of the left magnet), therefore, the right-hand rule indicates that the resulting magnetic force vector $\vec{F}_B$ direction is out of the page.

11. c. The direction of the magnetic field formed around a current-carrying wire can be determined using the right-hand rule.

13. a. The ranking of the magnetic field strength in the three regions can be deduced from the radii of the curved trajectories corresponding to each region. The magnetic force $\vec{F}_B$ is the centripetal force which curves the charged particle's trajectories:

$$F_B = F_C \Rightarrow qvB = m\frac{v^2}{r} \Rightarrow r = \frac{mv}{qB} \overset{q/m=const.}{\Rightarrow} r \sim \frac{1}{B}$$

Therefore, $B_{III} > B_I > B_{II}$

15. c. The radius of the electron's curved trajectory r is inverse proportional to the magnetic field B:

$r \sim 1/B$

It follows that if the magnetic field is decreased by a factor of 2, the radius of the electron is increased by a factor of 2:

$$\frac{B'}{B} = \frac{1}{2} \overset{r \sim \frac{1}{B}}{\Rightarrow} \frac{r'}{r} = 2$$

However, the deflection decreases, but not by a factor of 2 since the deflection is produced only by a fraction of the electron's curved trajectory, or, in other words, the radius and deflection are not directly proportional.

17. For each diagram from Figure 23-57, the magnitude of the magnetic field $\left|\vec{B}\right|$ can be calculated as follows:

(a) $\vec{B} = 0$

(b) $\vec{B} = 2B_0$

(c) $\vec{B} = 0$

(d) $\vec{B} = 2B_0$

(e) $\vec{B} = 2\sqrt{2}B_0$

(f) $\vec{B} = 0$

In the results written above, $B_0 = \dfrac{I}{2\pi\mu_0 r}$, where r is the distance from the centre of the square to any of the wires positioned in the corners of the square. Therefore, the ranking, from the highest to the lowest of the magnitude of the magnetic field $\left|\vec{B}\right|$ is as follows:

$$(e) > (b) = (d) > (a) = (c) = (f)$$

19. The force on the electron $\vec{F}$ is given by

$$\vec{F} = (-e)\vec{v} \times \vec{B} \tag{1}$$

$$-(\vec{v} \times \vec{B}) = \begin{vmatrix} \hat{i} & \hat{j} & \hat{k} \\ -4 & 0 & 8 \\ 0.8 & 0.6 & 0 \end{vmatrix} \times 10^4 = \left(-4.8\hat{i} + 6.4\hat{j} - 2.4\hat{k}\right) \times 10^4 \, \frac{\text{m} \cdot \text{T}}{\text{s}} \tag{2}$$

Using the results of equations (1) and (2),

$$\vec{F} = \left(1.6 \times 10^{-19} \ \text{C}\right) \times \left(8 \times 10^4 \times \left(-0.6\hat{i} + 0.8\hat{j} - 0.3\hat{k}\right) \, \frac{\text{m} \cdot \text{T}}{\text{s}}\right)$$

$$\vec{F} = 12.8 \times \left(-0.6\hat{i} + 0.8\hat{j} - 0.3\hat{k}\right) \times 10^{-15} \ \text{N} = 12.8 \times \left(-0.6\hat{i} + 0.8\hat{j} - 0.3\hat{k}\right) \ \text{fN}$$

Also,

$$\left|\vec{F}\right| = \left(12.8 \times \sqrt{0.6^2 + 0.8^2 + 0.3^2}\right)\text{fN} = 13.4 \, \text{fN}$$

21. The magnetic force $\vec{F}$ acting on the particle is in the direction of $\hat{j}$ and its magnitude $\left|\vec{F}\right|$ is given by

$$\left|\vec{F}\right| = qvB = \left(10 \times 10^{-9} \ \text{C}\right)\left(3 \times 10^8 \ \frac{\text{m}}{\text{s}}\right)\left(1.5 \times 10^{-3} \ \text{T}\right) = 4.5 \ \text{mN}$$

23. (a) The radius of the circular motion associated with a magnetic field $\vec{B}$ acting on a particle of charge q can be determined by recognizing that the magnetic force $\vec{F_B}$ is the centripetal force $\vec{F_r}$:

$$F_B = F_r \Leftrightarrow qvB = \frac{mv^2}{r} \Rightarrow r = \frac{mv}{qB} \tag{1}$$

Since the speeds of the particles and the magnetic field are the same in all three cases, the mass-to-charge ratio $\frac{m}{q}$ will affect the ranking. The ranking of the electron, proton, and α-particle, respectively, is as follows:

$$\left|-\frac{m_e}{e}\right| < \frac{m_p}{e} < \frac{4m_p}{2e}$$

Therefore, the ranking of the radii r_e, r_p, and r_α is as follows:

$$r_\alpha > r_p > r_e$$

Notice, larger radius means that the particle is less affected by the magnetic field.

(b) The period of the circular motion T is given by

$$T = \frac{2\pi r}{v} \overset{eq.\ (1)}{\Rightarrow} T = \frac{2\pi mv}{vqB} = \frac{2\pi m}{qB}$$

Based on the mass-to-charge ratio $\frac{m}{q}$ ranking given in the solution of part (a), the ranking of the periods T_e, T_p, and T_α is as follows:

$$T_\alpha > T_p > T_e$$

25. (a) The proton's speed v_p can be calculated using the following equation:

$$v_p = \frac{eBr}{m} = \frac{\left(1.60\times10^{-19}\,C\right)\left(0.9\ T\right)\left(0.1\times10^{-3}\,m\right)}{1.67\times10^{-27}\ kg} = 8.6\times10^3\,m/s = 8.6\ km/s$$

(b) The proton's angular frequency ω_p is

$$\omega_p = \frac{v_p}{r} = \frac{eB}{m} = \frac{\left(1.60\times10^{-19}\,C\right)\left(0.9\,T\right)}{1.67\times10^{-27}\ kg} = \frac{8.6\times10^3\ m/s}{0.1\times10^{-3}\ m} = 8.6\times10^7\ rad/s$$

27. The radius r of the trajectory of a particle of charge q in a uniform magnetic field B is given by

$$r = \frac{mv}{qB} \tag{1}$$

The speed of a particle accelerated by an electric potential difference V can be calculated classically as follows:

$$qV = \frac{mv^2}{2} \Rightarrow v = \sqrt{\frac{2qV}{m}} \tag{2}$$

Substituting the speed v given by equation (2) into equation (1), the following equation is obtained:

$$r = \frac{m\sqrt{\frac{2qV}{m}}}{qB} = \frac{1}{B}\sqrt{\frac{2mV}{q}} = \frac{\sqrt{2V}}{B}\sqrt{\frac{m}{q}} \tag{3}$$

The masses and charges of the deuteron (a nucleus of deuterium – heavy hydrogen, an isotope of hydrogen that has one proton and one neutron), α-particle, and proton are as follows:

$$m_d = 2m_p; q_d = e$$

$$m_\alpha = 4m_p; q_\alpha = 2e$$

$$m_p = m_p; q_p = e$$

Based on equation (3), the following ratios amongst the three radii of the deuteron, α-particle, and proton r_d, r_α, and r_p, respectively, can be written:

$$\frac{r_d}{r_p} = \sqrt{\frac{m_d}{m_p} \cdot \frac{q_p}{q_d}} = \sqrt{2} \Rightarrow r_d = \sqrt{2}\, r_p$$

$$\frac{r_\alpha}{r_d} = \sqrt{\frac{m_\alpha}{m_p} \cdot \frac{q_p}{q_\alpha}} = \sqrt{2} \Rightarrow r_\alpha = \sqrt{2}\, r_p$$

29. Let v_1, v_2, v_3 denote the speeds of the electron at the surface of the neutron star equal to 5%, 10%, and 20% of the speed of light in vacuum, respectively. The corresponding electron radii r_1, r_2, and r_3 can be calculated as follows:

$$r_1 = \frac{m}{e} \cdot \frac{v_1}{B} = \frac{m}{e} \cdot \frac{0.05c}{B} = 0.05\frac{mc}{eB} = 0.05\frac{\left(9.11 \times 10^{-31}\,\text{kg}\right)\left(3 \times 10^8\,\text{m/s}\right)}{\left(1.60 \times 10^{-19}\,\text{C}\right)\left(3 \times 10^7\,\text{T}\right)}$$

$$= 0.05 \times 5.69375 \times 10^{-11} = 0.28 \times 10^{-11}\,\text{m} = 2.8 \times 10^{-12}\,\text{m}$$

$$r_2 = \frac{m}{e} \cdot \frac{v_2}{B} = 0.1\frac{mc}{eB} = 0.1\frac{\left(9.11\times10^{-31}\,\text{kg}\right)\left(3\times10^8\,\text{m/s}\right)}{\left(1.60\times10^{-19}\,\text{C}\right)\left(3\times10^7\,\text{T}\right)} = 5.7\times10^{-12}\,\text{m}$$

$$r_3 = \frac{m}{e} \cdot \frac{v_3}{B} = 0.2\frac{mc}{eB} = 0.2\frac{\left(9.11\times10^{-31}\,\text{kg}\right)\left(3\times10^8\,\text{m/s}\right)}{\left(1.60\times10^{-19}\,\text{C}\right)\left(3\times10^7\,\text{T}\right)} = 1.14\times10^{-11}\,\text{m}$$

The corresponding magnitudes of magnetic forces $\overrightarrow{F_{B_1}}$, $\overrightarrow{F_{B_2}}$, and $\overrightarrow{F_{B_3}}$ are equal to

$$F_{B_1} = ev_1B = \left(1.60\times10^{-19}\,\text{C}\right)0.05\left(3\times10^8\,\text{m/s}\right)\left(3\times10^7\,\text{T}\right) = 72\,\mu\text{N}$$

$$F_{B_2} = ev_2B = \left(1.60\times10^{-19}\,\text{C}\right)0.1\left(3\times10^8\,\text{m/s}\right)\left(3\times10^7\,\text{T}\right) = 144\,\mu\text{N}$$

$$F_{B_3} = ev_3B = \left(1.60\times10^{-19}\,\text{C}\right)0.2\left(3\times10^8\,\text{m/s}\right)\left(3\times10^7\,\text{T}\right) = 288\,\mu\text{N}$$

31. The magnetic field $\vec{B}_\perp$ perpendicular to the electron's velocity $\vec{v} = v\hat{i}$ is

$$\vec{B}_\perp = \left(0\,\mu T, 20\,\mu T, 18\,\mu T\right) \Rightarrow \left|\vec{B}_\perp\right| = B_\perp = \sqrt{20^2 + 18^2}\,\mu T = 26.9\,\mu T \tag{1}$$

The radius of the electron's trajectory R is given by

$$R = \frac{mv}{eB_\perp} = \frac{\left(9.11\times10^{-31}\,\text{kg}\right)\left(6\times10^7\,\text{m/s}\right)}{\left(1.60\times10^{-19}\,\text{C}\right)\sqrt{20^2 + 18^2}\,\left(10^{-6}\,\text{T}\right)} = 12.696\,\text{m} = 12.7\,\text{m} \tag{2}$$

The geometry of the electron's trajectory is depicted in the figure shown below. The radius of curvature in the yz-plane and the electron's deflection are denoted by R and Δ, respectively.

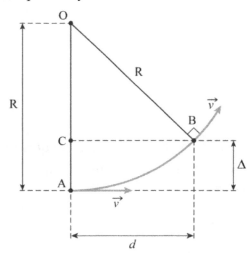

Applying Pythagoras's theorem in the right triangle OCB:

$$OC^2 + CB^2 = BO^2 \Rightarrow \left(R - \Delta\right)^2 + d^2 = R^2 \tag{3}$$

From equation (3), the deflection Δ is given by

$$\Delta = R - \sqrt{R^2 - d^2} = 12.696 - \sqrt{12.696^2 - 0.5^2} = 9.8 \text{ mm}$$

33. (a) The magnetic force per unit length F/l is given by

$$\frac{F}{l} = BI = (1.8 \text{ T})(10 \text{ A}) = 18 \text{ N/m}$$

(b) The magnetic force per unit length F'/l after the magnetic field is halved and the current is doubled is

$$\frac{F'}{l} = B'I' = \frac{B}{2}2I = BI = \frac{F}{l} = 18 \text{ N/m}$$

Therefore, the magnetic force per unit length remains the same.

35. The magnetic field at the centre O of the circular loop $\vec{B}$ is equal to

$$\vec{B} = \vec{B}_{upper-half} + \vec{B}_{lower-half} = \vec{B}_{upper-half} + \left(-\vec{B}_{upper-half}\right) = 0 \text{ T}$$

37. The magnetic field created by the lightning bolt can be approximated by the formula:

$$B = \frac{\mu_0 I}{2\pi r}$$

The magnitudes of the magnetic field at $r_1 = 10 \text{ m}$, $r_2 = 100 \text{ m}$, and $r_3 = 1 \text{ km}$ are

$$B(r_1) = \frac{\mu_0 I}{2\pi r_1} = \frac{\left(4\pi \times 10^{-7} \text{ T} \cdot \text{m/A}\right)\left(10^4 \text{ A}\right)}{2\pi(10 \text{ m})} = 200 \ \mu\text{T}$$

$$B(r_2) = \frac{\mu_0 I}{2\pi r_2} = \frac{\left(4\pi \times 10^{-7} \text{ T} \cdot \text{m/A}\right)\left(10^4 \text{ A}\right)}{2\pi(100 \text{ m})} = 20 \ \mu\text{T}$$

$$B(r_3) = \frac{\mu_0 I}{2\pi r_3} = \frac{\left(4\pi \times 10^{-7} \text{ T} \cdot \text{m/A}\right)\times\left(10^4 \text{ A}\right)}{2\pi(1000 \text{ m})} = 2 \ \mu\text{T}$$

By comparison, Earth's magnetic field is roughly $50 \ \mu T$.

39. (a) The magnetic field magnitude $B(r)$ produced by a wire carrying current I with diameter d at a distance r from the centre of the wire ($r \le d/2 = 5 \text{ mm}$) is given by

$$B(r)_{inside} = \frac{\mu_0 I_{enc}}{2\pi r} = \frac{\mu_0}{2\pi r} \cdot \frac{I\pi r^2}{\pi d^2/4} = \frac{2\mu_0 I}{\pi d} \cdot \frac{r}{d} = \frac{2\mu_0 I}{\pi d^2} \cdot r \qquad (1)$$

(Notice when $r = d/2 = 5 \text{ mm}$, equation (1), as expected, becomes $B\left(\dfrac{d}{2}\right) = \dfrac{\mu_0 I}{\pi d}$)

Based on equation (1),

$$B(2\,\mathrm{mm}) = \frac{2\left(4\pi\times10^{-7}\,\mathrm{T\cdot m/A}\right)\left(25\,\mathrm{A}\right)}{\pi\left(10\times10^{-3}\,\mathrm{m}\right)^2}\left(2\times10^{-3}\,\mathrm{m}\right) = \left(0.2\,\mathrm{T/m}\right)\left(2\times10^{-3}\,\mathrm{m}\right) = 0.4\,\mathrm{mT}$$

$$B(5\,\mathrm{mm}) = \frac{2\left(4\pi\times10^{-7}\,\mathrm{T\cdot m/A}\right)\left(25\,\mathrm{A}\right)}{\pi\left(10\times10^{-3}\,\mathrm{m}\right)^2}\left(5\times10^{-3}\,\mathrm{m}\right) = \left(0.2\,\mathrm{T/m}\right)\left(5\times10^{-3}\,\mathrm{m}\right) = 1.0\,\mathrm{mT}$$

Notice, only the magnitude of the magnetic field $B(r)$ at a radial distance r *within* the cable $r \le d/2$ is given by equation (1). The magnitude of the magnetic field $B(r)$ at a radial distance r outside the cable ($r = 7$ mm) is given by equation (2):

$$B(r)_{outside} = \frac{\mu_0 I}{2\pi r} \tag{2}$$

Given equations (1) and (2) and the numerical values of current I, one can summarize the expressions for $B(r)$ as

$$B(r) = \begin{cases} 0.2\ \dfrac{\mathrm{T}}{\mathrm{m}}\times r\,[\mathrm{m}] = 0.2\ \dfrac{\mathrm{mT}}{\mathrm{mm}}\times r\,[\mathrm{mm}], & r \le 5\,\mathrm{mm} \\[3mm] \dfrac{5.0\ \dfrac{\mathrm{T}}{\mathrm{m}}}{r\,[\mathrm{m}]} = \dfrac{5.0\ \dfrac{\mathrm{mT}}{\mathrm{mm}}}{r\,[\mathrm{mm}]}, & r > 5\,\mathrm{mm} \end{cases}$$

Therefore, $B(7\ \mathrm{mm}) = 0.71$ mT

(b) The graph of the above function $B = B(r)$ is shown below.

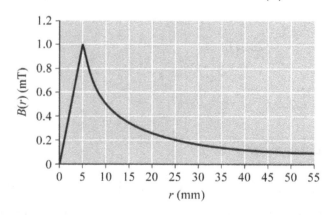

(c) Applying Gauss's law for a uniformly charged insulating sphere of radius R and charge Q, the electric field $E(r)$ inside and outside the sphere is given by

$$E(r) = \begin{cases} \dfrac{Q}{4\pi\varepsilon_0 R^2} \cdot \dfrac{r}{R} = \dfrac{Q}{4\pi\varepsilon_0 R^3} \cdot r, & r \le R \\[4mm] \dfrac{Q}{4\pi\varepsilon_0 r^2}, & r > R \end{cases} \qquad (3)$$

Applying Ampere's law for a long wire with radius R and carrying current I, the magnetic field $B(r)$ inside and outside the wire is given by

$$B(r) = \begin{cases} \dfrac{\mu_0 I}{2\pi R} \cdot \dfrac{r}{R} = \dfrac{\mu_0 I}{2\pi R^2} \cdot r, & r \le R \\[4mm] \dfrac{\mu_0 I}{2\pi r}, & r > R \end{cases} \qquad (4)$$

Relationships (3) and (4) are similar, except for the more rapid ($\sim 1/r^2$) decrease of the electric field $E(r)$ outside the uniformly charged sphere.

41. This problem will be revised for the second edition. An answer is not provided here.

43. (a) The magnetic moment of the loop $\vec{\mu}_{loop}$ of radius R and carrying current I is

$$\vec{\mu}_{loop} = I\vec{A} = I\pi R^2 \left(-\hat{k}\right) = \pi \times 10 A \times (0.1m)^2 \times \left(-\hat{k}\right) = \left(0.314 \text{ A} \cdot \text{m}^2\right)\left(-\hat{k}\right)$$

(b) The magnetic moment of the coil $\vec{\mu}_{coil}$ is given by

$$\vec{\mu}_{coil} = N\vec{\mu}_{loop} = 100 \times \left(0.314 \text{ A} \cdot \text{m}^2\right)\left(-\hat{k}\right) = \left(31.4 \text{ A} \cdot \text{m}^2\right)\left(-\hat{k}\right)$$

(c) The torque experienced by the coil $\vec{\tau}_{coil}$ is given by

$$\vec{\tau}_{coil} = \vec{\mu}_{coil} \times \vec{B} = \begin{vmatrix} \hat{i} & \hat{j} & \hat{k} \\ 0 & 0 & -\mu_{coil} \\ 0 & B & 0 \end{vmatrix} = \mu_{coil} B \hat{i} = \left(31.4 \text{ A} \cdot \text{m}^2\right)(1 \text{ T})\hat{i} = \left(31.4 \text{ N} \cdot \text{m}\right)\hat{i}$$

(d) All the directions of the answers as compared to the original question will get reversed, while the magnitudes will remain the same:

(a) $\vec{\mu}_{loop} = \left(0.314 \text{ A} \cdot \text{m}^2\right)\hat{k}$

(b) $\vec{\mu}_{coil} = \left(31.4 \text{ A} \cdot \text{m}^2\right)\hat{k}$

(c) $\vec{\tau}_{coil} = \left(31.4 \text{ N} \cdot \text{m}\right)\left(-\hat{i}\right)$

(e) If $\vec{B} = (1.0 \text{ T})\hat{i}$, the results of parts (a)–(c) change as follows:

(a) The original result is not changed.

(b) The original result is not changed.

(c) The torque experienced by the coil $\vec{\tau}_{coil}$ is given by

$$\vec{\tau}_{coil} = \vec{\mu}_{coil} \times \vec{B} = \begin{vmatrix} \hat{i} & \hat{j} & \hat{k} \\ 0 & 0 & -\mu_{coil} \\ B & 0 & 0 \end{vmatrix} = -\mu_{coil}B\hat{i} = (31.4 \text{ N}\cdot\text{m})\left(-\hat{i}\right)$$

If $\vec{B} = (1.0 \text{ T})\hat{k}$, the results of parts (a)–(c) change as follows:

(a) The original result is not changed.

(b) The original result is not changed.

(c) The torque experienced by the coil $\vec{\tau}_{coil}$ is given by

$$\vec{\tau}_{coil} = \vec{\mu}_{coil} \times \vec{B} = \begin{vmatrix} \hat{i} & \hat{j} & \hat{k} \\ 0 & 0 & -\mu_{coil} \\ 0 & 0 & B \end{vmatrix} = \vec{0} \text{ N}\cdot\text{m} \text{ , or } \tau_{coil} = 0 \text{ N}\cdot\text{m}$$

(f) By definition, the potential energy of the coil is given by

$$U \overset{def}{=} -\vec{\mu}\cdot\vec{B}$$

Based on the equation written above, the initial and final potential energy values U_i and U_f, respectively, are given by

$U_i = \mp\mu_{coil}B$, $\qquad \vec{\mu} \parallel \vec{B}$ (vectors $\vec{\mu}$ and $\vec{B}$ can be aligned parallel or antiparallel),

$U_f = 0$, $\qquad \vec{\mu} \perp \vec{B}$

Therefore, the change in the potential energy of the coil ΔU is given by

$$\Delta U \overset{def}{=} U_f - U_i = \pm\mu_{coil}B \cong \pm 31.4 \text{ J}$$

45. (a) The direction of the magnetic force $\vec{F}_B$ on the electron is perpendicular and away from the wire, as indicated in the diagram on the left in the figure shown below. Its magnitude F_B can be calculated as follows:

$$F_B = evB = ev\frac{\mu_0 I}{2\pi d} = \frac{\left(1.6\times10^{-19}\text{C}\right)\left(1.5\times10^7\text{ m/s}\right)\left(4\pi\times10^{-7}\text{ T}\cdot\text{m/A}\right)\left(5.0 \text{ A}\right)}{2\pi\left(2\times10^{-2}\text{ m}\right)}$$

$$= 1.2\times10^{-16} = 0.12\,\text{fN}$$

(b) The direction of the magnetic force $\vec{F}_B$ on the proton is perpendicular and toward the wire, as indicated in the right-hand side diagram from the figure shown below. Its magnitude F_B can be calculated as follows:

$$F_B = evB = ev\frac{\mu_0 I}{2\pi d} = \frac{\left(1.6\times10^{-19}\,\mathrm{C}\right)\left(1.5\times10^{7}\,\mathrm{m/s}\right)\left(4\pi\times10^{-7}\,\mathrm{T\cdot m/A}\right)\left(5.0\,\mathrm{A}\right)}{2\pi\left(2\times10^{-2}\,\mathrm{m}\right)}$$

$$= 1.2\times10^{-16}\,\mathrm{N} = 0.12\,\mathrm{fN}$$

The value of magnetic force on the proton is exactly the same as the value of magnetic force on the electron, as the particle move with the same velocity and have equal and opposite electric charges.

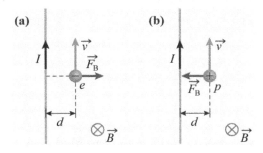

(c) The radius r of the circular trajectory induced by the magnetic force acting on a particle with mass-to-charge ratio m/q and moving with speed v perpendicular to a uniform magnetic field of magnitude B, is given by

$$r = \frac{m}{q}\cdot\frac{v}{B} \tag{1}$$

The magnetic field $B(d)$ created by a long wire carrying current I at distance d is given by

$$B(d) = \frac{\mu_0 I}{2\pi d} \tag{2}$$

Combining equations (1) and (2), the following equation is obtained:

$$r = \frac{m}{q}\cdot\frac{2\pi dv}{\mu_0 I} \tag{3}$$

Equation (3) indicates that the radius r of the charged particle's trajectory for which $\vec{v} \perp \vec{B}$, is directly proportional to the distance from the wire d. Qualitatively, the trajectories for electron and proton are shown in the figure on the next page.

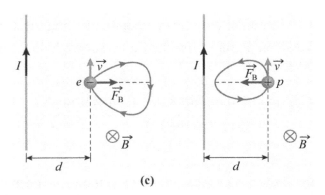

(c)

47. Let R_{R_1} and R_{R_2} denote the resistance of the upper and lower arcs, respectively. Also, let I_1 and I_2 denote the currents carried by the upper and lower arcs, respectively. Since the upper and lower wires have the same resistivity and cross-sectional area, their resistances will be proportional to their respective lengths:

$$\frac{R_{R_1}}{R_{R_2}} = \frac{\rho \dfrac{l_1}{A}}{\rho \dfrac{l_2}{A}} = \frac{l_1}{l_2} = 2$$

The following equations can be written by applying Kirchhoff's and Ohm's laws:

$$I = I_1 + I_2 \tag{1}$$

$$\frac{I_1}{I_2} = \frac{R_{R_2}}{R_{R_1}} = \frac{1}{2} \tag{2}$$

Using equations (1) and (2), it can be derived:

$$I_1 = \frac{I}{3} \tag{3}$$

$$I_2 = \frac{2I}{3} \tag{4}$$

The magnetic field magnitudes created by the two semicircles are given by the following equations. Let us denote the radius of the smaller semi-circle as R, then $R_1 \equiv 2R; R_2 \equiv R$.

$$B_1 = \frac{1}{2} \cdot \frac{\mu_0 I_1}{2R_1} \overset{eq.\,(3)}{\Rightarrow} B_1 = \frac{\mu_0 I}{6R_1} = \frac{\mu_0 I}{12R} \tag{5}$$

$$B_2 = \frac{1}{2} \cdot \frac{\mu_0 I_2}{2R_2} \overset{eq.\,(4)}{\Rightarrow} B_2 = \frac{\mu_0 I}{12R_2} = \frac{\mu_0 I}{12R} \tag{6}$$

Let $\hat{k}$ denote the unit vector perpendicular on the plane of the two arcs and directed into the page. Then, the total magnetic field $\vec{B}$ created by the two arcs is equal to

$$\vec{B} = (B_1 - B_2)\hat{k} \overset{eqs.\,(5)\&(6)}{\Rightarrow} \vec{B} = \frac{\mu_0 I}{6}\left(\frac{1}{R_1} - \frac{1}{2R_2}\right)\hat{k} = \frac{\mu_0 I}{12}\left(\frac{1}{R} - \frac{1}{R}\right)\hat{k} = 0$$

49. (a) Assuming a constant magnetic field throughout the length of the solenoid and applying Ampère's law, the magnitude of the magnetic field in the centre of the solenoid B_0 can be determined as follows:

$$B_0 L = \mu_0 N I \Rightarrow B_0 = \frac{\mu_0 N I}{L} = \frac{\left(4\pi \times 10^{-7}\,\text{T} \cdot \text{m/A}\right)\left(50 \times 1\,\text{A}\right)}{40 \times 10^{-2}\,\text{m}} = 5\pi\left(10^{-5}\,\text{T}\right) = 157\,\mu\text{T}$$

(b) Let us assume an infinitesimal element of length dx along the solenoid. The infinitesimal number of windings dN corresponding to this infinitesimal element is equal to

$$dN = N dx / L \tag{1}$$

For convenience, let the x-axis be the central axis of the solenoid and assume that the centre of the solenoid is at $x = 0$. The infinitesimal magnetic field $dB(x)$ created at distance x by the infinitesimal solenoidal element of length dx and radius R can be calculated by employing the Biot-Savart law (see Example 23-8):

$$dB(x) = \frac{\mu_0 I R^2\, dN}{2\left(x^2 + R^2\right)^{3/2}} \overset{eq.\ (1)}{\Rightarrow} dB(x) = \frac{\mu_0 N I R^2}{2L} \cdot \frac{dx}{\left(x^2 + R^2\right)^{3/2}} \tag{2}$$

Let $B(x)$ denote the magnetic field along the central axis of the solenoid as a function of distance x from its centre. The function $B(x)$ can be determined by integrating the value of $dB(x)$ from equation (2) as follows:

$$B(x) = \int_{-L/2}^{L/2} dB(x' - x) = \frac{\mu_0 N I R^2}{2L} \int_{l_1}^{l_2} \frac{dx'}{\left[x'^2 + R^2\right]^{3/2}} \tag{3}$$

The limits of integration in equation (3) are $l_1 = -(L/2 + x)$ and $l_2 = L/2 - x$. The following mathematical identity can be demonstrated by using the trigonometric substitution $x = R\tan(\theta)$:

$$\int \frac{dx}{\left(x^2 + R^2\right)^{3/2}} = \frac{1}{R^2} \cdot \frac{x}{\sqrt{x^2 + R^2}} + C \tag{4}$$

Where C denotes an arbitrary constant in equation (4). Hence, the magnetic field $B(x)$ from equation (3) is equal to

$$B(x) = \frac{\mu_0 N I R^2}{2L} \frac{1}{R^2} \frac{x'}{\sqrt{x'^2 + R^2}} \Bigg|_{l_1}^{l_2} = \frac{\mu_0 N I}{2L} \left[\frac{L + 2x}{\sqrt{(L + 2x)^2 + 4R^2}} + \frac{L - 2x}{\sqrt{(L - 2x)^2 + 4R^2}} \right] \tag{5}$$

It can be noticed that $|x| < L/2$ if one is to calculate the magnetic field magnitude inside the solenoid. The value of the magnetic field magnitude in the centre of the solenoid $B(0)$ can be calculated using equation (5):

$$B(0) = \frac{\mu_0 NI}{\sqrt{L^2 + 4R^2}} = \frac{\mu_0 NI}{L} \cdot \frac{1}{\sqrt{1 + \left(\dfrac{2R}{L}\right)^2}} \tag{6}$$

From an inspection of equation (6), it can be seen that $B_0 = B(0) = \mu_0 NI / L$ in the limit $L \gg R$.

(c) The plot of the function $B(x)$ from equation (5) is shown below.

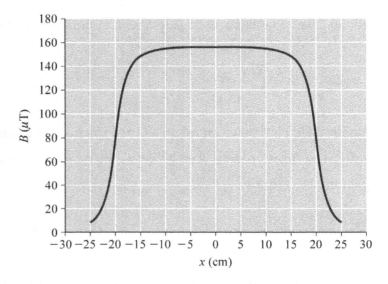

(d) A 5% decrease from the central value $B(0)$ is equal to

$$0.95 \times B(0) = 0.05 \times \frac{\mu_0 NI}{\sqrt{L^2 + 4R^2}} = \frac{0.95 \times \left(4\pi \times 10^{-7}\,\text{T} \cdot \text{m/A}\right) \times \left(50 \times 1\,\text{A}\right)}{\sqrt{40^2 + 5^2} \times 10^{-2}\,\text{m}} = 148\,\mu\text{T}$$

From the graph of $B(x)$ shown above, the 5% decrease value of the central magnetic field magnitude corresponds to a distance of approximately 15 cm from the centre of the solenoid.

51. Let the subscript 1 denote the left-hand wire carrying current $2I$ and let the subscript 2 denote the right-hand wire carrying current I. Also, let r_{1A} denote the distance between the left-hand wire and point A, let r_{2A} denote the distance between the right-hand wire and point A, and so on. The distance between the two wires is denoted by $d = 1$ m. Based on the diagram from Figure 23-66, the following distances can be estimated:

$$r_{1A} = \sqrt{(0.2\,\text{m})^2 + (0.3\,\text{m})^2} = 0.36\,\text{m} \ \text{and} \ r_{2A} = \sqrt{(0.8\,\text{m})^2 + (0.3\,\text{m})^2} = 0.85\,\text{m}$$

$$r_{1B} = r_{2B} = \frac{d}{2} = 0.5\,\text{m}$$

$$r_{1C} = r_{2C} = \sqrt{(0.5\,\text{m})^2 + (0.3\,\text{m})^2} = 0.58\,\text{m}$$

$$r_{1D} = 0.8\,\text{m} \ \text{and} \ r_{2D} = 0.2\,\text{m}$$

The corresponding magnetic field magnitudes can be calculated as follows:

$$B_{1A} = \frac{\mu_0 2I}{2\pi r_{1A}} = \frac{\mu_0 I}{\pi r_{1A}} = \frac{\left(4\pi \times 10^{-7}\,\text{T}\cdot\text{m/A}\right)(10\,\text{A})}{\pi(0.36\,\text{m})} = 11.11\,\mu\text{T}$$

$$B_{2A} = \frac{\mu_0 I}{2\pi r_{2A}} = \frac{\left(4\pi \times 10^{-7}\,\text{T}\cdot\text{m/A}\right)(10\,\text{A})}{2\pi(0.85\,\text{m})} = 2.35\,\mu\text{T}$$

$$B_{1B} = \frac{\mu_0 2I}{2\pi r_{1B}} = \frac{\mu_0 I}{\pi r_{1B}} = \frac{\left(4\pi \times 10^{-7}\,\text{T}\cdot\text{m/A}\right)(10\,\text{A})}{\pi(0.5\,\text{m})} = 8\,\mu\text{T}$$

$$B_{2B} = \frac{\mu_0 I}{2\pi r_{2B}} = \frac{\left(4\pi \times 10^{-7}\,\text{T}\cdot\text{m/A}\right)(10\,\text{A})}{2\pi(0.5\,\text{m})} = 4\,\mu\text{T}$$

$$B_{1C} = \frac{\mu_0 2I}{2\pi r_{1C}} = \frac{\mu_0 I}{\pi r_{1C}} = \frac{\left(4\pi \times 10^{-7}\,\text{T}\cdot\text{m/A}\right)(10\,\text{A})}{\pi(0.58\,\text{m})} = 6.9\,\mu\text{T}$$

$$B_{2C} = \frac{\mu_0 I}{2\pi r_{2C}} = \frac{\left(4\pi \times 10^{-7}\,\text{T}\cdot\text{m/A}\right)(10\,\text{A})}{2\pi(0.58\,\text{m})} = 3.49\,\mu\text{T}$$

$$B_{1D} = \frac{\mu_0 2I}{2\pi r_{1D}} = \frac{\mu_0 I}{\pi r_{1D}} = \frac{\left(4\pi \times 10^{-7}\,\text{T}\cdot\text{m/A}\right)(10\,\text{A})}{\pi(0.8\,\text{m})} = 5.0\,\mu\text{T}$$

$$B_{2D} = \frac{\mu_0 I}{2\pi r_{2D}} = \frac{\left(4\pi \times 10^{-7}\,\text{T}\cdot\text{m/A}\right)(10\,\text{A})}{2\pi(0.2\,\text{m})} = 10\,\mu\text{T}$$

The orientation of the magnetic field vectors is indicated in the diagram on the next page.

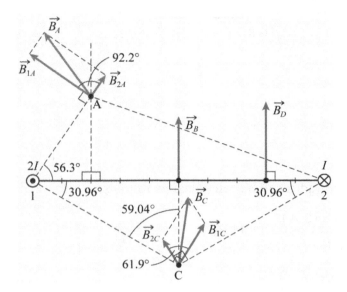

The magnitudes of the magnetic field in each point can be found using vector addition rules. Notice, for the magnitudes B_A and B_C we use the cosine rule. It is important to use the proper angle in the cosine law calculations. In the example shown below, the angle to be used is 110° and not the 70° angle.

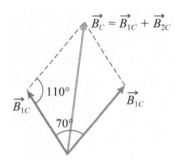

Therefore, the magnitudes of magnetic field are equal to

$$B_A = \sqrt{11.11^2 + 2.35^2 + 2 \times 11.11 \times 2.35 \times \cos(87.8°)}\ \mu T = 15.37\ \mu T$$

$$B_B = B_{1B} + B_{2B} = 12\ \mu T$$

$$B_D = B_{1D} + B_{2D} = 15\ \mu T$$

$$B_C = \sqrt{6.9^2 + 3.49^2 + 2 \times 6.9 \times 3.49 \times \cos(118.1°)}\ \mu T = 9.35\ \mu T$$

53. Let $d = 60$ cm, $a = 30$ cm, and $b = 20$ cm denote the distance between the loop and the wire, and the two dimensions of the loop, respectively. Let I_1 and I_2 denote the currents in the loop and in the very long wire, respectively.

For the vertical sides of the loop:

The magnetic field magnitudes B_1 and B_2 along the two vertical sides of the loop are equal and given by

$$B_1 = B_2 = \frac{\mu_0 I_2}{2\pi r} = \frac{\mu_0 I_2}{2\pi \sqrt{d^2 + (b/2)^2}} = \frac{\left(4\pi \times 10^{-7}\,\text{T·m/A}\right)(20\,\text{A})}{2\pi\sqrt{0.36 + 0.01}} = 6.58\,\mu\text{T}$$

The currents carried by the two vertical sides are equal but flowing in opposite directions. Therefore, the forces acting on the vertical wires (per unit length) can be found as follows (we consider here wires to be very long wires). The image below shows how these forces can be calculated:

$$F_B(r_1) = F_B(r_2) = BIL = (6.58\,\mu\text{T})(10\,\text{A})(0.3\,\text{m}) = 19.7\,\mu\text{N}$$

Notice, while these forces are equal in magnitude, they have different directions, thus the resultant of these two forces can be found using the cosine law:

$$(F_B)_R = \sqrt{2 \cdot 19.7^2 - 2 \cdot 19.7 \cos(18.9°)} = 6.47\,\mu\text{N}$$

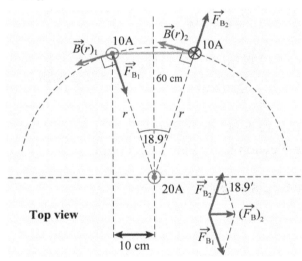

It is important to mention that in this case, the forces are acting on different sides of the loop, thus creating a torque. The magnitude of the torque can be calculated to be

$$\tau = F_B(r) \cdot \cos(9.5°)(b) = F_B(r) \cdot \cos(9.5°)(0.2\,\text{m}) = 1.28\,\mu\text{N·m}$$

Notice, this torque is very small, however, by increasing magnetic field, you can achieve a much larger torque values.

For the horizontal sides of the loop:

The total forces acting on the two horizontal sides are going to be zero. The reasoning lies within the symmetry of the loop. In the centre of the two horizontal sides the magnetic force is zero because the magnetic fields are parallel to the current. The infinitesimal magnetic forces $d\vec{F}_B$ acting on infinitesimal lengths dx positioned symmetrically with respect to the centre cancel each other due to the symmetry of the magnetic field, as indicated in the vector diagram shown below.

In summary, the net force acting on the loop is directed as shown in the image above for vertical sides of the loop and its magnitude is 6.47 μN.

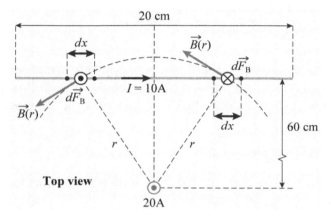

Top view

55. (a) $\dfrac{F_{AB}}{l} = I_B B_A = I_B \dfrac{\mu_0 I_A}{2\pi d} = \dfrac{\mu_0 I_A I_B}{2\pi d} = \dfrac{\left(4\pi\times10^{-7}\,\text{T}\cdot\text{m/A}\right)(10\ \text{A})(15\ \text{A})}{2\pi(1\ \text{m})} = 30\ \mu\text{N/m}$

(b) $\dfrac{F_{BA}}{l} = I_A B_B = I_A \dfrac{\mu_0 I_B}{2\pi d} = \dfrac{\mu_0 I_B I_A}{2\pi d} = \dfrac{\left(4\pi\times10^{-7}\,\text{T}\cdot\text{m/A}\right)(15\ \text{A})(10\ \text{A})}{2\pi(1\ \text{m})} = 30\ \mu\text{N/m}$

(c) The forces calculated in parts (a) and (b) are equal in magnitudes ($F_{AB} = F_{BA}$) and opposite in direction, as expected from Newton's third law of motion.

57. (a) The magnetic force acting on the proton $\vec{F}_B$ is given by

$$\vec{F}_B = e\vec{v}\times\vec{B} = e\begin{vmatrix} \hat{i} & \hat{j} & \hat{k} \\ v_x & 0 & v_z \\ 0 & 0 & B \end{vmatrix} = ev_x B\hat{j}$$

$$F_B = ev_x B = \left(1.6\times10^{-19}\,\text{C}\right)\left(2\times10^{5}\,\text{m/s}\right)(0.45\ \text{T}) = 1.4\times10^{-14}\ \text{N}$$

The gravitational force acting on the proton is equal to

$$F_G = mg = \left(1.67\times10^{-27}\,\text{kg}\right)\left(9.8\ \text{m/s}^2\right) = 1.6\times10^{-26}\ \text{N}$$

The gravitational force is about 12 orders of magnitude smaller than the magnetic force.

(b) The trajectory of the proton is an upward (in the positive z direction) helix of constant radius and pitch. The circular motion, as seen from above, is clockwise, or the helicity is positive.

(c) $\omega = \dfrac{eB}{m} = \dfrac{\left(1.6\times10^{-19}\,\text{C}\right)\left(0.45\,\text{T}\right)}{1.67\times10^{-27}\,\text{kg}} = 4.3\times10^{7}\,\text{rad/s}$

$T = \dfrac{2\pi}{\omega} = 14.6\times10^{-8}\,\text{s} = 0.15\,\mu\text{s}$

(d) $R = \dfrac{mv_x}{eB} = \dfrac{\left(1.67\times10^{-27}\,\text{kg}\right)\left(2\times10^{5}\,\text{m/s}\right)}{\left(1.6\times10^{-19}\,\text{C}\right)\left(0.45\,\text{T}\right)} = 4.6\,\text{mm}$

$Pitch = v_z T = \left(1.5\times10^{5}\,\text{m/s}\right)\left(0.15\times10^{-6}\,\text{s}\right) = 23\,\text{mm}$

(e) If the proton is replaced by an electron the following results are obtained:

(a) $\vec{F}_B = (-e)\vec{v}\times\vec{B} = (-e)\begin{vmatrix} \hat{i} & \hat{j} & \hat{k} \\ v_x & 0 & v_z \\ 0 & 0 & B \end{vmatrix} = -ev_x B\hat{j}$

$F_B = ev_x B = \left(1.6\times10^{-19}\,\text{C}\right)\left(2\times10^{5}\,\text{m/s}\right)\left(0.45\,\text{T}\right) = 1.4\times10^{-14}\,\text{N}$

$F_G = mg = \left(9.1\times10^{-31}\,\text{kg}\right)\left(9.8\,\text{m/s}^2\right) = 8.9\times10^{-30}\,\text{N}$

The gravitational force is about 15 orders of magnitude smaller than the magnetic force.

(b) The trajectory of the electron is an upward spiral of constant radius and pitch. The circular motion, as seen from above, is counter-clockwise, or the helicity is negative.

(c) $\omega = \dfrac{eB}{m} = \dfrac{\left(1.6\times10^{-19}\,\text{C}\right)\left(0.45\,\text{T}\right)}{9.1\times10^{-31}\,\text{kg}} = 7.9\times10^{10}\,\text{rad/s}$

$T = \dfrac{2\pi}{\omega} = 14.6\times10^{-8}\,\text{s} = 7.94\times10^{-11}\,\text{s} = 79\,\text{ps}$

(d) $R = \dfrac{mv_x}{eB} = \dfrac{\left(9.1\times10^{-31}\,\text{kg}\right)\left(2\times10^{5}\,\text{m/s}\right)}{\left(1.6\times10^{-19}\,\text{C}\right)\left(0.45\,\text{T}\right)} = 2.5\,\mu\text{m}$

$Pitch = v_z T = \left(1.5\times10^{5}\,\text{m/s}\right)\left(7.94\times10^{-11}\,\text{s}\right) = 12\,\mu\text{m}$

59. (a) Let $B_N = 18\,\mu\text{T}$ and $B_{down} = 48\,\mu\text{T}$ denote the two components of the Earth's magnetic field. Both components are perpendicular to the electron's velocity $\vec{v}$. The magnitude of the total magnetic field B_t is then given by

$B_t = \sqrt{B_N^2 + B_{down}^2} = 51\,\mu\text{T}$

Then the electron will move upwards on a circular trajectory in a plane perpendicular to the total magnetic field vector $\vec{B}_t$. The angle θ between the total magnetic field vector $\vec{B}_t$ and the North direction is equal to

$$\theta = \tan^{-1}\left(\frac{B_{down}}{B_N}\right) = \tan^{-1}\left(\frac{48}{18}\right) = 69°$$

(b) The radius R of the electron's trajectory is equal to

$$R = \frac{mv}{eB_t} = \frac{m\sqrt{\frac{2K}{m}}}{eB_t} = \frac{\sqrt{2mK}}{eB_t} = \frac{\sqrt{2\left(9.1\times10^{-31}\,\text{kg}\right)\left(20\times1.6\times10^{-19}\,\text{J}\right)}}{\left(1.6\times10^{-19}\,\text{C}\right)\left(51.3\times10^{-6}\,\text{T}\right)} = 0.29 \text{ m}$$

(c) The gravitational force can be neglected due to the very small mass of the electron.

61. (a) Let $B_{Earth} = 55\,\mu T$ and I_{max} denote the Earth's magnetic field and the maximum current intensity to be calculated, respectively. Then, I_{max} can be calculated as follows:

$$B_{Earth} = \frac{\mu_0 I_{max}}{2\pi d} \Rightarrow I_{max} = \frac{2\pi d B_{Earth}}{\mu_0} = \frac{2\pi\left(30\times10^{-2}\,\text{m}\right)\left(55\times10^{-6}\,\text{T}\right)}{4\pi\times10^{-7}\,\text{T}\cdot\text{m/A}} = 82.5\,\text{A} \approx 83 \text{ A}$$

(b) $I'_{max} = \dfrac{I_{max}}{2} = \dfrac{82.5\,\text{A}}{2} = 41\,\text{A}$

63. Let us assume that the magnetic field of the Earth $\vec{B}_{Earth}$ is entirely horizontal, pointing toward North, and with a magnitude $B_{Earth} = 55\,\mu T$. The net magnetic field $\vec{B}_{net}$ will then be horizontal and pointing toward North:

$$\vec{B}_{net} = \left(B + B_{Earth}\right)\left(-\hat{i}\right) = \left(5\times10^{-3} + 55\times10^{-6}\right)\left(-\hat{i}\right)\text{T} = -5.055\,\text{mT}\hat{i}$$

The distance from the wire r at which the total magnetic field is zero can be calculated by equating the net magnetic field magnitude B_{net} with the magnetic field magnitude created by the wire at this distance $B_w(r)$:

$$B_w(r) = B_{net} = \frac{\mu_0 I}{2\pi r} \Rightarrow r = \frac{\mu_0 I}{2\pi B_{net}} = \frac{\left(4\pi\times10^{-7}\,\text{T}\cdot\text{m/A}\right)\left(50 \text{ A}\right)}{2\pi\left(5.055\times10^{-3}\text{ T}\right)} = 2\,\text{mm}$$

Therefore, the total magnetic field is zero at a distance 2 mm below the wire as indicated by the magnetic field vector diagram shown in the figure on the next page.

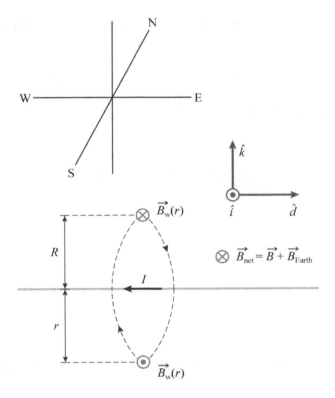

65. In the limiting case, that the length of the solenoid is significantly larger than its diameter, the magnitude of the magnetic field B inside the solenoid can be determined employing Ampère's law:

$$B = \frac{\mu_0 N I}{L} = \mu_0 n I = \left(4\pi \times 10^{-7}\,\text{T} \cdot \text{m/A}\right)\left(10^3\,\text{m}^{-1}\right)\left(10\,\text{A}\right) = 4\pi\,\text{mT} = 13\,\text{mT}$$

The magnetic permeability of iron μ_{iron} is given by

$$\mu_{iron} = \mu_0 \left(\chi_{iron} + 1\right)$$

Hence, the magnetic field inside the solenoid with an iron core B_{iron} is given by

$$B_{iron} = B\left(\chi_{iron} + 1\right) = \left(4\pi \times 10^{-3}\,\text{T}\right)\left(4 \times 10^3 + 1\right) = 50\,\text{T}$$

Thus, the iron core can increase the magnetic field inside the solenoid by 4000 times, which explains why it is used in solenoids.

67. The relative percentage increase or decrease $\left(\Delta B / B_{air}\right) \times 100\%$ of the magnetic field magnitude for the three substances compared to the magnetic field magnitude in air B_{air} can be calculated as follows:

$$\frac{\Delta B}{B_{air}} \times 100\% = \frac{B_m - B_{air}}{B_{air}} \times 100\% = \frac{\left(K_m - K_{air}\right)B_0}{K_{air}B_0} \times 100\% = \left(K_m - 1\right) \times 100\%$$

In the previous equation, the magnetic field magnitude in vacuum is denoted by B_0 and the approximation $K_{air} \approx 1$ is made.

For the three substances at a temperature of 20°C, the following results are obtained:

(a) $\dfrac{\Delta B_{water}}{B_{air}} \times 100\% = \left(K_{water} - 1 \right) \times 100\% = \left(0.999991 - 1 \right) \times 100\% = -0.0009\%$

(b) $\dfrac{\Delta B_{oxygen}}{B_{air}} \times 100\% = \left(K_{oxygen} - 1 \right) \times 100\% = \left(1.0000019 - 1 \right) \times 100\% = 0.00019\%$

(c) $\dfrac{\Delta B_{vacuum}}{B_{air}} \times 100\% \approx 0\%$

69. The Biot-Savart law is

$$d\vec{B} = \frac{\mu_0}{4\pi} \cdot \frac{I\,d\vec{l} \times \vec{r}}{r^3} \tag{1}$$

In the polar coordinates (R, ϕ) characterized by perpendicular unit vectors $(\hat{r}, \hat{\phi})$, the cross product in equation (1) can be written as

$$I\,d\vec{l} \times \vec{r} = Idl\,\hat{\phi} \times R\hat{r} = IRdl\,\hat{n} = IR^2\,d\phi\,\hat{n} \tag{2}$$

In equation (2), $\hat{n}$ denotes the unit vector perpendicular to the unit vectors $\hat{r}$ and $\hat{\phi}$ as indicated in the vector diagram from the figure shown below. Combining equations (1) and (2), the infinitesimal magnetic field $d\vec{B}$ from the Biot-Savart law is

$$d\vec{B} = \frac{\mu_0}{4\pi} \cdot \frac{IR^2\,d\phi}{R^3}\,\hat{n} = \frac{\mu_0 I}{4\pi R}\,d\phi\,\hat{n} \tag{3}$$

The total magnetic field at the point C, $B(C)$, can be obtained by integrating over the angle ϕ the infinitesimal magnetic field magnitude dB resulting from equation (3):

$$B(C) = \int_0^\phi \frac{\mu_0 I}{4\pi R}\,d\phi = \frac{\mu_0 I \phi}{4\pi R}$$

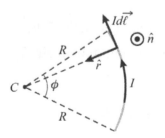

Copyright © 2015 by Nelson Education Ltd.

71. (a) Using the notations from the vector diagram from the figure shown below, applying the Biot-Savart law, it follows that

$$\left|d\vec{B}\right| = \frac{\mu_0 I}{4\pi r^3}\left|d\vec{x}\times\vec{r}\right| = \frac{\mu_0 I}{4\pi r^3} r\sin(\theta)\,dx = \frac{\mu_0 I}{4\pi}\cdot\frac{\sin(\theta)\,dx}{r^2} \qquad (1)$$

Also, one can write:

$$r = \frac{R}{\sin(\theta)} \qquad (2)$$

$$x = \frac{R}{\tan(\theta)} \Rightarrow dx = -\frac{R}{\left[\sin\theta\right]^2}\,d\theta \qquad (3)$$

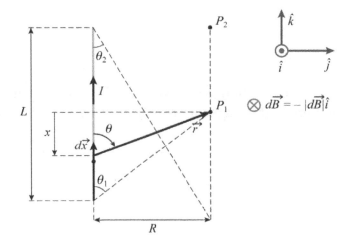

Using the results of equations (2) and (3), equation (1) becomes

$$dB = -\frac{\mu_0 I}{4\pi R}\sin(\theta)\,d\theta \qquad (4)$$

The total magnetic field magnitude at point P_1, $B(P_1)$, is equal to twice the integral of equation (4) from $\pi/2$ to angle θ_1 indicated in the diagram:

$$B(P_1) = 2\left(-\frac{\mu_0 I}{4\pi R}\right)\int_{\pi/2}^{\theta_1}\sin(\theta)\,d\theta = \frac{\mu_0 I}{2\pi R}\cos(\theta_1) = \frac{\mu_0 I}{2\pi R}\cdot\frac{L}{\left(L^2+4R^2\right)^{1/2}} \qquad (5)$$

(b) Based on equation (5),

$$B(P_1) = \frac{\mu_0 I}{2\pi R}\cdot\frac{L}{L\left[1+\left(\dfrac{2R}{L}\right)^2\right]^{\frac{1}{2}}} = \frac{\mu_0 I}{2\pi R}\cdot\frac{1}{\left[1+\left(\dfrac{2R}{L}\right)^2\right]^{\frac{1}{2}}} \qquad (6)$$

For an infinitely long wire $L \to \infty$, equation (6) simplifies to

$$B(P_1) = \frac{\mu_0 I}{2\pi R} \tag{7}$$

The simplified expression from equation (7) is the known formula of the magnetic field magnitude created at distance R by a wire carrying current I.

(c) Based on the solution of part (a), the magnitude of the magnetic field at point P_2, $B(P_2)$, is given by

$$B(P_2) = \frac{\mu_0 I}{4\pi R} \cos(\theta_2) = \frac{\mu_0 I}{4\pi R} \cdot \frac{L}{\left(L^2 + R^2\right)^{1/2}} \tag{8}$$

(d) It is easy to verify that

$$B(P_1) = 2B(P_2)$$

if the length L is replaced by $L/2$ in equation (8) as expected. That is, the magnetic field at point P_1 is the algebraic sum of the magnetic fields generated by the two halves of the wire.

73.

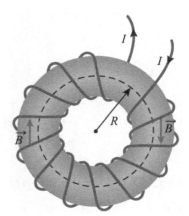

(a) The magnetic field generated inside the toroid B_{in} can be calculated by applying Ampère's law using a circular loop of radius R located inside to toroid:

$$2\pi R B_{\text{inside toroid}} = \mu_0 NI \Rightarrow B_{\text{inside toroid}} = \frac{\mu_0 NI}{2\pi R} \tag{1}$$

The magnetic field outside the toroid $B_{\text{outside toroid}}$ is zero, since the enclosed current I_{enc} is zero.

(b) The magnitudes of the magnetic fields inside and outside, $B_{\text{inside solenoid}}$ and $B_{\text{outside solenoid}}$, for the solenoid with N windings and length L are given by

$$B_{\text{inside solenoid}} = \frac{\mu_0 N I}{L} \tag{2}$$

$$B_{\text{outside solenoid}} = 0\,\text{T} \tag{3}$$

The magnetic field magnitude of a toroid of radius R, where R is the average between the inner and outer radii of the toroid, as shown in the figure is equal to that of a solenoid bent into a circle. Then,

$$L = 2\pi R \tag{4}$$

Substituting the length L from equation (4) into equation (2), the result from equation (1) is obtained.

(c) Based on the comparison and analysis from part (c), it can be inferred that a toroid is a solenoid bent into a circular shape.

(d) $B_{\text{inside toroid}} = \dfrac{\mu_0 N I}{2\pi R} = \dfrac{\left(4\pi \times 10^{-7}\,\text{T} \cdot \text{m/A}\right)\left(10^3\right)\left(15\,\text{A}\right)}{2\pi\left(0.5\text{ m}\right)} = 6\,\text{mT}$

(e) $B_{\text{inside solenoid}} = \dfrac{\mu_0 N I}{L} = \dfrac{\left(4\pi \times 10^{-7}\,\text{T} \cdot \text{m/A}\right)\left(10^3\right)\left(15\text{A}\right)}{1\text{ m}} = 6\pi\,\text{mT}$

$$\frac{B_{\text{inside solenoid}}\left(e\right)}{B_{\text{inside toroid}}\left(d\right)} = \pi$$

The ratio $B_{\text{inside solenoid}}\left(e\right) / B_{\text{inside toroid}}\left(d\right)$ written above is equal to the toroid-solenoid length ratio. This result is expected since the length of the solenoid is equal to the diameter of the toroid.

75. It is advisable for people who have pacemakers to keep their electronic devices (such as iPods) as far away from their pacemakers as possible.

1. h

3. e

5. a

7. d. The induced voltage ε in the case described in Figure 24-41 in which $l = 1\,\mathrm{m}$ denotes the length of the metal bar and $B = 1\,\mathrm{T}$ is the magnitude of the uniform and constant horizontal magnetic field, is given by

$$\varepsilon = \left|\frac{d\Phi_B}{dt}\right| = \frac{Bldx}{dt} = Blv \Rightarrow v_{min} = \frac{\varepsilon_{min}}{Bl} = \frac{10\ \mathrm{V}}{(1\ \mathrm{T})(1\ \mathrm{m})} = 10\ \mathrm{m/s}$$

9. d. The following relationships are valid:

$$\frac{N_2}{N_1} = \frac{V_2}{V_1} = \frac{I_1}{I_2} = \frac{25\,000}{25} = 1000$$

It follows that

$$V_1 = \frac{V_2}{1000} = \frac{120\,000\,\mathrm{V}}{1000} = 120\,\mathrm{V}$$

$$I_1 = I_2 \times 1000 = \left(10 \times 10^{-3}\,\mathrm{A}\right)10^3 = 10\ \mathrm{A}$$

11. b. The magnetic flux through the loop is decreasing; therefore, the induced current is such that magnetic field in the loop is in the same direction as the one generated by the current-carrying wire. Hence, the current runs clockwise and there is an attractive force between the loop and the wire.

13. a. The magnetic flux through the ring increases until it reaches about half of the permanent magnet's length; therefore, the induced current is such the magnetic field generated in the ring is opposite to the magnetic field of the permanent magnet. Hence, the current runs clockwise during this time: as the ring approaches the centre of the magnet, this current gradually decreases. The current is zero right when the ring is in the centre of the magnet and then, the direction of the current changes to the counterclockwise direction.

15. The inductance of the solenoid $L_{solenoid}$ is given by the following expression:

$$L_{solenoid} = \mu\left(\frac{N}{L}\right)^2 V = \mu\left(\frac{N}{L}\right)^2 \pi R^2 L = \frac{\pi\mu(NR)^2}{L}$$

Based on the equation above, the following solenoid inductance values can be calculated:

(a) $\quad L_{solenoid} = \dfrac{\pi\mu(N \times 2R)^2}{L} = 4\dfrac{\pi\mu(NR)^2}{L}$

(b) $L_{solenoid} = \dfrac{\pi\mu(2N \times 2R)^2}{L} = 16\dfrac{\pi\mu(NR)^2}{L}$

(c) $L_{solenoid} = \dfrac{\pi\mu(2N \times R)^2}{L} = 4\dfrac{\pi\mu(NR)^2}{L}$

(d) $L_{solenoid} = \dfrac{\pi\mu(2N \times R)^2}{2L} = 2\dfrac{\pi\mu(NR)^2}{L}$

(e) $L_{solenoid} = \dfrac{\pi\mu(N \times R)^2}{2L} = \dfrac{1}{2}\dfrac{\pi\mu(NR)^2}{L}$

Hence, the ranking of the inductance from the largest to the smallest is as follows:

$(b) > (a) = (c) > (d) > (e)$

17. The energy density u_B stored in a solenoid with N turns, of length L, radius R, and through which runs a current I is given by

$$u_B \overset{def}{=} \frac{U_B}{V} = \frac{\dfrac{L_{solenoid}I^2}{2}}{\pi R^2 L} = \frac{\pi\mu N^2 R^2 I^2}{2L} \cdot \frac{1}{\pi R^2 L} = \frac{\mu N^2 I^2}{2L^2}$$

The following results are obtained:

(a) $u_B = \dfrac{\mu N^2 I^2}{2L^2}$

(b) $u_B = \dfrac{\mu(2N)^2 I^2}{2L^2} = 4\dfrac{\mu N^2 I^2}{2L^2}$

(c) $u_B = 4\dfrac{\mu N^2 I^2}{2L^2}$

(d) $u_B = \dfrac{\mu(2N)^2 I^2}{2(2L)^2} = \dfrac{\mu N^2 I^2}{2L^2}$

(e) $u_B = \dfrac{\mu N^2 I^2}{2(2L)^2} = \dfrac{1}{4} \cdot \dfrac{\mu N^2 I^2}{2L^2}$

The ranking of the energy density u_B is as follows:

$(b) = (c) > (a) = (d) > (e)$

19. When the current in the RL circuits drops to 75% of its initial value ($I(t) = 0.75I_{max}$), the current as a function of time $I(t)$ can be described as follows:

$$I(t) = I_{max}\left(1 - e^{-\frac{t}{L/R}}\right) \Rightarrow e^{-\frac{t}{L/R}} = 1 - \frac{I(t)}{I_{max}} = 0.25 \Rightarrow t = \frac{L}{R}\ln(4)$$

Therefore, the ranking of the time follows the ranking of the L/R time constant.

Circuit 1: $\dfrac{L}{R} = \dfrac{10 \times 10^{-3}\,\text{H}}{10\ \Omega} = 1\,\text{ms}$

Circuit 2: $\dfrac{L}{R} = \dfrac{20 \times 10^{-3}\,\text{H}}{20\ \Omega} = 1\,\text{ms}$

Circuit 3: $\dfrac{L}{R} = \dfrac{20 \times 10^{-3}\,\text{H}}{20\ \Omega} = 1\,\text{ms}$

Circuit 4: $\dfrac{L}{R} = \dfrac{30 \times 10^{-3}\,\text{H}}{50\ \Omega} = 0.6\,\text{ms}$

Circuit 5: $\dfrac{L}{R} = \dfrac{30 \times 10^{-3}\,\text{H}}{50\ \Omega} = 0.6\,\text{ms}$

The ranking of the time t, from the largest to the smallest, is

$$(1) = (2) = (3) > (4) = (5)$$

21. c. When the iron bar is inserted into the air-filled inductor, its inductance L increases due to an increase in the magnetic permeability μ. Hence, the energy stored in the coil increases, which, in turn, reduces the available power on the bulb.

23. (a) The magnetic flux as a function of time t, $\Phi_B(t)$, is given by

$$\Phi_B(t) \overset{def}{=} \vec{B}(t) \cdot \vec{A} = -B(t)A = -B(t)w(l - vt) = B(t)w(vt - l) \qquad (1)$$

The rate of change of the magnetic flux $d\Phi_B/dt$ is then given by

$$\frac{d\Phi_B}{dt} = wvB(t) + wvt\frac{dB(t)}{dt} - wl\frac{dB(t)}{dt} = w\left[vB(t) + (vt - l)\frac{dB(t)}{dt}\right] \qquad (2)$$

The time t' during which $3/4$ of the length l is out of the magnetic field is equal to

$$t' = \frac{3l}{4v} = \frac{3 \times 0.5\,\text{m}}{4 \times 1.5\,\text{m/s}} = 0.25\,\text{s}$$

Numerically, equation (2) gives

$$\frac{d\Phi_B}{dt} = (0.25\ \text{m})\left[\left(1.5\ \frac{\text{m}}{\text{s}}\right)\left(0.002\ \frac{\text{T}}{\text{s}}(0.25\ \text{s}) + 0.001\ \text{T}\right)\right.$$

$$\left. + \left(\left(1.5\ \frac{\text{m}}{\text{s}}\right)(0.25\ \text{s}) - 0.5\ \text{m}\right)\left(0.002\ \frac{\text{T}}{\text{s}}\right)\right] = 0.5\ \frac{\text{mWb}}{\text{s}}$$

(b) The time t'' during which the entire length l is out of the magnetic field is equal to

$$t'' = \frac{l}{v} = \frac{0.5\ \text{m}}{1.5\ \text{m/s}} = \frac{1}{3}\ \text{s} = 0.333\ \text{s}$$

Numerically, equation (2) gives

$$\frac{d\Phi_B}{dt} = (0.25\ \text{m})\left[\left(1.5\ \frac{\text{m}}{\text{s}}\right)(0.002\ \text{T} \times 0.333 + 0.001\ \text{T})\right.$$

$$\left. + \left(\left(1.5\ \frac{\text{m}}{\text{s}}\right)(0.333\ \text{s}) - 0.5\ \text{m}\right)\left(0.002\ \frac{\text{T}}{\text{s}}\right)\right] = 0.625\ \frac{\text{mWb}}{\text{s}}$$

$$\approx 0.63\ \frac{\text{mWb}}{\text{s}}$$

(c) Since now there are 20 times the loops as in part (a) and (b), the effective areas and the rates of change of magnetic flux are also 20 times higher than calculated in parts (a) and (b).

For part (a): $\dfrac{d\Phi_B}{dt} = \left(0.5\ \dfrac{\text{mWb}}{\text{s}}\right)(20\ \text{loops}) = 10.0\ \dfrac{\text{mWb}}{\text{s}}$

For part (b): $\dfrac{d\Phi_B}{dt} = \left(0.625\ \dfrac{\text{mWb}}{\text{s}}\right)(20\ \text{loops}) = 12.5\ \dfrac{\text{mWb}}{\text{s}}$

25. The rate of change of the magnetic flux through the inner (denoted by index 1) and outer (denoted by index 2) solenoids is the same. Therefore, it can be written that

$$\frac{d\Phi_B}{dt} = N_1 A_1 \left(\frac{dB}{dt}\right)_1 = N_2 A_2 \left(\frac{dB}{dt}\right)_2$$

Hence,

$$\left(\frac{dB}{dt}\right)_2 = \frac{N_1 A_1}{N_2 A_2} \cdot \left(\frac{dB}{dt}\right)_1 = \frac{N_1}{N_2} \cdot \frac{R_1^2}{R_2^2} \cdot \frac{\mu_0 N_1}{l} \cdot \left(\frac{dI}{dt}\right)_1 = \frac{100}{200} \cdot \frac{4}{9} \cdot \frac{(4\pi \times 10^{-7}\ \text{T} \cdot \text{m/A}) \cdot 100}{0.5\ \text{m}} \cdot 0.5\ \frac{\text{A}}{\text{s}}$$

$$= 28\ \frac{\mu\text{T}}{\text{s}}$$

27. The induced emf ε is

$$\varepsilon = -\frac{\Delta\Phi_B}{\Delta t} = -\frac{\left(8\times10^{-4} - 6\times10^{-2}\right)\text{Wb}}{0.1\ \text{s}} = 0.6\ \text{V}$$

29. The magnitude of the induced emf ε is given by

$$\varepsilon = \left|-\frac{d\Phi_B}{dt}\right| = \left|-\frac{NA(0-B)}{\Delta t}\right| = \frac{NAB}{\Delta t}$$

Therefore, the magnitude of the magnetic field B is equal to

$$B = \frac{\varepsilon\,\Delta t}{NA} = \frac{\left(45\times10^{-3}\ \text{V}\right)\left(0.02\ \text{s}\right)}{\left(20\ \text{turns}\right)\left(6\times10^{-4}\ \text{m}^2\right)} = 75\ \text{mT}$$

31. According to Faraday's law of electromagnetic induction the induced emf ε in a coil with N loops and resistance R is given by

$$\varepsilon = -N\frac{\Delta\Phi_B}{\Delta t} = IR = \frac{\Delta Q}{\Delta t}R \Rightarrow \Delta Q = -N\frac{\Delta\Phi_B}{R}$$

33. (a) Employing Lenz's law, the magnetic field generated by the induced current I flowing through the rectangular loop has to be in the same direction (upwards) as the applied decreasing magnetic field. Then, the induced current I runs counterclockwise, as seen from above. Hence, the magnetic force $\vec{F}_B$ is in the direction of the x-axis as indicated in the diagram shown below.

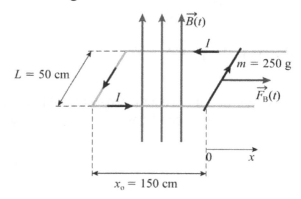

The magnitude of the magnetic force as a function of time $F_B(t)$ can be determined as follows:

$$F_B(t) = B(t)LI(t) = B(t)L\frac{\varepsilon(t)}{R} = \frac{B(t)L}{R}\left(-\frac{d\Phi_B}{dt}\right) = -\frac{B(t)L}{R}\cdot\frac{d(BA)}{dt} \quad (1)$$

Further,

$$\frac{d(BA)}{dt} = A(t)\frac{dB}{dt} + B(t)\frac{dA}{dt} \quad (2)$$

At time t in which the bar moved to the right a distance $x(t)$, the rectangular loop area $A(t)$ is given by

$$A(t) = L[x_0 + x(t)] \qquad (3)$$

In equation (3), $x_0 = 150$ cm. Using equation (3), one can write that

$$\frac{dA}{dt} = L\frac{dx}{dt} \qquad (4)$$

The magnitude of the magnetic field dependence on time $B(t)$ is given by

$$B(t) = B_0 - Ct \qquad (5)$$

In equation (5), $B_0 = B(t = 0 \text{ s}) = 15$ mT, and $C = 0.3$ mT/s.

Hence,

$$\frac{dB}{dt} = -C \qquad (6)$$

Substituting the results of equations (3)–(6) into equation (2),

$$\frac{d(BA)}{dt} = -LC[x_0 + x(t)] + L(B_0 - Ct)v(t) \qquad (7)$$

Equation (1) can be re-written as

$$F_B(t) = -\frac{(B_0 - Ct)L^2}{R}\{(B_0 - Ct)v(t) - C[x_0 + x(t)]\}$$

$$= -\frac{(B_0 - Ct)L^2}{R}\{B_0 v(t) - C[x_0 + x(t) + tv(t)]\} \qquad (8)$$

At time $t = 0$ s: $x(0) = 0$ m and $v(0) = 0$ m/s. Therefore, equation (8) gives

$$F_B(0) = -\frac{B_0 L^2}{R}(-Cx_0) = \frac{B_0 CL^2 x_0}{R}$$

$$= \frac{(15 \times 10^{-3}\,\text{T})\left(0.3 \times 10^{-3}\,\dfrac{\text{T}}{\text{s}}\right)(0.5^2\,\text{m}^2)(1.5\,\text{m})}{2\,\Omega} = 0.84\,\mu\text{N}$$

(b) The acceleration at time $t = 0\,s$, $a(0)$ is equal to

$$a(0) = \frac{F_B(0)}{m} = \frac{B_0 CL^2 x_0}{mR} \overset{part\ (a)}{=} \frac{0.84\,\mu\text{N}}{0.25\,\text{kg}} = 3.4\,\frac{\mu\text{m}}{\text{s}^2}$$

35. (a) The magnitude of the induced emf ε in the coil is given by

$$\varepsilon = \left|-\frac{d\Phi_B}{dt}\right| = B\left|\frac{dA}{dt}\right| = B\left|\frac{d}{dt}\left[A\cos\left(2\pi ft + \alpha_0\right)\right]\right| = 2\pi fBA\left|\sin\left(2\pi ft + \alpha_0\right)\right|$$

Hence, the maximum induced emf ε_{max} is

$$\varepsilon_{max} = 2\pi fBA = 2\pi\left(60\text{ Hz}\right)\left(3\times10^{-2}\text{T}\right)\left(100\times10^{-4}\text{m}^2\right) = 113\text{ mV}$$

(b) When $N = 100$ loops are used:

$$\varepsilon_{max} = 2\pi NfBA = 2\pi\left(100\text{ loops}\right)\left(60\text{ Hz}\right)\left(3\times10^{-2}\text{ T}\right)\left(100\times10^{-4}\text{ m}^2\right) = 11.3\text{ V}$$

(c) The expression for output voltage as obtained in part (a) can be written as

$$\varepsilon(t) = \varepsilon_{max}\sin\left(2\pi ft + \alpha_0\right)$$

Since $\varepsilon(t) = \varepsilon_{max}$ at $t = 0 \Rightarrow \alpha_0 = \dfrac{\pi}{2}$

$$\varepsilon(t) = \varepsilon_{max}\sin\left(2\pi ft + \frac{\pi}{2}\right) = \varepsilon_{max}\cos\left(2\pi ft\right)$$

37. (a) The number of turns in the secondary coil N_2 can be determined by assuming that the magnetic flux variation $\dfrac{d\Phi_B}{dt}$ trough a single loop of either of the two coils is the same. Hence,

$$V_1 = N_1\frac{d\Phi_B}{dt} \Rightarrow \frac{V_1}{N_1} = \frac{d\Phi_B}{dt} \tag{1}$$

$$V_2 = N_2\frac{d\Phi_B}{dt} \Rightarrow \frac{V_2}{N_2} = \frac{d\Phi_B}{dt} \tag{2}$$

From equations (1) and (2), it follows that

$$\frac{V_1}{N_1} = \frac{V_2}{N_2} \Rightarrow N_2 = N_1\frac{V_2}{V_1} = 1000\times\frac{12\text{ V}}{120\text{ V}} = 100$$

(b) If the transformer is 100% efficient, conservation of energy leads to

$$I_1V_1 = I_2V_2 \tag{3}$$

$$I_2 = \frac{V_2}{R} = \frac{12\text{ V}}{15\Omega} = 0.8\text{ A} = 800\text{ mA}$$

Using equation (3),

$$I_1 = I_2\frac{V_2}{V_1} = 0.8\text{ A}\times\frac{12\text{ V}}{120\text{ V}} = 0.08\text{ A} = 80\text{ mA}$$

The rms values of the two currents are

$$I_{rms\,1} = \frac{I_1}{\sqrt{2}} = \frac{0.08\text{ A}}{\sqrt{2}} = 57\text{ mA}$$

$$I_{rms\,2} = \frac{I_2}{\sqrt{2}} = \frac{0.8\text{ A}}{\sqrt{2}} = 570\text{ mA}$$

(c) The primary current has the same value as calculated in part (b):

$$I_1 = 80\text{ mA}$$

$$I_{rms\,1} = \frac{I_1}{\sqrt{2}} = \frac{80\text{ mA}}{\sqrt{2}} = 57\text{ mA}$$

If the transformer works with 90% efficiency, equation (3) can be re-written as follows:

$$0.9 I_1 V_1 = I_2 V_2$$

Hence,

$$I_2 = \frac{0.9 I_1 V_1}{V_2} = \frac{0.9(0.08\text{ A})(120\text{ V})}{12\text{ V}} = 720\text{ mA}$$

$$I_{rms\,2} = \frac{I_2}{\sqrt{2}} = \frac{720\text{ mA}}{\sqrt{2}} = 510\text{ mA}$$

39. (a) The magnitude of the electric field induced outside the solenoid $E(r,t)$, $r > R$ can be derived using the following equation:

$$2\pi r E(r,t) = \varepsilon \Rightarrow E(r,t) = \frac{\varepsilon(t)}{2\pi r} \tag{1}$$

Further, the induced emf ε can be determined using Faraday's law of electromagnetic induction:

$$\varepsilon(t) = \left|-\frac{d\Phi_B}{dt}\right| = N\pi R^2 \frac{dB}{dt} = N\pi R^2 \frac{d}{dt}\left(\frac{\mu_0 NI}{L}\right) = \frac{2\mu_0 \pi^2 N^2 R^2 f I_{max} \cos(2\pi ft)}{L} \tag{2}$$

From equations (1) and (2),

$$E(r,t) = \frac{1}{r} \cdot \frac{\pi \mu_0 N^2 R^2 f I_{max} \cos(2\pi ft)}{L} \tag{3}$$

(b) The magnitude of the electric field induced inside the solenoid $E(r,t)$, $r < R$ can be derived from equation (3) where radius R is replaced by radius r:

$$E(r) = r \cdot \frac{\pi \mu_0 N^2 f I_{max} \cos(2\pi ft)}{L} \tag{4}$$

(c) The results of parts (a) and (b) can be summarized as follows:

$$E(r,t) = \begin{cases} E_{max}\dfrac{r}{R}\cos(2\pi ft), & r < R \\[2mm] E_{max}\dfrac{R}{r}\cos(2\pi ft), & r \ge R \end{cases} \tag{5}$$

In equation (5), E_{max} is given by

$$E_{max} = \frac{\pi\mu_0 N^2 Rf I_{max}}{L} \tag{6}$$

The plot of the function $E(r)/E_{max}$ from equation (5) at an arbitrary time t is shown below.

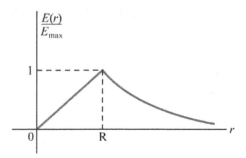

41. The mutual inductance of the two circuits, M, can be calculated as follows:

$$\varepsilon_2 = M\frac{dI_1}{dt} \Rightarrow M = \frac{\varepsilon_2}{\dfrac{dI_1}{dt}} = \frac{20\ \text{V}}{200\ \dfrac{\text{A}}{\text{s}}} = 0.1\ \text{H}$$

43. (a) The mutual inductance of the two solenoids M is given by

$$M = \sqrt{L_1 L_2} = \frac{\pi\mu_0 r^2}{l}N_1 N_2$$

$$= \frac{\pi\left(4\pi\times10^{-7}\ \text{T}\cdot\text{m/A}\right)(0.05\ \text{m})^2}{2\ \text{m}}(25000\ \text{turns})(5000\ \text{turns})$$

$$= \frac{4\pi^2\times10^{-7}\times0.05^2\times25\times10^3\times5\times10^2}{2}\ \text{H} = 62\ \text{mH}$$

(b) The induced emf in the coil ε_{coil} is

$$\varepsilon_{coil} = M\frac{dI_{solenoid}}{dt} = \left(61.7\times10^{-3}\ \text{H}\right)\left(10\ \frac{\text{A}}{\text{s}}\right) = 0.62\ \text{V}$$

45. (a) The energy stored in the solenoid from question 44 is equal to

$$U = \frac{1}{2}LI_{max}^2 = \frac{1}{2}\left(8.2\times10^{-3}\,\text{H}\right)\left(1.2^2\,\text{A}^2\right) = 5.9\ \text{mJ}$$

(b) The energy stored in the solenoid from question 44 when $I' = I_{max}/2$ is equal to

$$U' = \frac{1}{2}LI'^2 = \frac{1}{2}L\left(\frac{I_{max}}{4}\right)^2 = \frac{1}{4}U = \frac{5.9\,\text{mJ}}{4} = 1.5\,\text{mJ}$$

Another way to approach the problem:

$$U' = \frac{1}{2}LI'^2 = \frac{1}{2}\left(8.2\times10^{-3}\ \text{H}\right)\left(0.6\ \text{A}\right)^2 = 1.5\,\text{mJ}$$

(c) The energy stored in the solenoid when the current is zero ($I'' = 0$ A) is

$$U'' = \frac{1}{2}L\left(I''\right)^2 = 0$$

47. The infinitesimal induced emf $d\varepsilon$ in an infinitesimal element of length dx positioned at distance x from the non-rotating end of the rod, is given by

$$d\varepsilon = Bvdx$$

Here v is the tangential velocity of the element with respect to the centre of rotation. This is related to the frequency of rotation by $v = 2\pi fx$. Therefore,

$$d\varepsilon = Bvdx = B2\pi fxdx = 2\pi Bfxdx$$

In the equation written above, B and f denote the static magnetic field magnitude and the rotational frequency, respectively. The total emf induced in the rod, ε, is obtained by integration:

$$\varepsilon = 2\pi Bf\int_0^L xdx = 2\pi Bf\frac{L^2}{2} = \pi BfL^2 = \pi\left(0.035\ \text{T}\right)\left(5\ \text{Hz}\right)\left(0.1\ \text{m}\right)^2 = 5.5\,\text{mV}$$

49. (a) In this problem, the area is denoted as S. The magnitude of the induced emf, $\varepsilon(t)$, in the equilateral triangle loop of side $a = 10$ cm, is

$$\varepsilon(t) = \left|-\frac{d\Phi_B}{dt}\right| = B\left|\frac{dS(t)}{dt}\right| = B\left|\frac{d}{dt}\left\{\frac{\left[h(t)\right]^2}{\sqrt{3}}\right\}\right| = \frac{2}{\sqrt{3}}B\left|h(t)\frac{dh}{dt}(t)\right| \qquad (1)$$

In equation (1), $h(t)$ denotes the time dependence of the height of isosceles triangle, which is in the uniform and constant magnetic field $\vec{B}$ area. The initial height of the triangle when all of it is submerged in magnetic field is h_0. The geometry and the variable area $S(t)$ calculation are indicated in the diagram shown below. The expression of $h(t)$ in terms of the total height h_0 of the triangle is

$$h(t) = h_0 - vt = a\sin(60°) - 2t = 0.1\sqrt{\frac{3}{4}} - 2t = \left(\sqrt{0.0075} - 2t\right) \text{ m} \qquad (2)$$

$$\frac{dh(t)}{dt} = -v = -2 \ \frac{\text{m}}{\text{s}}$$

Based on equation (2), equation (1) can be re-written as

$$\varepsilon(t) = \frac{2}{\sqrt{3}} B \left| h(t) \frac{dh}{dt}(t) \right| = \frac{2}{\sqrt{3}} B \left| h(t)(-v) \right|$$

$$= \frac{2}{\sqrt{3}} Bv(h_0 - vt) = \frac{2}{\sqrt{3}}(0.045 \text{ T})(2 \text{ m/s})\left(\left(\sqrt{0.0075} - 2t\right) \text{ m}\right) \qquad (3)$$

$$= (0.009 - 0.208t) \text{ V}$$

From this equation we see that $\varepsilon(t = 0 \text{ s}) = 9.0 \text{ mV}$.

(b) The induced current runs counterclockwise as indicated in the diagram shown below.

(c) The maximum force can be calculated in two ways: (i) the mechanical force is equal and opposite to the net magnetic force acting on the loop, (ii) the mechanical power P_m is equal to the electrical power P_e generated in the circuit. The second option is the easiest because it does not involve calculation of the net magnetic force on the isosceles triangle loop:

$$P_m = P_e \Rightarrow F(t)v = \frac{\left[\varepsilon(t)\right]^2}{R} \Rightarrow$$

$$F_{max} = \frac{\varepsilon_{max}^2}{vR} = \frac{(0.009 \text{ V})^2}{(2 \text{ m/s})(1 \ \Omega)} = 40.5 \ \mu\text{N}$$

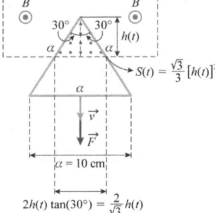

51. The induced emf in the band ε is

$$\varepsilon = -\frac{d\Phi_B}{dt} = -\frac{d}{dt}\left(B\pi r^2\right) = -2\pi r B \frac{dr}{dt} = -2\pi BC\left(r_i + Ct\right)$$

In the equation written above, $B = 50.0$ mT and $C = 0.025$ cm/s $= 2.5 \times 10^{-4}$ m/s. Hence,

$$\varepsilon = -2\pi\left(0.05 \text{ T}\right)\left(2.5 \times 10^{-4}\frac{\text{m}}{\text{s}}\right)\left(\left(5 + 0.025t\right) \times 10^{-2}\text{m}\right), \text{ where } t \text{ is measured in seconds.}$$

We can check the units, to see that the answer has proper dimensions:

$$[\varepsilon] = \text{T} \cdot \frac{\text{m}}{\text{s}} \cdot \text{m} = \frac{\text{T} \cdot \text{m}^2}{\text{s}} = \text{V}$$

$$\varepsilon = -3.9 \times \left(1 + 0.05t\right) \ \mu\text{V}$$

53.

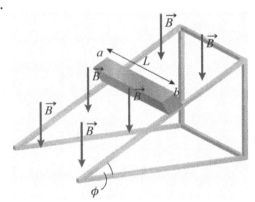

(a) The rod's downward motion will be slowed down by the induced emf (eddy currents). Since as the rod goes down the area of frame the current can flow through will be changing. Newton's second law on the downward axis in the inclined plane gives

$$ma = mg\sin\left(\phi\right) - F_B\cos\left(\phi\right) \Rightarrow a = g\sin\left(\phi\right) - \frac{F_B}{m}\cos\left(\phi\right) \qquad (1)$$

(b) The magnitude of the induced emf in the rod $\varepsilon(v)$ is

$$\varepsilon\left(v\right) = \left|-\frac{d\Phi_B}{dt}\right| = B\left|\frac{dA}{dt}\right|\cos\left(\phi\right) = BLv\cos\left(\phi\right) \qquad (2)$$

(c) The magnitude of the magnetic force F_B is given by

$$F_B = BIL = B\frac{\varepsilon}{R}L \overset{eq. (2)}{\Rightarrow} F_B = \frac{B^2L^2v\cos\left(\phi\right)}{R} \qquad (3)$$

Using equation (3), equation (1) can be re-written as

$$a = \frac{dv}{dt} = g\sin(\phi) - \frac{\left[BL\cos(\phi)\right]^2}{mR}v \tag{4}$$

To solve differential equation (4), the following notations can be employed:

$$T = \frac{mR}{\left[BL\cos(\phi)\right]^2} = \frac{(0.3 \text{ kg})(2\ \Omega)}{\left[(0.5 \text{ T})(0.2 \text{ m})\sqrt{3}/2\right]^2} = 80 \text{ s} \tag{5}$$

$$a = g\sin(\phi) - \frac{v}{T} \Rightarrow \frac{da}{dt} = -\frac{1}{T}\frac{dv}{dt} = -\frac{a}{T} \tag{6}$$

With these notations from equations (5) and (6), equation (4) can be re-written as

$$a = -T\frac{da}{dt} \Rightarrow \frac{da}{a} = -\frac{dt}{T} \Rightarrow \ln(a) = -\frac{t}{T} \tag{7}$$

It follows that the value of quantity a at time t, $a(t)$, is given by

$$a(t) = a(0)e^{-\frac{t}{T}} \tag{8}$$

At time $t = 0$ s, $v = 0$ m/s. Hence,

$$a(0) = g\sin(\phi) \tag{9}$$

Going back to the original notation from equation (6),

$$g\sin(\phi) - \frac{v}{T} = g\sin(\phi)e^{-\frac{t}{T}} \Rightarrow v = gT\sin(\phi)\left(1 - e^{-\frac{t}{T}}\right) \tag{10}$$

Numerically,

$$v \cong \left(9.8\ \frac{\text{m}}{\text{s}^2}\right)(80 \text{ s})\frac{1}{2}\left(1 - e^{-\frac{t[s]}{80\,s}}\right) = \left(392\ \frac{\text{m}}{\text{s}}\right)\left(1 - e^{-\frac{t[s]}{80\,s}}\right) \tag{11}$$

(d) Using equations (2) and (10), the expression of $\varepsilon(t)$ is

$$\varepsilon(t) = BLgT\sin(\phi)\cos(\phi)\left(1 - e^{-\frac{t}{T}}\right)$$

Numerically,

$$\varepsilon(t) = (0.5 \text{ T})(0.2 \text{ m})\left(9.8\ \frac{\text{m}}{\text{s}^2}\right)(80 \text{ s})\frac{1}{2}\frac{\sqrt{3}}{2}\left(1 - e^{-\frac{t[s]}{80\,s}}\right) = 34\cdot\left(1 - e^{-\frac{t[s]}{80\,s}}\right)\text{V}$$

(e) From equation (11), terminal velocity v_f is equal to

$$v_f = 392 \text{ m/s}$$

55. (a) Using Ampère's law:

$$2\pi r B = \mu N I \Rightarrow B = \frac{\mu N I}{2\pi r} = \frac{\left(1.2 \times 10^{-3}\ \text{Hm}^{-1}\right)\left(10^4\ \text{turns}\right)\left(10\ \text{A}\right)}{2\pi\left(0.3\ \text{m}\right)} = 64\ \text{T}$$

(b) The magnetic flux through the toroid Φ_B is given by

$$\Phi_B = BAN = \frac{\mu N^2 A}{2\pi r} I$$

Therefore, the inductance of the toroid, L, is equal to

$$L = \frac{\Phi_B}{I} = \frac{\mu N^2 A}{2\pi r} = \frac{\left(1.2 \times 10^{-3}\,\text{Hm}^{-1}\right)\left(10^4\ \text{turns}\right)^2\left(10 \times 15 \times 10^{-6}\ \text{m}^2\right)}{2\pi\left(0.3\ \text{m}\right)} = 9.6\ \text{H}$$

(c) $$B = \frac{\mu N I}{2\pi r} = \frac{\left(1.2 \times 10^{-3}\ \text{Hm}^{-1}\right)\left(10^4\ \text{turns}\right)\left(20\ \text{A}\right)}{2\pi\left(0.3\ \text{m}\right)} = 130\ \text{T}$$

$$L = \frac{\mu N^2 A}{2\pi r} = \frac{\left(1.2 \times 10^{-3}\ \text{Hm}^{-1}\right)\left(10^4\ \text{turns}\right)^2\left(10 \times 15 \times 10^{-6}\,\text{m}^2\right)}{2\pi\left(0.3\,\text{m}\right)} = 9.6\ \text{H}$$

57. (a) The current I the car engine draws when running at is operating speed is

$$I = \frac{V - \varepsilon}{R} = \frac{12\ \text{V} - 10\ \text{V}}{0.4\ \Omega} = 5.0\ \text{A}$$

(b) During the start-up there is no induced back emf ε. Therefore, the current I during this time is equal to

$$I = \frac{V}{R} = \frac{12\ \text{V}}{0.4\ \Omega} = 30\ \text{A}$$

59. (a) The sketch of the oscilloscope trace that can be observed is in the plot (a) from the figure on the next page. Two pulses corresponding to the magnetic flux passing through the two coils will be formed.

(b) If the speed is doubled the time between pulses will be halved as long and the amplitude is twice as large, as sketched in the plot (b) in the figure shown on the next page.

(c) If the number of loops in the left coil were to double, the amplitude of the first pulse will double as sketched in plot (c) on the next page.

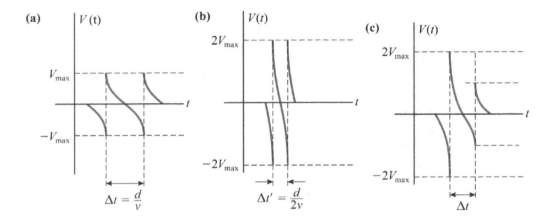

(a) $V(t)$, V_{max}, $-V_{max}$, $\Delta t = \dfrac{d}{v}$

(b) $V(t)$, $2V_{max}$, $-2V_{max}$, $\Delta t' = \dfrac{d}{2v}$

(c) $V(t)$, $2V_{max}$, $-2V_{max}$, Δt

(d) The measured speed v is equal to

$$v = \frac{\Delta x}{\Delta t} = \frac{2\text{ m}}{3.5 \times 10^{-3}\text{ s}} = 570\text{ m/s}$$

(e) For higher speeds the two pulses merge; a high speed can reduce the time separation between pulses below the width of the pulse (i.e., the two pulses cannot be resolved). Increasing the distance poses two problems: (i) the projectile trajectory is affected by gravitation over longer distance, hence, the projectile might not pass through the centre of the second coil and lead to an inaccurate measurement, (ii) the speed of the projectile is affected by air friction over long distances leading, again to inaccurate speed measurements.

61. Let ε_1 and ε_2 denote the emfs induced in the two loops of radii $r_1 = 4$ cm and $r_2 = 2$ cm, respectively. Then,

$$\varepsilon_1 = \pi r_1^2 \frac{dB}{dt} \tag{1}$$

$$\varepsilon_2 = \pi r_2^2 \frac{dB}{dt} \tag{2}$$

The ratio of the loop resistances R_1 / R_2 is

$$\frac{R_1}{R_2} = \frac{r_1}{r_2} \Rightarrow R_1 = \frac{r_1}{r_2} R_2 = 2(0.5\ \Omega) = 1\ \Omega$$

The individual loop currents run in opposite directions, therefore, the net current in the wire I is

$$I = \frac{\varepsilon_1 - \varepsilon_2}{R_1 + R_2} \overset{eqs.\ (1)\&(2)}{\Rightarrow} I = \frac{\pi \left(r_1^2 - r_2^2\right)\dfrac{dB}{dt}}{R_1 + R_2} = \frac{\pi\left((16-4)\times 10^{-4}\text{m}^2\right)\left(0.5 \times 10^{-3}\ \dfrac{\text{T}}{\text{s}}\right)}{(1+0.5)\ \Omega} = 1.3\ \mu\text{A}$$

The net current flows counterclockwise in the large loop of radius r_1, and clockwise in the small loop of radius r_2.

Copyright © 2015 by Nelson Education Ltd.

63. (a) The current time dependence in the series RL circuit, $I(t)$, is given by

$$I(t) = \frac{\varepsilon}{R}\left(1 - e^{-\frac{Rt}{L}}\right) = \frac{12 \text{ V}}{15 \text{ }\Omega}\cdot\left(1 - e^{-\frac{15}{3}t[s]}\right) = 0.8\cdot\left(1 - e^{-5t[s]}\right) \text{ A} \tag{1}$$

Then,

$$V_R(t) = I(t)R = 12\cdot\left(1 - e^{-5t[s]}\right) \text{ V} \tag{2}$$

$$V_L(t) = \varepsilon e^{-\frac{Rt}{L}} = 12\cdot e^{-5t[s]} \text{ V} \tag{3}$$

Based on equations (1)–(3), the values in the Table 24-3 are as given below.

Time (s)	V_R (V)	V_L (V)	I (A)	ε (V)
0.001	0.06	11.94	0.004	12
0.01	0.59	11.41	0.040	12
0.1	4.72	7.28	0.315	12
0.2	7.59	4.41	0.506	12
0.4	10.38	1.62	0.692	12

(b) The following relationships can be used to check the values from the table above at any given time t:

$$V_R(t) + V_L(t) = \varepsilon = 12 \text{ V} \tag{4}$$

$$I(t) = \frac{V_R(t)}{R} \tag{5}$$

(c) If the battery is not ideal, the resistance of the battery r has to be added to the circuit resistance R.

65. (a) The external force F_{ext} needed to move the bar to the left is

$$F_{ext} = F_B = BIl = B\frac{\varepsilon}{R}l = B\frac{Blv}{R}l = \frac{B^2l^2v}{R}$$

(b) $F_{ext} = \frac{B^2l^2v}{R} = \frac{\left(3\times10^{-3} \text{ T}\right)^2(0.6\text{m})^2(1.5 \text{ m/s})}{5 \text{ }\Omega} = 0.972 \text{ }\mu\text{N} = 1 \text{ }\mu\text{N}$

(c) The energy delivered to the resistor U_R in time t is given by

$$U_R = P\cdot t = \frac{\varepsilon^2}{R}t = \frac{B^2l^2v^2t}{R}$$

(d) $U_R = \dfrac{B^2 L^2 v^2 t}{R} = \dfrac{\left(3 \times 10^{-3}\,\text{T}\right)^2 \left(0.6\,\text{m}\right)^2 \left(1.5\,\text{m/s}\right)^2 \left(3\,\text{s}\right)}{5\,\Omega} = 4\,\mu\text{J}$

67. (a) The time to reach 99% of the maximum current I_{max} can be calculated using the known time dependence of the current, $I(t)$:

$$I(t) = I_{max}\left(1 - e^{-\frac{Rt}{L}}\right) \Rightarrow t = \dfrac{L}{R}\ln\left[\left(1 - \dfrac{I(t)}{I_{max}}\right)^{-1}\right] = \dfrac{40 \times 10^{-3}\,\text{H}}{5\,\Omega}\ln(100) = 37\,\text{ms}$$

The value of this current I is

$$I = 0.99 \times I_{max} = 0.99 \times \dfrac{12\,\text{V}}{5\,\Omega} = 2.4\,\text{A}$$

(b) The voltage across the inductor V_L is

$$V_L = \varepsilon e^{-\frac{Rt}{L}} = 12\,\text{V} \times e^{-\frac{(5\,\Omega)(0.037\ \text{s})}{0.040\ \text{mH}}} = 0.12\,\text{V}$$

(c) The time dependence of the current $I(t)$ when the switch is in the position b is an exponential time decay from the initial maximum value I_{max}, as shown below:

$$L\dfrac{dI}{dt} + IR = 0 \Rightarrow \dfrac{dI}{I} = -\dfrac{R}{L}dt \Rightarrow I = I_{max}e^{-\frac{Rt}{L}}$$

The time t' needed to decrease the current to 50% of the maximum value is, then, given by

$$t' = \dfrac{L}{R}\ln(2) = \dfrac{40 \times 10^{-3}\,\text{H}}{5\,\Omega}\ln(2) = 5.5\,\text{ms}$$

69. (a) The time dependence of the emf $\varepsilon(t)$ can be derived as follows:

$$\varepsilon(t) = -\dfrac{d\Phi_B}{dt} = -B\dfrac{d}{dt}\left[NA\cos(\omega t)\right] = NBA\omega\sin(\omega t) = 2\pi NBlwf\sin(2\pi ft)$$

(b) The maximum emf ε_{max} is equal to

$$\varepsilon_{max} = NBlw\omega = 10^3\,(2.5\,\text{T})(0.3\,\text{m})(0.6\,\text{m})2\pi\left(20\dfrac{\text{rev}}{\text{s}}\right) = 57\,\text{kV}$$

where, $\omega = 2\pi f = 2\pi\left(1200\dfrac{\text{rev}}{\text{min}} \times \dfrac{1\,\text{min}}{60\,\text{s}}\right) = 2\pi\left(20\dfrac{\text{rev}}{\text{s}}\right)$

(c) The frequency of the emf is equal to the rotation frequency of the coil f. The amplitude of the generated emf can be increased by increasing the number of loops N, the magnitude of the magnetic field B, the loop area $l \times w$, or the rotation frequency of the coil f.

71. The magnetic and electric fields energy densities, u_B and u_E, respectively, are given by

$$u_B = \frac{B^2}{2\mu_0}$$

$$u_E = \frac{\varepsilon_0 E^2}{2}$$

The lowest values, $(u_B)_l$ and $(u_E)_l$, are

$$(u_B)_l = \frac{B_l^2}{2\mu_0} = \frac{\left(25 \times 10^{-6} \text{ T}\right)^2}{2\left(4\pi \times 10^{-7} \text{ T} \cdot \text{m/A}\right)} = 2.5 \times 10^{-4} \ \frac{\text{J}}{\text{m}^3}$$

$$(u_E)_l = \frac{\varepsilon_0 E_l^2}{2} = \frac{8.85 \times 10^{-12} \ \frac{\text{C}^2}{\text{N} \cdot \text{m}^2} \times \left(40 \frac{\text{V}}{\text{m}}\right)^2}{2} = 7.1 \times 10^{-9} \ \frac{\text{J}}{\text{m}^3}$$

Notice, you can check that the units for the above equations make sense:

$$\left[(u_B)_l\right] = \frac{\text{T}^2}{\text{T} \cdot \text{m/A}} = \frac{\text{T} \cdot \text{A}}{\text{m}} = \frac{\text{C} \cdot \frac{\frac{\text{N}}{\text{C} \cdot \frac{\text{m}}{\text{s}}} \cdot \frac{\text{C}}{\text{s}}}{}}{\text{m}} = \frac{\text{N}}{\text{m}^2} = \frac{\text{N} \cdot \text{m}}{\text{m}^3} = \frac{\text{J}}{\text{m}^3}$$

and

$$\left[(u_E)_l\right] = \frac{\text{C}^2}{\text{N} \cdot \text{m}^2}\left(\frac{\text{V}}{\text{m}}\right)^2 = \frac{\text{C}^2}{\text{C} \cdot \frac{\text{V}}{\text{m}} \text{m}^2} \cdot \frac{\text{V}^2}{\text{m}^2} = \frac{\text{C} \cdot \text{V}}{\text{m}^3} = \frac{\text{J}}{\text{m}^3}$$

Another way to perform dimensional analysis for the electric field energy density:

$$\left[(u_E)_l\right] = \frac{\text{F}}{\text{m}}\left(\frac{\text{V}}{\text{m}}\right)^2 = \frac{\text{FV}^2}{\text{m}^3} = \frac{\frac{\text{C}^2}{\text{J}} \cdot \frac{\text{J}^2}{\text{C}^2}}{\text{m}^3} = \frac{\text{J}}{\text{m}^3}$$

$$\frac{\text{C}^2}{\text{N} \cdot \text{m}^2} = \frac{\text{C}^2}{\text{N} \cdot \text{m} \cdot \text{m}} = \frac{\text{C}^2}{\text{J} \cdot \text{m}} = \frac{\text{C}^2/\text{J}}{\text{m}} \equiv \frac{\text{F}}{\text{m}}$$

The highest values: $(u_B)_h$ and $(u_E)_h$, are

$$(u_B)_h = \frac{B_h^2}{2\mu_0} = \frac{\left(57 \times 10^{-6} \text{ T}\right)^2}{2\left(4\pi \times 10^{-7} \text{ T} \cdot \text{m/A}\right)} = 1.3 \times 10^{-3} \ \frac{\text{J}}{\text{m}^3}$$

$$(u_E)_h = \frac{\varepsilon_0 E_h^{\ 2}}{2} = \frac{\left(8.85 \times 10^{-12}\ \dfrac{F}{m}\right)\left(150\ \dfrac{V}{m}\right)^2}{2} = 1.0 \times 10^{-7}\ \frac{J}{m^3}$$

The magnetic field energy density at the surface of Earth is more than 10 000 times larger than the electric field energy density.

73. The smallest voltage ε_{min} induced in the coil of the breathing monitor can be calculated employing Faraday's law as follows. In this calculation we realize that the relationship between the time interval between two consequent baby breaths and the frequency of his breathing is $\Delta t = \dfrac{1}{f}$. Therefore, the minimum emf can be calculated as follows:

$$\varepsilon_{min} = \frac{B_{Earth} N (\Delta A)_{min}}{\Delta t} = \frac{B_{Earth} N (\Delta A)_{min}}{1/f} = f B_{Earth} N (\Delta A)_{min}$$

$$\varepsilon_{min} = \left(40\ \frac{times}{min} \times \frac{1\ min}{60\ s}\right)(30 \times 10^{-6}\ T)(300\ turns)(5 \times 10^{-4}\ m^2) = 3\ \mu V$$

75. (a) The current intensity can be calculated as follows:

$$I = \frac{P}{V} = \frac{10^9\ W}{2 \times 10^5\ V} = 5 \times 10^3\ A = 5\ kA$$

The magnetic field at the surface of the inner conductor, B_{inner}, is given by

$$B_{inner} = \frac{\mu_0 I}{2\pi a} = \frac{\left(4\pi \times 10^{-7}\ T \cdot m/A\right)\left(5 \times 10^3\ A\right)}{2\pi\left(2.5 \times 10^{-2}\ m\right)} = 0.040\ T = 40\ mT$$

(b) The magnetic field at the inner surface of the outer conductor, $B_{outer-in}$, is

$$B_{outer-in} = \frac{\mu_0 I}{2\pi (a+b)} = \frac{\left(4\pi \times 10^{-7}\ T \cdot m/A\right)\left(5 \times 10^3\ A\right)}{2\pi(2.5+5.0) \times 10^{-2}\ m} = \frac{4}{3} \times 10^{-2}\ T = 13\ mT$$

(c) The magnetic field at the outer surface of the outer conductor, $B_{outer-out}$, is

$$B_{outer-out} = 0\ T$$

The result expressed by the equation above is due to the fact the current through a circular loop enclosing the superconducting coaxial cable is zero: $I_{encl} = 0\ A$.

(d) The infinitesimal energy of the magnetic field dU_B in the space between the two conductors of the coaxial cable is given by

$$dU_B = \frac{[B(r)]^2}{2\mu_0} dV = \frac{[B(r)]^2}{2\mu_0} \cdot 2\pi r l dr = \frac{\left(\dfrac{\mu_0 I}{2\pi r}\right)^2}{2\mu_0} \cdot 2\pi r l dr = \frac{\mu_0 I^2 l dr}{4\pi r}$$

Integrating the equation on the previous page, the total magnetic energy U_B is

$$U_B = \frac{\mu_0 I^2 l}{4\pi} \int_a^b \frac{dr}{r} = \frac{\mu_0 I^2 l}{4\pi} \cdot \ln\left(\frac{b}{a}\right)$$

Hence, the magnetic energy density, u_B, is given by

$$u_B = \frac{U_B}{V} = \frac{\dfrac{\mu_0 I^2 l}{4\pi} \cdot \ln\left(\dfrac{b}{a}\right)}{\pi\left(b^2 - a^2\right)l} = \frac{\mu_0 I^2 \ln\left(\dfrac{b}{a}\right)}{4\pi^2\left(b^2 - a^2\right)} = \frac{\left(4\pi \times 10^{-7}\,\text{T}\cdot\text{m/A}\right)\left(25 \times 10^6\,\text{A}^2\right)\ln(2)}{4\pi^2\left(5^2 - 2.5^2\right) \times 10^{-4}\,\text{m}^2}$$

$$= 0.92\ \frac{\text{kJ}}{\text{m}^3}$$

(e) The magnetic energy stored in the space between the conductors U_B over a distance $l = 1500\,km$ is equal to

$$U_B\left(l = 1500\,\text{km}\right) = \frac{\mu_0 I^2 l}{4\pi} \cdot \ln\left(\frac{b}{a}\right) = \frac{\left(4\pi \times 10^{-7}\,\text{T}\cdot\text{m/A}\right)\left(25 \times 10^6\,\text{A}^2\right)\left(15 \times 10^5\,\text{m}\right)\ln(2)}{4\pi}$$

$$= 2.6\,\text{MJ}$$

(f) The pressure exerted on the outer conductor p as a result of the power transmission can be calculated as follows:

$$p = \frac{dF_B}{dA} = \frac{B_{outer-in}Idl}{2\pi b dl} = \frac{B_{outer-in}I}{2\pi b} = \frac{\left(\dfrac{4}{3} \times 10^{-2}\,\text{T}\right)\left(5 \times 10^3\,\text{A}\right)}{2\pi\left(5 \times 10^{-2}\,\text{m}\right)} = \frac{20}{3\pi} \times 10^2\,\text{Pa} = 0.2\text{ kPa}$$

77. (a) The voltages across each inductors connected in series are given by

$$V_{L1} = -L_1 \frac{dI}{dt}$$

$$V_{L2} = -L_2 \frac{dI}{dt}$$

$$\vdots$$

The total voltage across all inductors, V_L, is

$$V_L = V_{L1} + V_{L2} + \cdots = -L_{eq}\frac{dI}{dt} \Rightarrow L_{eq} = L_1 + L_2 + \cdots$$

(b) The voltages across each inductor connected in parallel are equal to

$$V_{L1} = V_{L2} = \cdots = -L_1 \frac{dI_1}{dt} = -L_2 \frac{dI_2}{dt} = \cdots$$

The sum of the currents is

$$I = I_1 + I_2 + \cdots$$

$$V = -L_{eq}\frac{dI}{dt} = -L_{eq}\left(\frac{dI_1}{dt} + \frac{dI_2}{dt} + \cdots\right) = \left(-L_{eq}\right)\left(-V\right)\left(\frac{1}{L_1} + \frac{1}{L_2} + \cdots\right)$$

Hence,

$$L_{eq} = \left(L_1^{-1} + L_2^{-1} + \cdots\right)^{-1}$$

79. (a) The voltage across the inductor, V_L, in a series RL circuit is given by

$$V_L = \varepsilon e^{-\frac{t}{\tau}} \Rightarrow \ln\left(V_L\right) = \ln\left(\varepsilon\right) - \frac{t}{\tau}$$

Hence, the semi-logarithmic representation of the function $V_L\left(t\right)$ is a line where the slope, b, and intercept, a, are given by

$$b = -\frac{1}{\tau}$$

$$a = \ln\left(\varepsilon\right)$$

The data from Table 24-4 gives the following linear graph with $a = 2.485$ and $b \cong -0.333\,\text{ms}$

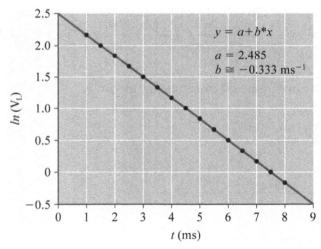

It follows that $\tau = -b \cong 3.0\,\text{ms}$

Further,

$$\tau = \frac{L}{R} \Rightarrow L = \tau R = \left(3.0 \times 10^{-3}\,\text{s}\right)\left(2\,\Omega\right) = 6.0\,\text{mH}$$

(b) $\tau = -b = 3.0\,\text{ms}$

(c) The electric current $I(t)$ is given by

$$I(t) = \frac{\varepsilon}{R}\left(1 - e^{-\frac{t}{\tau}}\right) = \frac{e^{2.485}\ V}{2\ \Omega}\left(1 - e^{-\frac{t[ms]}{3}}\right) = 6.0\left(1 - e^{-\frac{t[ms]}{3}}\right)A$$

The following table can be generated using a spreadsheet program. The last column gives the current values for each time according to the equation written above.

Time (ms)	V_L (V)	ln V_L	I (A)
1.0	8.598	2.152	1.701
1.5	7.278	1.985	2.361
2.0	6.161	1.818	2.919
2.5	5.215	1.652	3.392
3.0	4.414	1.485	3.793
3.5	3.737	1.318	4.132
4.0	3.163	1.152	4.418
4.5	2.678	0.985	4.661
5.0	2.267	0.818	4.867
5.5	1.919	0.652	5.041
6.0	1.624	0.485	5.188
6.5	1.375	0.318	5.313
7.0	1.164	0.152	5.418
7.5	0.985	−0.015	5.507
8.0	0.834	−0.182	5.583

1. The RLC circuit is above resonance.

3. The inductive reactance X_L is directly proportional to the AC angular frequency ω:

    $$X_L \overset{def}{=} \omega L$$

    Therefore, a low frequency ω translates into a low inductive reactance X_L. Qualitatively, the inductor's emf depends on how fast the magnetic flux varies. Therefore, the induced emf is directly proportional to the AC frequency ω.

5. The capacitive reactance X_C is, by definition, given by

    $$X_C \overset{def}{=} \frac{1}{\omega C}$$

    Therefore, capacitive reactance decreases with frequency.

7. No, the rms value of a voltage can never exceed its maximum value. An rms value of a voltage equals its maximum value only for a constant voltage.

9. The phase angle approaches the value $\pi/2$ as the frequency increases significantly.

11. The following equations can be written:

    $$\varepsilon_{rms} \overset{def}{=} \sqrt{\left(\varepsilon^2\right)_{avg}} = \sqrt{\frac{1}{T}\int_0^T \left[\varepsilon(t)\right]^2 dt} = \sqrt{\frac{1}{T} 4 \int_0^{T/4} \left[\varepsilon(t)\right]^2 dt}$$

    Further,

    $$\varepsilon(t) = \frac{4\varepsilon_{max}}{T}t \Rightarrow \left[\varepsilon(t)\right]^2 = \frac{16\left(\varepsilon_{max}\right)^2}{T^2}t^2 \Rightarrow \int_0^{T/4}\left[\varepsilon(t)\right]^2 dt = \frac{\left(\varepsilon_{max}\right)^2 T}{12}$$

    Hence,

    $$\varepsilon_{rms} = \sqrt{\frac{4}{T}\cdot\frac{T\left(\varepsilon_{max}\right)^2}{12}} = \frac{\varepsilon_{max}}{\sqrt{3}}$$

13. (a) $i_{rms} = \dfrac{P}{\varepsilon_{rms}} = 0.5\,\text{A}$

    (b) $R = \dfrac{\varepsilon_{rms}^2}{P} = 240\,\Omega$

15. (a) $X_L = \omega L = 10^3 \dfrac{\text{rad}}{\text{s}} \times 10^{-3}\,\text{H} = 1\,\Omega$

(b) $i_{max} = \dfrac{\varepsilon_{max}}{X_L} = \dfrac{\sqrt{2}\,\varepsilon_{rms}}{X_L} = \dfrac{\sqrt{2}\times 20\,\text{V}}{1\,\Omega} = 30\,\text{A}$

(c) $i(t) = i_{max}\sin\left(\omega t + \dfrac{\pi}{2}\right) = 30\sin\left(1000t + \dfrac{\pi}{2}\right)$

17. (a) $X_C = \dfrac{1}{\omega C} = \dfrac{1}{10^2\dfrac{\text{rad}}{\text{s}}\times 10^{-9}\,\text{F}} = 10^7\,\Omega = 10\,\text{M}\Omega$

(b) $i_{max} = \dfrac{\sqrt{2}\,\varepsilon_{rms}}{X_C} = \dfrac{\sqrt{2}\times 20\,\text{V}}{10^7\,\Omega} = 2\sqrt{2}\times 10^{-6}\,\text{A} = 2.8\,\mu\text{A}$

(c) $i(t) = i_{max}\sin\left(\omega t - \dfrac{\pi}{2}\right) = 0.0000028\sin\left(100t - \dfrac{\pi}{2}\right)$

19. The inductance L can be derived as follows:

$$2\pi f = \dfrac{1}{\sqrt{LC}} \Rightarrow L = \dfrac{1}{4\pi^2 f^2 C} = \dfrac{1}{4\pi^2 \times \left(10^4\,\text{Hz}\right)^2 \times 10^{-5}\,\text{F}} = 25\,\mu\text{H}$$

21. The factor by which the resonance frequency changes is denoted by $f_{2C}\,/\,f_C$, and is equal to

$$\dfrac{f_{2C}}{f_C} = \sqrt{\dfrac{C}{2C}} = \dfrac{1}{\sqrt{2}} = 0.707$$

23. (a) $\sin(\theta) = 0 \Rightarrow 5000t = k\pi \Rightarrow t = \dfrac{k\pi}{5000}$ s, $\quad k = 0,\,1,\,2,\cdots$

(b) $\sin(\theta) = 1 \Rightarrow 5000t = \dfrac{(2m+1)\pi}{2} \Rightarrow t = \dfrac{(2m+1)\pi}{10000}$ s, $\quad m = 0,2,4,\cdots$

(c) $\sin(\theta) = -1 \Rightarrow 5000t = \dfrac{(2n+1)\pi}{2} \Rightarrow t = \dfrac{(2n+1)\pi}{10000}$ s, $\quad n = 1,3,5,\cdots$

25. (a) $X_C = \dfrac{1}{\omega C} = 6.06\ \text{k}\Omega$

(b) $X_L = \omega L = 660\ \Omega$

(c) $\quad Z = \sqrt{R^2 + \left(X_L - X_C\right)^2} = 5400\,\Omega$

(d) $\quad \phi = \tan^{-1}\left(\dfrac{X_L - X_C}{R}\right) = -89.9°$

(e) $\quad i_{max} = \dfrac{\varepsilon_{max}}{Z} = \dfrac{25\,\text{V}}{5400\ \Omega} = 4.63\,\text{mA}$

27. (a) $\quad \tan(\phi) = \dfrac{X_L - X_C}{R} \Rightarrow X_L = R\tan\phi + \dfrac{1}{\omega C} = 22\,\text{k}\Omega$

(b) $\quad L = \dfrac{X_L}{\omega} = \dfrac{22.1\times10^3\,\Omega}{9000\,\text{rad/s}} = 2.5\,\text{H}$

(c) $\quad i_{max} = \dfrac{\varepsilon_{max}}{Z} = \dfrac{\varepsilon_{max}}{R\sqrt{1 + \left[\tan(\phi)\right]^2}} = \dfrac{250\,\text{V}}{100\,\Omega\sqrt{1 + \left[\tan(-0.79)\right]^2}} = 1.8\,\text{A}$

29. The inductance L can be calculated as follows:

$$Q = \dfrac{1}{R}\sqrt{\dfrac{L}{C}} \Rightarrow L = C\left(RQ\right)^2 = 10^{-7}\,\text{F}\times\left(10^2\,\Omega\times15\right)^2 = 0.23\,\text{H}$$

31. The maximum number of bulbs that can go in the circuit N is equal to

$$N = \dfrac{\varepsilon_{rms}i_{max}}{P_{bulb}} = \dfrac{(110\ \text{volts})(15\ \text{amps})}{100\ \text{W/bulb}} = 16.5\ \text{bulbs}$$

33. Yes, the emf delivers an initial amount of energy to get the oscillation going. This would be equal to the energy that any instant is stored in the electric field of the capacitor and the magnetic field of the inductor.

35. $\quad L = \dfrac{X_L}{\omega} = \dfrac{\dfrac{1}{\omega C} + \sqrt{\left(\dfrac{\varepsilon_{max}}{i_{max}}\right)^2 - R^2}}{\omega} = 140\,\text{mH}$

$\quad \phi = \tan^{-1}\left(\dfrac{X_L - X_C}{R}\right) = 89.9°$

37. $\quad C = \dfrac{1}{\omega X_C} = 3.3\,\mu\text{F}$

39. The inductance of the motor L can be calculated as follows:

$$i_{rms} = \frac{\varepsilon_{rms}}{\sqrt{R^2 + 4\pi^2 f^2 L^2}} \Rightarrow L = \frac{\sqrt{\left(\frac{\varepsilon_{rms}}{i_{rms}}\right)^2 - R^2}}{2\pi f} = \frac{\sqrt{\left(\frac{110\,V}{2\,A}\right)^2 - (30\,\Omega)^2}}{2\pi \times 60\,Hz} = 120\,mH$$

41. $L = \dfrac{1}{4\pi^2 f^2 C} = 6.3\,\mu H$

$$i_{max} = \frac{q_{max}}{\sqrt{LC}} = 0.25\,A$$

43. $V_R(t = 10\,s) = V_{Rmax} \sin(2\pi f t - \phi) = 10\sqrt{2}\,V = 14\,V$

45. The resistance R and the capacitance C can be calculated as follows:

$$i_{max} = \frac{\varepsilon_{max}}{\sqrt{R^2 + X_C^2}} \Rightarrow R^2 + X_C^2 = \left(\frac{\varepsilon_{max}}{i_{max}}\right)^2 = 45 \times 10^3\,\Omega \qquad (1)$$

$$\tan(\phi) = -\frac{X_C}{R} = -\tan(0.7) \Rightarrow X_C = R\tan(0.7) \qquad (2)$$

Using equations (1) and (2),

$$R = \sqrt{\frac{45 \times 10^3}{1 + \left[\tan(0.7)\right]^2}} = 160\,\Omega$$

$$C = \frac{1}{\omega X_C} = 1.5\,\mu F$$

47. $P_R = max \mapsto i_{rms} = max \mapsto Z = min \Rightarrow X_L = X_C \Rightarrow \phi = 0°$ (resonance)

49. $R = \dfrac{P_R}{i_{rms}^2} = 20\,\Omega$

$$\phi = \tan^{-1}\sqrt{\left(\frac{\varepsilon_{rms} / i_{rms}}{R}\right)^2 - 1} = 25°$$

51. Using equation (25-39e) derived in Section 25-6:

$$\Delta f = \frac{R}{2\pi L} \Rightarrow R = 2\pi L \Delta f = 2\pi \times 10^{-8}\,H \times 40 \times 10^3\,Hz = 2.5\,m\Omega$$

$$C = \frac{1}{4\pi^2 f_0^2 L} = \frac{1}{4\pi^2 \times \left(101.5 \times 10^6 \, \text{Hz}\right)^2 \times 10^{-8} \, \text{H}} = 0.25 \, \text{nF}$$

53. We will exploit the fact that if we sit on the side of the resonance peak, we can get very large changes in amplitude of the current in an RLC circuit. We will drive the circuit with an emf with a fixed frequency ω_f. We will select our inductor so that at the maximum capacitance value we expect, the resonance frequency of the circuit will be just above ω_f. This is shown in the figure below.

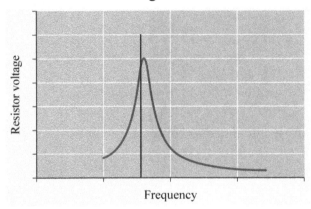

The black vertical line represents the emf frequency and the blue line gives us the amplitude of the voltage across the resistor.

As the capacitance decreases, the resonance frequency of the circuit will increase and the drive frequency will move away form resonance. As a result, the amplitude of the voltage across the resistor will decrease. This is shown in the next figure.

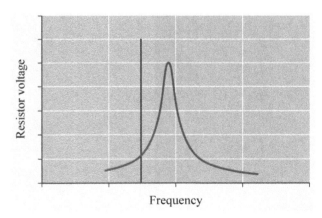

We are now driving the circuit well below resonance, and as a result the amplitude of the resistor voltage is much lower.

We can make the circuit more sensitive by adjusting the Q factor. However, as we do this we will lose some of our range.

Chapter 26—ELECTROMAGNETIC WAVES AND MAXWELL'S EQUATIONS

1. It explains the generation of varying magnetic fields from varying electric fields.

3. (a) $\vec{E}$ and $\vec{B}$ waves are on the same axis.

   (c) $\vec{E}$ and $\vec{B}$ waves have different phases.

   (d) $\vec{E}$ and $\vec{B}$ waves have the same amplitude.

5. (i) Sound waves are not fields; they do not propagate in vacuum.

   (ii) Sound waves are not transverse waves.

   (iii) Microscopically, massless particles called photons are associated with electromagnetic waves; quasi-particles called phonons are associated with sound waves.

7. Deceleration and acceleration in a straight line, a curved trajectory, or oscillations.

9. No, it cannot because of the energy loss that is associated with the emission of electromagnetic radiation.

11. The radiation pressure is twice as large on a surface that is a perfect reflector than on a surface that is a perfect absorber. The difference is due to the change of total momentum of the system (radiation + surface), which is equal to twice the initial momentum of the radiation for the perfect reflector surface, and equal to the initial momentum for the perfect absorber surface.

13. $I_d = \varepsilon_0 \dfrac{d\Phi_E}{dt} = \varepsilon_0 \pi r^2 \dfrac{dE}{dt} = 8.85 \times 10^{-12} \dfrac{\text{F}}{\text{m}} \times \pi \times 10^{-4}\,\text{m}^2 \times 10^2 \dfrac{\text{V}}{\text{m} \cdot \text{s}} = 0.3\,\text{pA}$

15. (a) $B = \dfrac{\mu_0 I \left(\dfrac{r}{R}\right)^2}{2\pi r} = 0.320\,\mu\text{T}$

    (b) $B = \dfrac{\mu_0 I}{2\pi r} = 2.00\,\mu\text{T}$

17. (a) $I(t = 0\,\text{s}) = \dfrac{\varepsilon}{R} = 12\,\text{A}$

    (b) $I(t = 60\,\text{s}) = \dfrac{\varepsilon}{R} e^{-\frac{t}{RC}} \approx 0\,\text{A}$

(c) $\dfrac{dE}{dt}(t=0\,\mathrm{s}) = \dfrac{I(t=0\,\mathrm{s})}{\varepsilon_0 \pi r^2} = 43 \times 10^{12}\ \dfrac{\mathrm{V}}{\mathrm{m} \cdot \mathrm{s}}$

$\dfrac{dE}{dt}(t=60\,\mathrm{s}) = \dfrac{I(t=60\,\mathrm{s})}{\varepsilon_0 \pi r^2} \approx 0\,\dfrac{\mathrm{V}}{\mathrm{m} \cdot \mathrm{s}}$

(d) $B(t=0\,\mathrm{s}) = \dfrac{\mu_0 I(t=0\,\mathrm{s})}{2\pi r} = 24\,\mu\mathrm{T}$

$B(t=60\,\mathrm{s}) = \dfrac{\mu_0 I(t=60\,\mathrm{s})}{2\pi r} = 0\,\mu\mathrm{T}$

19. (a) $\vec{S}$ is in the y-direction.

   (b) $\vec{S}$ is in the $(-z)$-direction.

   (c) $\vec{S}$ is in the z-direction.

   (d) $\vec{S}$ is in the y-direction.

21. $B_0 = \dfrac{E_0}{c} = 1.7 \times 10^{-11}\,\mathrm{T} = 17\,\mathrm{pT}$

   $\vec{E}$ can be oriented in any direction in the y-z plane.

23. (a) $\lambda = \dfrac{2\pi}{k} = 1.57 \times 10^{-9}\,\mathrm{m} = 1.57\,\mathrm{nm}$

   $f = \dfrac{c}{\lambda} = 1.91 \times 10^{17}\,\mathrm{Hz}$

   (b) $E_0 = cB_0 = 1500\ \mathrm{V/m}$

   (c) Since the magnetic field is described by a function of the form $\sin(kx - \omega t)$ the direction of motion of the electromagnetic wave is along the x-axis, toward increasing x. As the direction of the electric field must be perpendicular to the orientation of the magnetic field and the direction of motion of the wave, it must be oriented along the y-axis. Therefore,

   $$\vec{E}(x,t) = (1500V / m)\,\sin\left(4.00 \times 10^9\, x - \omega t\right)\hat{y}$$

   (d) The wave travels in the x-direction.

25. $B(t) = \dfrac{E(t)}{c} = 3.3 \times 10^{-9}$ T in the $(-z)$-direction

27. Microwaves < infrared light < yellow light < blue light < X-rays < gamma rays

29. (a) $f = \dfrac{c}{\lambda} = 6.67 \times 10^{14}$ Hz

(b) $f = \dfrac{c}{\lambda} = 5.63 \times 10^{14}$ Hz

(c) $f = \dfrac{c}{\lambda} = 4.29 \times 10^{14}$ Hz

31. (a) $\overline{S} = \dfrac{P}{4\pi r^2} = \dfrac{100}{4\pi(10)^2} = 7.96 \times 10^{-2}$ W/m^2

$E_0 = \sqrt{2\mu_0 c \overline{S}} = \sqrt{2\mu_0 c \dfrac{P}{4\pi r^2}} = 7.8$ V/m

$B_0 = \dfrac{E_0}{c} = 2.6 \times 10^{-8}$ T

(b) $\overline{S} = \dfrac{P}{4\pi r^2} = \dfrac{100}{4\pi(100)^2} = 7.96 \times 10^{-4}$ W/m^2

$E_0 = \sqrt{2\mu_0 c \overline{S}} = \sqrt{2\mu_0 c \dfrac{P}{4\pi r^2}} = 0.78$ V/m

$B_0 = \dfrac{E_0}{c} = 2.6 \times 10^{-9}$ T

(c) $\overline{S} = \dfrac{P}{4\pi r^2} = \dfrac{100}{4\pi(10^5)^2} = 7.96 \times 10^{-10}$ W/m^2

$E_0 = \sqrt{2\mu_0 c \overline{S}} = \sqrt{2\mu_0 c \dfrac{P}{4\pi r^2}} = 7.8 \times 10^{-4}$ V/m

$B_0 = \dfrac{E_0}{c} = 2.6 \times 10^{-12}$ T

33. $\overline{S} = \dfrac{0.10 \text{ J/m}^2}{5.0 \text{ s}} = 0.02$ W/m^2

35. $E_0 = \sqrt{\dfrac{2\overline{u}}{\varepsilon_0}} = 0.1\,\text{V/m}$

   $B_0 = \dfrac{E_0}{c} = 3\times10^{-10}\,\text{T}$

37. Let $\overline{S}_1$ be the intensity at a radius R_1 and $\overline{S}_2$ be the intensity at radius R_2. Then,

   $\dfrac{\overline{S}_2}{\overline{S}_1} = \left(\dfrac{R_1}{R_2}\right)^2$

   $\overline{S}_1 = 2.0\,\text{W/m}^2$, $R_1 = 1.0\,\text{m}$ and $R_2 = 5.0\,\text{m}$. Therefore,

   $\overline{S}_2 = 0.08\,\text{W/m}^2$

39. The average Lorentz force ($\overline{F}_B$) is zero due to various orientations of the electron speed with respect to the magnetic field of the Sun's light:

   $\overline{F}_B = e v \overline{B}_0 \overline{\sin(\vec{v}, \vec{B})} = 0$

   $\overline{F} = e\overline{E}_0 + e v \overline{B}_0 \overline{\sin(\vec{v}, \vec{B})} = 2 e \mu_0 c \overline{S} = 6.0\times10^{-14}\,\text{N}$

41. Let $\eta = 0.15$ denote the efficiency of the photovoltaic cells. Then for incident intensity $\overline{S}$, energy absorbed by an area A in time t is

   $E = \eta \overline{S} A t$

   $= (0.15)(200)(16)(365\times24) = 4000\,\text{kW}\cdot\text{h}$

43. (a) $\overline{S} = \dfrac{P}{A} = \dfrac{P}{\pi r^2} = 1.59\times10^3\,\text{W/m}^2$

   (b) $P_{rad} = 2\dfrac{\overline{S}}{c} = 10.6\times10^{-6}\,\text{Pa}$

   $F = P_{rad}A = \dfrac{2P}{c} = 3.33\times10^{-11}\,\text{N}$

45. The electromagnetic wave is polarized along the y-axis.

47. From Malus' law: $I = 0.25\,I_0 = I_0\left(\cos\theta\right)^2 \Rightarrow \theta = 60°$

   Therefore, the angle with respect to the vertical is $90° - \theta = 30°$.

49. $I_2 = I_1\left(\cos 45°\right)^2 = \dfrac{I_0}{2}\left(\cos 45°\right)^2 = 0.25\,\text{W/m}^2$

51. (a) Let I_0 be the initial intensity, I_1 the intensity after passing through the first polarizer, and I_2 the intensity after passing through the second analyzer. The light emerging from the first polarizer is polarized along the direction of its transmission axis, which is perpendicular to the transmission axis of the second polarizer. Therefore,

$$I_2 = I_1 (\cos 90°)^2 = 0 \text{ W/m}^2$$

(b) The light emerging from the first polarizer is polarized along the direction of its transmission axis, which is perpendicular to the transmission axis of the second polarizer. In this case, $I_1 = I_0$ and

$$I_2 = I_1 (\cos 90°)^2 = 0 \text{ W/m}^2$$

(c) In this case no light passes through the first polarizer. Hence, $I_1 = 0$ and hence, $I_2 = 0$.

(d) The light emerging from the first polarizer is polarized along the direction of its transmission axis, which is perpendicular to the transmission axis of the second polarizer. In this case, $I_1 = I_0 / 2$ and

$$I_2 = I_1 (\cos 90°)^2 = 0 \text{ W/m}^2$$

53. The initial light is unpolarized. So the orientation of the first polarizer does not matter and only half of the initial intensity will pass through the first polarizer.

$$I_A = \frac{I_0}{2} = 0.5 I_0$$

$$I_B = I_A \left[\cos(45° - 30°) \right]^2 = 0.5 I_0 (\cos 15°)^2 = 0.47 I_0$$

$$I_C = I_B (\cos 45°)^2 = 0.47 I_0 \times 0.5 = 0.23 I_0$$

55. When the unpolarized light is incident on the first polarizer oriented at angle θ with respect to the vertical, the transmitted wave will decrease its intensity by a factor equal to $(\cos 45°)^2$ and its polarization plane is oriented at angle θ with respect to the vertical. The second polarizer is oriented at an angle θ with respect to the first polarizer, so the intensity is decreased by a factor equal to $(\cos \theta)^2$, and so on. Therefore, the following equation can be written:

$$\frac{I_l}{I_0} = (\cos 45°)^2 (\cos \theta)^2 (\cos \theta)^2 (\cos \theta)^2 (\cos \theta)^2 = \frac{1}{10}$$

$$\Rightarrow (\cos \theta)^8 = \frac{1}{5} \Rightarrow \theta = \cos^{-1} \left[\left(\frac{1}{5} \right)^{1/8} \right] = 35°$$

57. (a) $F_g = F_{rad} \Rightarrow \rho \dfrac{4\pi r^3}{3} g_{Sun} = P_{rad}\pi r^2 = 2\dfrac{\overline{S}}{c}\pi r^2$

$$\Rightarrow r = \dfrac{3\overline{S}}{2\rho c g_{Sun}} = \dfrac{3\dfrac{P_{Sun}}{4\pi R^2}}{2\rho c G \dfrac{M_{Sun}}{R^2}} = \dfrac{3P_{Sun}}{8\pi \rho c\, G\, M_{Sun}} \approx 0.5\,\mu m$$

Thus the diameter is: $d = 2r \cong 1\,\mu m$. The following numerical values were used in the equation written above:

$G = 6.7 \times 10^{-11}\ \mathrm{m^3 \cdot kg^{-1} \cdot s^{-2}}$ (universal gravitational constant)

$P_{Sun} = 3.9 \times 10^{26}\ \mathrm{W}$ (radiation power of the Sun)

$M_{Sun} = 2.0 \times 10^{30}\ \mathrm{kg}$ (mass of the Sun)

(b) For any radius of the particle r, the ratio between the two forces is given by

$$\dfrac{F_g}{F_{rad}} = \dfrac{r}{r_{(a)}} \Rightarrow r < r_{(a)} \Rightarrow F_{rad} > F_G$$

Therefore, if the particles have a diameter smaller than 1 μm, the force due to the radiation pressure of the Sun will exceed its gravitational attraction force.

Yes, the tail will be longer as the comet approaches the Sun. For particles below the radius value calculated in part (a), $r_{(a)}$, the radiation pressure will exceed the gravitational attraction of the Sun. Thus, these particles will be pushed further away from the core of the comet and further away from the Sun, forming a longer tail.

1. The distance between an object and its image in a plane mirror is equal to twice the object-mirror distance. Therefore, if the object-mirror distance is increased by Δx, then the object-image distance will increase by $2\Delta x$:

$$d_{oi}{}' - d_{oi} = 2d_o{}' - 2d_o = 2\left(d_o{}' - d_o\right) = 2(0.5 \text{ m}) = 1 \text{ m}$$

3. A plane mirror inverts images in a direction perpendicular to the plane of the mirror; i.e., plane mirrors perform depth inversion. Plane mirrors do not interchange left and right; see the discussion in the "Making Connections" box on page 732 of the text.

   The word **AMBULANCE** is written in a mirror script, as ƎƆИA⅃UꓭMA (flipped horizontally) on the front of ambulances so that drivers in cars moving ahead of the ambulance can easily read it in their rear-view mirrors. Notice, if the word **AMBULANCE** was written backwards (from right to left), it would have been **ECNALUBMA** and not ƎƆИA⅃UꓭMA

5. c

7. The side-view mirrors of cars are convex mirrors where the image of the object is smaller than the object itself (magnification M < 1). Since the objects that the driver is looking at are relatively far away (compared to the mirror's focal length, which is about $f = -10$ cm), the distance from the object to the mirror is much larger than the focal distance of the mirror: $d_o \gg f$. Moreover, $d_o > 2f$. This means it is the case described in Table 27-1 in the textbook on the bottom row (right column). The image in that case is upright, reduced, and closer to the mirror than the object. As a result, the objects in the rear-view mirror appear closer than they really are.

$$M = \frac{h_i}{h_o} = -\frac{d_i}{d_o} < 1 \Rightarrow |d_i| < |d_o|$$

   The convex mirrors are used because the image is upright and because the magnification of less than one allows the driver a larger field of view of the road behind the car.

9. (a) Yes, an image of the rose will be formed.

   (b) To see the image of the rose in the mirror the observer has to be located in the lower left region of the rose-mirror assembly as shown in the figure on the next page. For a more detailed explanation, see Figure 27-10(b).

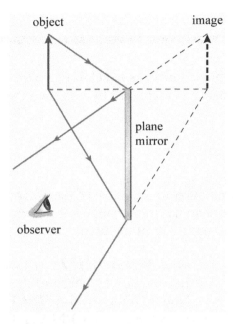

object   image

plane
mirror

observer

11. The maximum number of images that could be viewed is 5. The number of images is equal to the number of times a ray of light parallel to the axis of symmetry between the two mirrors is reflected. The number of images is given by

$$N = \frac{360°}{\theta} - 1 = \frac{360°}{60°} - 1 = 5$$

For the derivation of this equation, see also the solution and the ray diagram included in the solution of question 30.

13. The answer of question 12 is based on the fact that a reflected ray of light from the toe of the observer has to reach his/her eye as shown in the ray diagram from Figure 27-12. The correct answer for a spherical mirror is (f), based on the image properties of spherical mirrors given in Table 27-1.

15. In order to decide what mirror one has to choose, we have to realize that the image has to be an upright, magnified, and virtual image that will appear behind the mirror. It is clear that it will be a little inconvenient to use a mirror for make-up that produces an inverted image. Consulting Table 27-1 in the textbook, we can find only one option for a magnified upright image: to use a concave mirror, and place an object between the focal point and the mirror: $d_o < f$.

The ray diagram for this case is shown in Figure 27-19 (b) in the textbook:

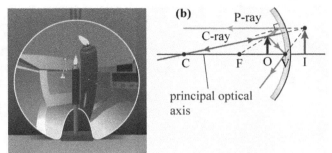

Figure 27-19 (b): A ray diagram for image formation for an object located between the focal point and the concave mirror. We recommend you to draw ray diagrams for different scenarios or experiment with an online simulation, such as http://physics.bu.edu/~duffy/java /Opticsa1.html (this simulation is based on the Physlet simulations built by Wolfgang Christian from Davidson College).

17. All three rays of light emitted simultaneously arrive at the screen in the same time following Fermat's principle of least time. Fermat principle states that the path taken between two points by a ray of light is the path that can be traversed in the least amount of time. The two rays of light undergoing the longer paths (rays A and C) have a shorter path through the lens than the central ray of light B has. Since the speed of light is slower through the lens, than through the air, the time taken by rays A, B, and C is the same—it is the least possible for light time to traverse between the two points. Therefore, the rays arrive at the screen simultaneously.

19. The result of painting the bottom half of the converging lens is an entire but a fainter image because half of the rays of light making up the image are blocked. You can easily test this result with a magnifying glass or another lens you might have on hand.

21. The focal point for a parallel beam of light incident at an angle $\theta \neq 0$ with respect to the optical axis is indicated in two ray diagrams shown below which correspond to a convergent lens (upper diagram) and a concave mirror (lower diagram).

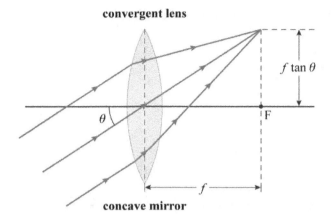

convergent lens

concave mirror

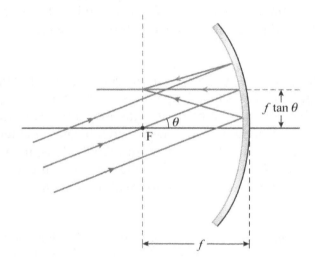

The parallel beam of light striking a concave mirror or a converging lens will converge in a point located on a focal place of the mirror or a lens respectively. The focal plane is a plane perpendicular to the principal optical axis of a mirror or a lens and it crosses the optical principal axis at the focal point F. A mirror has one optical plane, while the lens has two optical planes.

23. Two similar triangles can be identified in Figure 27-46 associated with this question. The following relationships can be derived:

$$\frac{3 \text{ m}}{(3+100) \text{ m}} = \frac{1 \text{ m}}{h} \Rightarrow h = \frac{103}{3} \text{ m} \simeq 34 \text{ m}$$

The following assumptions were made:

(i) The Sun, the tree, and the gnomon are coplanar (located in the same plane).

(ii) The shadow of the gnomon is formed by a parallel sunlight beam. This assumption implies that the size of the gnomon is larger than the image of the Sun in the sky.

(iii) The length of the shadow was measured just when the Sun disappears behind the tree such that the tip of the gnomon, the upper part of the Sun, and the tip of the tree are collinear.

This is a very tall tree—its height is about the height of a 10-story building. For comparison, the giant sequoia, the world's most massive tree, can reach the height of 87 m, which means it can be as tall as a 26-story building. Another California native tree—California incense cedar—can reach the heights of 40–60 m. Canada is home to some of the oldest and tallest trees in the world. The Carmanah Walbran Provincial Park in British Columbia is home to a 95-metre Sitka spruce tree.

25. (a) Observers A will see a total solar eclipse (the Sun is entirely eclipsed by the Moon), observers B will see a partial solar eclipse (the Sun is partially eclipsed by the Moon), and observers C will not see an eclipse.

    (b) Earth rotates about its axis from West to East in geographical terms. Mathematically, the rotation is counterclockwise in the Cartesian reference frame of Earth with its z-axis oriented from South to North. An observer on Earth perceives this motion through the position of the Sun in the sky from sunrise (East) to sunset (West). During the time of a total solar eclipse, the area in which this astronomical event can be observed will move from West to East in geographical terms.

    To learn more about eclipses and use interactive simulations to model these interesting phenomena, visit Stellarium—a free open-source planetarium for your computer: http://www.stellarium.org/.

27. Because the angle of incidence and the angle of reflection are equal, it follows that the angle of incidence was initially equal to $\dfrac{45°}{2}$; after the rotation of the mirror it is equal to $\dfrac{45°}{2} - 7° = 15.5°$. Therefore, the angle between the incident and the reflected beams is equal to $2\left(\dfrac{45°}{2} - 7°\right) = 31°$.

29. When the mirror is tilted upwards by an angle of 4° with respect to the initial horizontal position, the angles of incidence and reflection increase by 4° each (see question 27). Using the geometry from Figure 27-50, and neglecting the elevation of the mirror, the difference in height Δh between the spots formed by the beams in the two cases is given by

    $$\Delta h = (2.5 \text{ m})(\tan(34°) - \tan(30°)) = 0.243 \text{ m} = 24 \text{ cm}$$

31. (a) The image is not inverted as indicated by the ray diagram shown below.

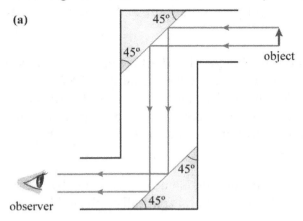

(b) See the ray diagram shown below.

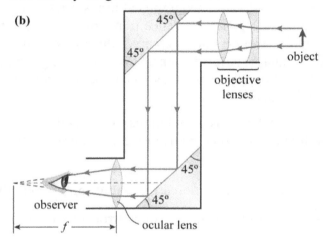

(c) Using curved mirrors would not be a viable design because these mirrors distort the object-observer distance due to varying magnification. The object-observer distance is important in many applications of the periscope.

An interesting modern application of a periscope is climbing belay glasses. The person who is belaying a climber has to look up. This often causes neck problems. These glasses are used to help alleviate neck pain from belaying. These belay glasses allow you to look straight ahead, but see upward.

33. (a) In the case described in this question the object-mirror distance d_o is equal to the image-mirror distance d_i: $d_o = d_i = 7$ cm. Using the spherical mirror equation the focal length of the mirror f can be calculated:

$$\frac{1}{d_o} + \frac{1}{d_i} = \frac{1}{f} \overset{d_i = d_o}{\Rightarrow} f = \frac{d_o}{2} = \frac{d_i}{2} = \frac{7\,\text{cm}}{2} = 3.5 \text{ cm}$$

(b) The image is real and inverted.

Copyright © 2015 by Nelson Education Ltd.

35. (a) In the right triangle OAB from the diagram shown below one can write the following equation:

$$\sin(80°) = \frac{AB}{OA} = \frac{AB}{R} = \frac{(12/2)\ \text{in}}{R} \Rightarrow R = \frac{6}{\sin(80°)}\ \text{in} = \frac{6}{\sin(80°)}\ \text{in} \times \frac{2.54\ \text{cm}}{1\ \text{in}} = 15\ \text{cm}$$

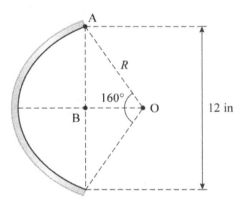

The focal length of the convex spherical mirror f is then given by

$$f = -\frac{R}{2} = -7.7\ \text{cm}$$

(b) Knowing the object-mirror distance $d_o = 2$ m, the image-mirror distance d_i can be calculated using the spherical mirror equation:

$$\frac{1}{d_o} + \frac{1}{d_i} = \frac{1}{f} \Rightarrow d_i = \frac{fd_o}{d_o - f} \cong \frac{(-7.7\ \text{cm})(200\ \text{cm})}{(200 + 7.7)\ \text{cm}} = -7.4\ \text{cm}$$

The magnification M is given by

$$M = -\frac{d_i}{d_o} = -\frac{-7.4\ \text{cm}}{200\ \text{cm}} = 0.04$$

The image is formed behind the mirror, is virtual, upright, and reduced. This corresponds to the description located in Table 27-1 for the case of convex mirror when $d_o \gg f$.

37. The following simplifying assumptions were made to solve this question:

(i) The flashlight is perpendicular to the bottom of the pool; hence, the centre line of the light beam is perpendicular on the surface of the water.

(ii) The light beam is conical with angle equal to the critical total reflection angle θ_c.

(iii) The length of the flashlight is negligible when compared to the pool's depth. The critical total reflection angle at the water-air boundary θ_{cr} is given by

$$\theta_{cr} = \tan^{-1}\left(\frac{1}{1.33}\right) \cong 36.9°$$

The light spot area A is given by the following equation:

$$A = \pi R^2 = \pi \left[(2\,\text{m}) \tan(\theta_{cr}) \right]^2 = \frac{4\pi}{1.33^2} \,\text{m}^2 = 7\,\text{m}^2$$

39. From the geometry indicated in the ray diagram the following relationships can be derived:

$$\tan(i) = \frac{R}{h} \tag{1}$$

$$\tan(r) = \frac{R}{h_a} \tag{2}$$

From equations (1) and (2) it follows that

$$\frac{h_a}{h} = \frac{\tan(i)}{\tan(r)} \approx \frac{\sin(i)}{\sin(r)} = \frac{1}{n} \Rightarrow h_a \approx \frac{h}{n} = 60 \text{ cm}$$

The approximation used above is valid for small angles. The observation angles were considered small because the radius of the coin is much smaller than the water depth: $R \ll h$.

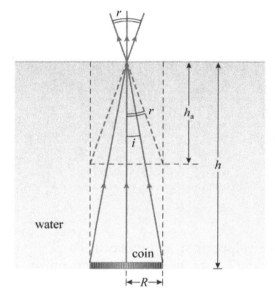

41. (a) The condition for total internal reflection is

$$n \sin(i) = n_{air} \sin(90°) = 1$$

In the above equation, the index of refraction for air n_{air} is considered to be 1. Hence, the minimum value of the prism's index of refraction n is given by

$$n = \frac{1}{\sin(i)} = \frac{1}{\sin(45°)} = \sqrt{2} = 1.41$$

(b) If the prism is submerged in oil ($n_{oil} = 1.47$), then,

$$n = \frac{n_{oil}}{\sin(45°)} = \frac{1.47}{1/\sqrt{2}} = 2.08$$

43. (a) The distance film-lens is equal to the distance image-lens in order to obtain a sharp image of the object. Using the thin lens equation, one can obtain:

$$\frac{1}{d_o} + \frac{1}{d_i} = \frac{1}{f} \Rightarrow d_i = \frac{fd_o}{d_o - f} = \frac{(45 \text{ mm}) \times (6 \text{ m})}{6 \text{ m} - 45 \text{ mm}} = \frac{45 \times 6000}{6000 - 45} \text{ mm} = 45 \text{ mm}$$

(b) The magnification on the film M is given by

$$M = -\frac{d_i}{d_o} = -\frac{\dfrac{45 \times 6000}{5955} \text{ mm}}{6000 \text{ mm}} = -0.008$$

45. (a) Using the thin lens equation the following results are obtained:

$$\frac{1}{d_o} + \frac{1}{d_i} = \frac{1}{f} \Rightarrow d_i = \frac{fd_o}{d_o - f} = \frac{(-4 \text{ cm})(25 \text{ cm})}{25 \text{ cm} - (-4 \text{ cm})} = -3.4 \text{ cm}$$

$$M = -\frac{d_i}{d_o} = -\frac{-(100/29) \text{ cm}}{25 \text{ cm}} = 0.14$$

(b) It is a divergent lens ($f < 0$)

(c) The image is formed in front of the lens, is virtual, upright, and reduced. See also the ray diagram below.

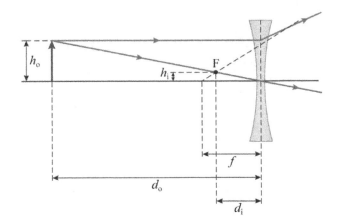

47. (a) Using lens maker's equation, the focal length of the thin concave-convex lens in water f_w can be calculated as follows:

$$\frac{1}{f_w} = \frac{n_{lens} - n_w}{n_w}\left(\frac{1}{R_1} - \frac{1}{R_2}\right)$$

$$\Rightarrow f_w = \frac{n_w}{n_{lens} - n_w} \cdot \frac{R_1 R_2}{R_2 - R_1} = \frac{1.33}{1.52 - 1.33} \cdot \frac{(-12 \text{ cm})(14 \text{ cm})}{14 \text{ cm} - (-12 \text{ cm})} = -45.2 \text{ cm}$$

The magnification of the lens in water M_w is then given by

$$M_w = -\frac{d_i}{d_o} = -\frac{d_o f_w / (d_o - f_w)}{d_o} = \frac{f_w}{f_w - d_o} = \frac{-45.2 \text{ cm}}{-45.2 \text{ cm} - 12 \text{ cm}} = 0.8$$

(b) Similarly to the solution in part (a), the focal length of the thin concave-convex lens in air f_{air} can be calculated as follows:

$$f_{air} = \frac{n_{air}}{n_{lens} - n_{air}} \cdot \frac{R_1 R_2}{R_2 - R_1} = \frac{1}{1.52 - 1} \times \frac{(-12 \text{ cm})(14 \text{ cm})}{14 \text{ cm} - (-12 \text{ cm})} = -12.4 \text{ cm}$$

The magnification of the lens in air M_{air} is then given by

$$M_{air} = -\frac{d_i}{d_o} = -\frac{d_o f_{air} / (d_o - f_{air})}{d_o} = \frac{f_{air}}{f_{air} - d_o} \cong \frac{-12.4 \text{ cm}}{-12.4 \text{ cm} - 12 \text{ cm}} = 0.5$$

The ray diagram shown below indicates qualitatively that a larger focal length for a divergent thin lens corresponds to a larger magnification as demonstrated by the calculations of the concave-convex glass lens in water and air.

Concave-complex Lens in Water

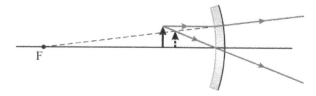

Concave-complex Lens in Air

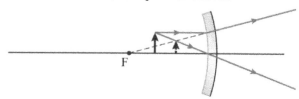

49. The smallest radius of the disc such that no light exits the surface of the water is given when the incident angle of the ray of light at the edge of the disc is equal to the critical angle θ_{cr} for the water-air boundary. It is assumed that the centre of the disc and the light source are

collinear along the normal to the water surface as shown in the diagram below. Mathematically, this fact can be expressed by two equations:

$$n_w \sin(\theta_{cr}) = n_{air} \sin(90°) \Rightarrow \sin(\theta_{cr}) = \frac{1}{n_w} \qquad (1)$$

$$\sin(\theta_{cr}) = \frac{R}{\sqrt{R^2 + H^2}} \qquad (2)$$

From equations (1) and (2), it follows that

$$\frac{R^2}{R^2 + H^2} = \frac{1}{n_w^2} \Rightarrow R^2 \left(n_w^2 - 1\right) = H^2 \Rightarrow R = \frac{H}{\sqrt{n_w^2 - 1}} = \frac{H}{\sqrt{1.33^2 - 1}} \cong 1.14H$$

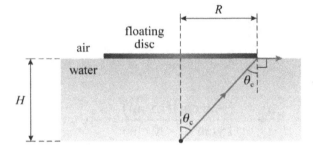

51. If the index of refraction of glass is denoted by n and the index of refraction of air is taken to be one, the law of refraction is

$$\sin(i) = n \sin(r) \Rightarrow r = \sin^{-1}\left[\frac{\sin(i)}{n}\right] = \sin^{-1}\left(\frac{0.5}{1.52}\right) = 19.2°$$

The angle of deflection δ between the incident and the refracted rays of light is given by

$$\delta = (90° - i) + 90° + r \cong 90° - 30° + 90° + 19.2° = 169°$$

For an angle of incidence $i = 45°$,

$$r = \sin^{-1}\left(\frac{\sqrt{2}/2}{1.52}\right) = 27.7° \Rightarrow \delta = 163°$$

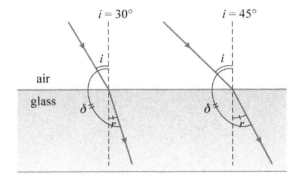

53. The distance image-lens can be calculated using the thin lens equation:

$$\frac{1}{d_o} + \frac{1}{d_i} = \frac{1}{f} \Rightarrow d_i = \frac{fd_o}{d_o - f} = \frac{(-30\text{ cm})(15\text{ cm})}{15\text{ cm} - (-30\text{ cm})} = -10\text{ cm}$$

The image is formed in front of the divergent lens; it is virtual, upright, and reduced. The image of the object cannot be seen on a screen because it is a virtual image.

From the ray diagram shown below the following relationships can be derived:

$$\Delta CBI \sim \Delta CAO \Rightarrow \frac{d_i}{d_o} = -\frac{h_i}{h_o} \qquad (\text{notice, } d_i < 0, \text{ since the image is virtual}) \qquad (1)$$

$$\Delta FBI \sim \Delta FMC \Rightarrow \frac{|f| - |d_i|}{|f|} = \frac{h_i}{h_o}, \text{ where } f < 0 \text{ since the lens is a diverging lens} \qquad (2)$$

Since both f and d_i are negative $\Rightarrow \dfrac{-f - (-d_i)}{-f} = \dfrac{h_i}{h_o}$

From equations (1) and (2) it follows that

$$-\frac{d_i}{d_o} = \frac{-f - (-d_i)}{-f} \Rightarrow -\frac{d_i}{d_o} = \frac{-f + d_i}{-f} \Rightarrow d_i = \frac{fd_o}{d_o - f} = \frac{(-30\text{ cm})(15\text{ cm})}{15\text{ cm} - (-30\text{ cm})} = -10\text{ cm}$$

The negative signs d_i in the thin lens equation solution indicates that the image is virtual (the image is on the same side as the object). The negative sign for the focal length indicates that it is a divergent lens—the focal point is a virtual focus.

These distances are on the left-hand side of the lens and are opposite to the direction of the light rays.

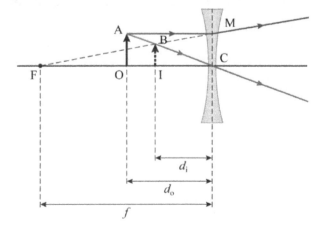

Copyright © 2015 by Nelson Education Ltd.

55. (a) For the image to be the largest possible the magnification of the camera lens has to be $|M| = 0.1$. The thin lens and magnification equations can be used to determine the object-lens distance d_o:

$$\frac{1}{d_o} + \frac{1}{d_i} = \frac{1}{f} \qquad (1)$$

$$|M| = \left| -\frac{d_i}{d_o} \right| = \frac{d_i}{d_o} \text{ (We assumed the case where } d_o > 0 \text{ and } d_i > 0) \qquad (2)$$

It follows that

$$\frac{1}{d_o} + \frac{1}{|M| d_o} = \frac{1}{f} \Rightarrow d_o = \frac{|M| + 1}{|M|} f = \frac{1.1}{0.1} (5 \text{ cm}) = 55 \text{ cm}$$

(b) The distance between the lens and the CCD sensor is equal to the lens-image distance d_i:

$$d_i = |M| d_o = 0.1 (55 \text{ cm}) = 5.5 \text{ cm}$$

57. (a) If $n = 1.52$ is the index of refraction of the glass prism, the refraction law on the left-hand side face of the prism and the ray geometry indicated in the diagram shown below yield the following equations:

$$\sin(i) = n \sin(r) \Rightarrow r = \sin^{-1} \left[\frac{\sin(i)}{n} \right] = \sin^{-1} \left(\frac{\sqrt{3}/2}{1.52} \right) = 34.7° \qquad (1)$$

$$i' + r \overset{eq.\ (1)}{=} 90° \Rightarrow i' = 90° - 34.7° = 55.3° \qquad (2)$$

The critical angle θ_{cr} for the glass-air boundary is

$$\theta_{cr} = \sin^{-1} \left(\frac{1}{1.52} \right) = 41.1°$$

Since the incident angle on the right-hand side face of the prism $i' \cong 55.3°$ is larger than the critical angle of the glass-air boundary, it follows that the light ray undergoes a total reflection at this interface and exits at the base of the prism.

(b) The angle between the emergent beam and the horizontal is denoted by α in the ray diagram. Using triangle $\triangle NOC$ the following relationship can be written:

$$\alpha = (90° - r') + 45° = 135° - r' = 135° - i' = 135° - 55.3° = 80°$$

(c) The light ray exits through the base of the prism, so the height of this point is zero.

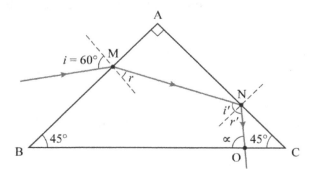

59. (a) The focal lengths f_1 and f_2 of the lenses with optical powers 4.0 D and 5.0 D, respectively, are

$$f_1 = \frac{1}{4} \text{ m} = 25 \text{ cm}$$

$$f_2 = \frac{1}{5} \text{ m} = 20 \text{ cm}$$

(b) The image-lens distance for the first mirror d_{i1} can be calculated from the thin lens equation:

$$d_{i1} = \frac{d_{o1} f_1}{d_{o1} - f_1} = \frac{(50 \text{ cm})(25 \text{ cm})}{50 \text{ cm} - 25 \text{ cm}} = 50 \text{ cm}$$

The image formed by the first lens is an object for the second lens:

$$d_{o2} = d - d_{i1} = 90 \text{ cm} - 50 \text{ cm} = 40 \text{ cm}$$

The image-lens distance for the second lens is

$$d_{i2} = \frac{d_{o2} f_2}{d_{o2} - f_2} = \frac{(40 \text{ cm})(20 \text{ cm})}{40 \text{ cm} - 20 \text{ cm}} = 40 \text{ cm}$$

Therefore, the image is formed 40 cm behind the second lens.

(c) No, the combined power of the two lenses cannot be used to solve this problem because the distance between them is large relatively to their focal lengths. The combined optical power describes a system where lenses are very close together.

61. Simplified notations for incident and refractive angles are given in the ray diagram shown below. Notice that $\theta_i = i$ in this new notation.

Using angle relationships in triangle MQN:

$$\delta = (i - r) + (r' - i') = i + r' - r - i' \tag{1}$$

Copyright © 2015 by Nelson Education Ltd.

The following angle relationships can be written:

$$\widehat{AMN} + \widehat{MNA} = (90° - r) + (90° - i') = 180° - \phi \Rightarrow r + i' = \phi \qquad (2)$$

The refraction law applied at the entry and exit boundaries of the prism gives

$$\sin(i) = n\sin(r) \Rightarrow r = \sin^{-1}\left[\frac{\sin(i)}{n}\right] \qquad (3)$$

$$n\sin(i') = \sin(r') \Rightarrow r' = \sin^{-1}\left[n\sin(i')\right] \overset{eq.\ (2)}{\Rightarrow} r' = \sin^{-1}\left[n\sin(\phi - r)\right] \qquad (4)$$

Using equations (1)–(4), the deflection angle δ is given by

$$\delta = i + r' - \phi = i - \phi + \left\{n\sin\left[\phi - \sin^{-1}\left(\frac{\sin(i)}{n}\right)\right]\right\} \qquad (5)$$

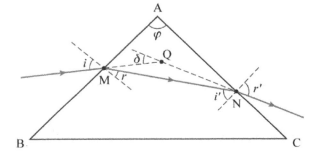

63. (a) The focal length of the convergent lens has to be equal to the sum of the distance between the lenses and the absolute value of the focal distance of the divergent lens as indicated in the ray diagram shown below. Hence, the distance between the lenses d is

$$d = |F_2| - |F_1| = 3F_1 - F_1 = 2F_1 \qquad (1)$$

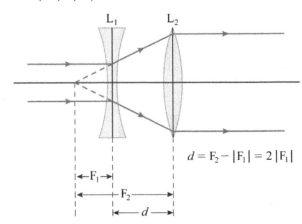

(b) The distance between the lenses is equal to the sum of the absolute values of the focal lengths of the two lenses as indicated in the ray diagram shown below.

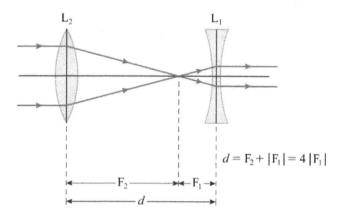

$$d = |F_2| + |F_1| = 3F_1 + F_1 = 4F_1$$

(2)

65. (a) The ray diagram is shown below.

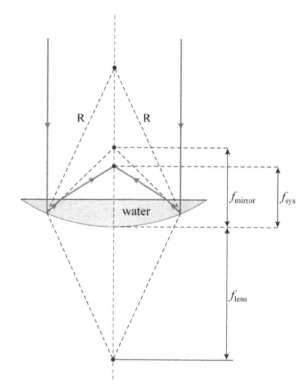

(b) The focal length of the mirror f_{mirror} is given by

$$f_{mirror} = \frac{R}{2} = 25 \text{ cm} \Rightarrow P_{mirror} = 4 \text{ D}$$

(1)

The focal length of the lens can be calculated by employing the lens maker's equation:

$$\frac{1}{f_{lens}} = (n_w - 1)\left(\frac{1}{R_1} - \frac{1}{R_2}\right) \tag{2}$$

The values of radii R_1 and R_2 are given by

$$R_1 = \infty \tag{3}$$

$$R_2 = -R = -50 \text{ cm} \tag{4}$$

Therefore,

$$f_{lens} = \frac{1}{1.33 - 1} \times 50 \text{ cm} = 1.5 \text{ m} \Rightarrow P_{lens} = 0.67 \, D$$

(c) The optical power of the system P_{sys} is equal to the sum of the optical powers of its components:

$$P_{sys} = P_{mirror} + P_{lens} = 4.0 \, D + 0.67 \, D = 4.7 \, D$$

67. The converging plano-convex lens produces a virtual image if $0 < d_o < f$.

The total time this condition is fulfilled t_{total} is the sum of the times before and after collision:

$$t_{total} = t_{before} + t_{after} \tag{1}$$

These times can be determined employing kinematical equations as follows:

$$t_{before} = \frac{f}{v_{marble}} = \frac{0.1 \text{ m}}{6 \text{ m/s}} = 0.017 \text{ s} = 17 \text{ ms} \tag{2}$$

$$t_{after} = \frac{f}{v_{after}} = \frac{f}{v'_{marble} + v'_{stand}} \tag{3}$$

Let us consider x axis to be horizontal and to be directed to the right. Conservation of momentum and energy yields the following equations, considering v is the velocity before the collision and v' is the velocity after the collision, M and m are the masses of the stand and the marble, respectively:

$$mv_{marble} = Mv'_{stand} + mv'_{marble} \tag{4}$$

$$\frac{mv_{marble}^2}{2} = \frac{mv'^2_{marble}}{2} + \frac{Mv'^2_{stand}}{2} \tag{5}$$

Solving equations (4) and (5) for the final velocities of the marble and the stand, the following results are obtained:

$$v'_{marble} = \frac{m-M}{M+m} v_{marble} = \frac{0.05 \text{ kg} - 0.4 \text{ kg}}{0.4 \text{ kg} - 0.05 \text{ kg}} (6 \text{ m/s}) = -\frac{14}{3} \text{ m/s} \qquad (6)$$

$$v'_{stand} = \frac{m}{M}(v_{marble} - v'_{marble}) = \frac{4}{3} \text{ m/s} \qquad (7)$$

Combining results of equations (6) and (7), we can calculate relative speed of the marble relatively to the stand before and after the collision:

$$\left|(v_{marble\text{-}stand})_{before}\right| = \left|v_{marble} - v_{stand}\right| = 6 \text{ m/s} - 0 \text{ m/s} = 6 \text{ m/s}$$

$$\left|(v_{marble\text{-}stand})_{after}\right| = \left|v'_{marble} - v'_{stand}\right| = \left|-\frac{14}{3} \text{ m/s} - \frac{4}{3} \text{ m/s}\right| = \frac{18}{3} \text{ m/s} = 6 \text{ m/s} \qquad (8)$$

The result of equation (8) indicates that the relative speed between the stand and the marble does not change after the elastic collision. Therefore,

$$t_{total} = 2(t_{before}) = 2\left(\frac{0.1 \text{ m}}{6 \text{ m/s}}\right) = 0.033 \text{ s} = 33 \text{ ms}$$

69. Let f_1 and f_2 be the focal lengths characterizing the two lenses. Since these are two thin lenses located very close to each other, we assume that the distance between them is negligible. For simplicity, let us assume that lens 1 is a convergent lens: $f_1 > 0$. Let us place an object in front of lens 1, at a distance larger than its focal length. According to Table 27-3 in the textbook, the image created by the first lens will be a real image, located on the other side of the lens. The location of this image can be found using the lens equation:

$$\frac{1}{f_1} = \frac{1}{d_{o1}} + \frac{1}{d_{i1}} \Rightarrow \frac{1}{d_{i1}} = \frac{1}{f_1} - \frac{1}{d_{o1}} \qquad (1)$$

The image created by the first lens will become an object for the second lens. However, the object-lens distance for the second lens will be negative as was explained in question 68 (b and c). Since the lenses are thin and the distance between them is negligible, we can write:

$$\frac{1}{f_2} = \frac{1}{d_{o2}} + \frac{1}{d_{i2}} \Rightarrow \frac{1}{d_{i2}} = \frac{1}{f_2} - \frac{1}{d_{o2}} \qquad (2)$$

Using the result obtained in equation (1) and remembering that $d_{o2} \equiv -d_{i1}$ we find:

$$\frac{1}{d_{i2}} = \frac{1}{f_2} + \frac{1}{d_{i1}} \Rightarrow \frac{1}{d_{i2}} = \frac{1}{f_2} + \frac{1}{f_1} - \frac{1}{d_{o1}}$$

Let us denote $\dfrac{1}{f_1} + \dfrac{1}{f_2} \equiv \dfrac{1}{f_{pair}}$ This is equivalent to $P_1 + P_2 = P_{pair}$

$$\dfrac{1}{d_{i2}} = \dfrac{1}{f_{pair}} - \dfrac{1}{d_{o1}} \Rightarrow \dfrac{1}{f_{pair}} = \dfrac{1}{d_{o1}} + \dfrac{1}{d_{i2}}$$

This derivation emphasizes the usefulness of the physical concept of optical power, P, measured in dioptres (D). The shorter the focal length, the more powerful the lens is and the optical power of two think lenses located very close to tech other equals to the sum of their optical powers. The optical power is also used by optometrists and ophthalmologists (eye doctors) who prescribe eye glasses and lenses.

$$P \equiv \dfrac{1}{f} \Rightarrow P_{pair} = P_1 + P_2$$

71. The condition to have the image of the hook on the floor is

$$d_o + d_i = H = 3 \text{ m} \qquad (1)$$

The thin lens equation can be written as

$$\dfrac{1}{f} = \dfrac{1}{d_o} + \dfrac{1}{d_i} \qquad (2)$$

Combining equations (1) and (2),

$$d_o^2 - Hd_o + Hf = 0 \qquad (3)$$

One of the solutions of quadratic equation (3) is not acceptable because it gives

$d_o > H$ which doesn't make sense. The remaining solution is

$$d_o = \dfrac{H - \sqrt{H^2 - 4Hf}}{2} \qquad (4)$$

The geometry and the balancing of forces yield the following equations as indicated in the diagram shown on the next page:

$$d_o = l\cos(\theta) \qquad (5)$$

$$R = l\sin(\theta) \qquad (6)$$

$$\tan(\theta) = \dfrac{m\omega^2 R}{mg} \stackrel{eq.\ (6)}{\Rightarrow} \tan(\theta) = \dfrac{\omega^2 l\sin(\theta)}{g} \Rightarrow \cos(\theta) = \dfrac{g}{\omega^2 l} \qquad (7)$$

Combining equations (5) and (7), the following result is obtained:

$$d_o = \dfrac{g}{\omega^2} \Rightarrow \omega = \sqrt{\dfrac{g}{d_o}} \qquad (8)$$

Using the result of equation (4),

$$\omega = \sqrt{\frac{2g}{H - \sqrt{H^2 - 4Hf}}} = \sqrt{\frac{2(9.81 \text{ m/s}^2)}{3 \text{ m} - \sqrt{(3 \text{ m})^2 - 4(3 \text{ m})(0.15 \text{ m})}}} = 8 \text{ rad/s}$$

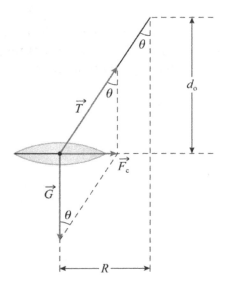

73. (a) From the ray diagram geometry of this optical system shown on the next page:

$$d_{i2} = -L/2 = -40 \text{ cm}/2 = -20 \text{ cm} \tag{1}$$

Using the thin lens equation for the second lens (L_2):

$$\frac{1}{f_2} = \frac{1}{d_{o2}} + \frac{1}{d_{i2}} \overset{eq. (1)}{\Rightarrow} d_{o2} = \frac{d_{i2}f_2}{d_{i2} - f_2} = \frac{(-20 \text{ cm})(30 \text{ cm})}{-20 \text{ cm} - 30 \text{ cm}} = 12 \text{ cm} \tag{2}$$

From the ray diagram geometry:

$$d_{i1} = L - d_{o2} = 40 \text{ cm} - 12 \text{ cm} = 28 \text{ cm} \tag{3}$$

Using the thin lens equation for the first lens (L_1):

$$\frac{1}{f_1} = \frac{1}{d_{o1}} + \frac{1}{d_{i1}} \Rightarrow d_{o1} = \frac{d_{i1}f_1}{d_{i1} - f_1} = \frac{(28 \text{ cm})(5 \text{ cm})}{28 \text{ cm} - 5 \text{ cm}} = \frac{140}{23} \text{ cm} = 6.1 \text{ cm} \tag{4}$$

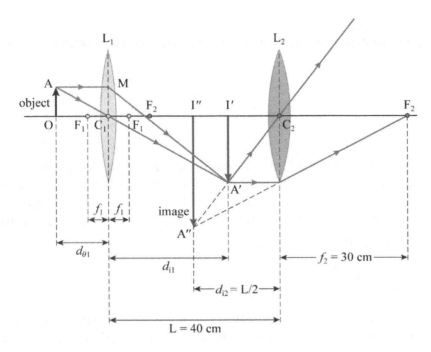

(b) The magnification of the optical system M is given by

$$M = M_1 M_2 = \frac{d_{i1}}{d_{o1}} \cdot \frac{d_{i2}}{d_{o2}} = -\frac{23}{3} = -7.7$$

(c) Yes, the compound microscope can produce an upright final image if the image produced by the first lens (L_1) is between the two lenses. That is,

$$f_1 < d_{i1} < L - f_2 \Rightarrow 5 \text{ cm} < d_{i1} < 10 \text{ cm}$$

75. (a) No, it is not possible to confine all the light from a large source into a significantly reduced image of the light source. In other words, this would mean that all the light of a source would have to be confined in a smaller volume. This would correspond to a decrease in entropy of an isolated system (no light loss or gain), which contradicts the second law of thermodynamics.

(b) Only an ideal system would project all the light from a smaller source into a magnified image. Such system does not exist due to light-matter interactions; there are no perfect mirrors and lenses. Real mirrors and lenses do not reflect or transmit, respectively, 100% of the incident light.

77. Answers may vary.

1. d

3. c. The constructive interference condition for the Michelson interferometer is

$$m\lambda = 2d \qquad (1)$$

In the equation (1), λ denotes the wavelength of the light used in the interference experiment, d is the distance the mirror was moved, and m is the number of constructive interference maxima or bright fringes. If monochromatic light with two wavelengths $\lambda_1 = 645\,\text{nm}$ and $\lambda_2 = 430\,\text{nm}$ are used separately, but in the same interference condition (the same value of d), then equation (1) leads to

$$m_1\lambda_1 = m_2\lambda_2 \Rightarrow m_2 = m_1 \frac{\lambda_1}{\lambda_2} = 64 \times \frac{645\,\text{nm}}{430\,\text{nm}} = 96$$

5. b. Using the constructive interference condition for the double-slit experiment and the small-angle approximation, it can be determined that the angle separation between the central spot and the first-order bright fringe on each side is approximately given by (see also equations (3) and (4) from the solution of question 4):

$$\theta \approx \frac{\lambda}{d}$$

Hence, for two different separations between the two slits d_1 and d_2, and the same wavelength λ, one can write that

$$\frac{\theta_1}{\theta_2} = \frac{d_2}{d_1} \Rightarrow d_2 = d_1 \frac{\theta_1}{\theta_2} = s\frac{1.2°}{2.4°} = \frac{s}{2}$$

7. d. The constructive interference condition for a diffraction grating is

$$d\sin(\theta) = m\lambda$$

Therefore, since d is constant, the bright fringes which are nearest to the central white spot correspond to the smallest wavelength λ. For the stellar visible light spectrum, the minimum and maximum wavelength values correspond to the colour violet and red, respectively.

9. b. The destructive interference condition for single-slit diffraction is

$$w\sin(\theta) = m\lambda$$

Therefore, for a constant slit width w, a larger wavelength λ will yield a larger angle θ satisfying the equation above; hence a more spread out diffraction pattern in both directions.

11. b. The first order of constructive interference for a light beam of wavelength λ in vacuum normally incident on the soap film with index of refraction n is

$$2t = \frac{\lambda}{2n} \Rightarrow \lambda = 4nt$$

Hence, as the soap film evaporates its thickness t decreases, and so does the wavelength of the light λ for which the interference condition holds.

13. b. The resolution improves if an electromagnetic wave with smaller wavelength λ is used.

15. Geometric optics can be used in cases where the electromagnetic radiation wavelength λ is much smaller than the size of the object to be investigated. The smallest dimension concerned in this case is 1 mm.

   (a) $\lambda = 532 \text{ nm} = 0.532 \times 10^{-3} \text{ mm} \ll 1 \text{ mm}$

   Geometric optics can be used.

   (b) $\lambda = 5 \mu\text{m} = 5 \times 10^{-3} \text{ mm} \ll 1 \text{mm}$

   Geometric optics can be used.

   (c) $\lambda = 180 \text{ nm} = 0.18 \times 10^{-3} \text{ mm} \ll 1 \text{ mm}$

   Geometric optics can be used.

   (d) $\lambda = \dfrac{c}{f} = \dfrac{3 \times 10^8 \text{ m/s}}{88 \times 10^6 \text{ Hz}} = 3.41 \text{ m} \gg 1 \text{ mm}$

   Geometric optics cannot be used.

17. (a) The phase difference between the two radio waves travelling the same distance is zero given that the initial phase difference between the two antennas is zero. Therefore, there is constructive interference which can be observed by a detector positioned at the halfway point between the two antennas.

   (b) The destructive interference occurs when the phase difference is equal to 180°. This corresponds to a wavelength difference equal to odd multiples of $\lambda/2$. If the detector is moved by a distance equal to $\lambda/4$ to the left or to the right, then the phase shift between the two detectors will be 180°, corresponding to destructive interference condition. The minimum distance d_{min} required is then equal to

   $$d_{min} = \frac{\lambda}{4} = \frac{c}{4f} = \frac{3 \times 10^8 \text{ m/s}}{4 \times 1500 \times 10^3 \text{ Hz}} = 50 \text{ m}$$

19. (a) The constructive interference condition for the double-slit interferometer is

   $$d \sin(\theta) = m\lambda$$

The first-order bright interference fringe occurs for $m = 1$ in the above equation. Hence,

$$\theta_1 = \sin^{-1}\left(\frac{\lambda}{d}\right) = \sin^{-1}\left(\frac{633 \times 10^{-9}\,\text{m}}{25 \times 10^{-5}\,\text{m}}\right) = \sin^{-1}(0.002532) = 0.145°$$

(b) From the geometry of the double-slit experiment shown in Figure 28-8 from the textbook, it follows that the distance between the first-order bright fringes denoted by d_1 is

$$d_1 = 2D\tan(\theta_1) \cong 2 \times 45\,\text{cm} \times \tan(0.145°) = 0.228\,\text{cm} = 2.28\,\text{mm}$$

21. (a) The distance from the centre of the $m = 1$ image $d_1 = 2.00$ cm is given by:

$$d_1 = D\tan(\theta_1)$$

In the equation written above D denotes the distance between the diffraction grating and screen. It follows that

$$\theta_1 = \tan^{-1}\left(\frac{d_1}{D}\right) = \tan^{-1}\left(\frac{2 \times 10^{-2}\,m}{0.750\,m}\right) \cong 1.53°$$

The number of lines per centimetre N can be calculated from the constructive interference condition:

$$\frac{1}{N}\sin(\theta_1) = \lambda \Rightarrow N = \frac{\sin(\theta_1)}{\lambda} = \frac{\sin(1.53°)}{700 \times 10^{-9}\,\text{m}} = 38082\,\text{m}^{-1} = 381\,\text{cm}^{-1}$$

(b) The distance d_1 for $\lambda = 400$ nm is equal to

$$d_1 = D\tan(\theta_1) \approx D\theta_1 \approx DN\lambda = 0.75\,\text{m} \times 38143\,\text{m}^{-1} \times 400 \times 10^{-9}\,\text{m} = 1.14\,\text{cm}$$

23. (a) Assuming a hard reflection on both interfaces by the antireflection film, the minimum film thickness t can be calculated by using the destructive interference condition:

$$t = \frac{\lambda}{4n} = \frac{1250 \times 10^{-9}\,\text{m}}{4 \times 1.38} = 226\,\text{nm}$$

(b) The constructive interference condition is

$$2t = m\frac{\lambda}{n} \Rightarrow \lambda = \frac{2nt}{m}$$

For the first-order of the constructive interference ($m = 1$), the above equation gives that

$$\lambda = 2 \times 1.38 \times 226\,\text{nm} = 624\,\text{nm}$$

25. The width of the slit w can be calculated using the destructive interference condition for a single-slit diffraction is

$$w \sin(\theta) = m\lambda \Rightarrow w = \frac{m\lambda}{\sin(\theta)} = \frac{1 \times 589 \times 10^{-9} \text{ m}}{\sin(1.15°)} = 29.3 \, \mu\text{m}$$

27. Only bright fringes within the central maximum corresponding to the single-slit diffraction pattern are counted. The first-order destructive interference condition for the single-slit diffraction gives

$$\sin(\theta) = \frac{\lambda}{w} \tag{1}$$

The constructive interference condition for the double-slit experiment gives:

$$\sin(\theta) = \frac{m\lambda}{d} \tag{2}$$

From equations (1) and (2), it follows that

$$\frac{\lambda}{w} = \frac{m\lambda}{d} \Rightarrow m = \frac{d}{w} = \frac{0.125 \text{ mm}}{0.0150 \text{ mm}} = 8$$

Eight bright fringes can be seen on each side. Therefore, including the central peak, $2 \times 8 + 1 = 17$ bright fringes are visible.

29. Using the Rayleigh criterion, the smallest angle θ_r separating two points that can be resolved with light of wavelength λ and using a circular aperture of diameter D is

$$\sin(\theta_r) = 1.22 \frac{\lambda}{D}$$

Assuming that the average light wavelength the eye is most sensitive to is $\lambda = 550$ nm and $D = 2.85$ mm, then

$$\theta_r = \sin^{-1}\left(1.22 \times \frac{550 \times 10^{-9} \text{ m}}{2.85 \times 10^{-3} \text{ m}}\right) = 0.0135°$$

If the distance between the observer and the car is l and the distance between the lights is $d = 1.45$ m, the following geometrical relationship is valid:

$$\tan\left(\frac{\theta_r}{2}\right) = \frac{d/2}{l} \Rightarrow l = \frac{d}{2 \tan\left(\frac{\theta_r}{2}\right)} = \frac{1.45 \text{ m}}{2 \times \tan(0.0135°/2)} = 6.16 \text{ km}$$

31. The bright fringes counted when going from evacuated to containing the gas are due to the phase difference between the two arms of the Michelson interferometer that is induced by the wavelength-reducing shift (λ/n) due to the light passing twice through the arm containing the gas with index of refraction n. If the length of each arm is denoted by l, this

means that the equivalent length of the arm filled with gas is equal to nl. Given that the only phase shift is due to the presence of the gas, the constructive interference condition is

$$m\lambda = 2l(n-1)$$

It follows that

$$n = 1 + \frac{m\lambda}{2l} = 1 + \frac{205 \times 633 \times 10^{-9} \text{ m}}{2 \times 0.365 \text{ m}} = 1.00018$$

33. The constructive interference condition for the two-slit experiment using light of wavelength λ and the distance d between the two slits, is

$$d\sin(\theta) = m\lambda$$

Based on the equation written above, the angular separation θ_1 between $m=0$ and $m=1$ spots is given by

$$\theta_1 = \sin^{-1}\left(\frac{\lambda}{d}\right) = \sin^{-1}\left(\frac{264 \times 10^{-9} \text{ m}}{0.1 \times 10^{-3} \text{ m}}\right) = 0.151°$$

35. If the angle between one $m=2$ maximum and the center is denoted by θ_2 then the angle between the two $m=2$ maxima is equal to $2\theta_2 = 1.30°$. Hence, $\theta_2 = 0.65°$. The spacing between the slits d can be calculated from the constructive interference condition:

$$d = \frac{2\lambda}{\sin(\theta_2)} = \frac{2 \times 480 \times 10^{-9} \text{ m}}{\sin(0.65°)} = 0.0846 \text{ mm} = 84.6\,\mu\text{m}$$

37. (a) Let θ_1 and θ_2 denote the deflection angles for the first-order interference fringes corresponding to the wavelengths $\lambda_1 = 350$ nm and $\lambda_2 = 1000$ nm, respectively. Using the constructive interference condition for the diffraction grating with N lines per unit length, one can calculate angles θ_1 and θ_2 as follows:

$$\theta_1 = \sin^{-1}(N\lambda_1) = \sin^{-1}\left[600 \times (10^{-3} \text{ m})^{-1} \times 350 \times 10^{-9} \text{ m}\right] = 12.1°$$

$$\theta_2 = \sin^{-1}(N\lambda_2) = \sin^{-1}\left[600 \times (10^{-3} \text{ m})^{-1} \times 1000 \times 10^{-9} \text{ m}\right] = 36.9°$$

(b) The effective width of the detector w is equal to

$$w = 2048 \text{ pixels} \times 14 \frac{\mu\text{m}}{\text{pixel}} = 28672\,\mu\text{m} = 28.7 \text{ mm}$$

(c) Given the detector width w calculated in part (b) and the grating-screen distance denoted by D, the following geometric relationship is valid:

$$w = D\left[\tan(\theta_2) - \tan(\theta_1)\right]$$

Hence, the distance D, in millimetres, is equal to

$$D = \frac{w}{\tan(\theta_2) - \tan(\theta_1)} = \frac{28.7 \text{ mm}}{\tan(36.9°) - \tan(12.1°)} = 53.5 \text{ mm}$$

39. The constructive interference condition for the reflected light rays between the two glass plates separated by distance t is given by

$$2t = (2k+1)\frac{\lambda}{2}, \ k = 0,1,2,\cdots \tag{1}$$

In the previous equation, a hard reflection of light on the glass plate equivalent to a 180° phase shift was assumed. Equation (1) can also be written as

$$t = \left(m - \frac{1}{2}\right)\frac{\lambda}{2}, \ m = 1,2,3,\cdots \tag{2}$$

The maximum distance between the glass plates is $t_{max} = 75 \,\mu m$. Hence, using equation (2), the maximum number of bright fringes that can be observed will be given by

$$m = \frac{1}{2} + \frac{2t_{max}}{\lambda} = \frac{1}{2} + \frac{2 \times 75 \times 10^{-6} \text{ m}}{532 \times 10^{-9} \text{ m}} = 282 \tag{3}$$

41. (a) The Rayleigh criterion for a circular aperture of diameter D and light wavelength λ gives the angular resolution θ_r as

$$\theta_r = \sin^{-1}\left(1.22\frac{\lambda}{D}\right)$$

Assuming $\lambda = 550 \,nm$, the value of θ_r for the France-Canada-Hawaii Telescope is

$$\theta_r = \sin^{-1}\left(1.22 \times \frac{550 \times 10^{-9} \text{ m}}{3.6 \text{ m}}\right) = 0.0000107°$$

(b) $\theta_r = 0.0000107° = 0.0000107° \times 3600\frac{1''}{1°} = 0.0385''$

43. The array approximates a telescope with a much larger effective diameter D. In this case, the diameter can be approximated as being equal to the distance between the two telescopes: $D \approx 3000 \,km$. The angular resolution θ_r given by the Rayleigh criterion is

$$\theta_r = \sin^{-1}\left(\frac{\lambda}{D}\right) = \sin^{-1}\left(\frac{c}{Df}\right) = \sin^{-1}\left(\frac{3 \times 10^8 \text{ m/s}}{3 \times 10^6 \text{ m} \times 1.42 \times 10^9 \text{ Hz}}\right) = \left(4 \times 10^{-6}\right)°$$

45. The Rayleigh criterion relating the angular resolution θ_r for a circular aperture of diameter D with the radiation wavelength λ is

$$\sin(\theta_r) = 1.22 \frac{\lambda}{D}$$

It follows that

$$D = \frac{1.22\lambda}{\sin(\theta_r)} = \frac{1.22 \times 500 \times 10^{-9}\,\text{m}}{\sin\left[(1.5/3600)^\circ\right]} = 0.0839\,\text{m} = 8.4\,\text{cm}$$

47. The destructive interference condition for the diffraction on the slit of width w is

$$w \sin(\theta) = m\lambda \tag{1}$$

Therefore, the angle θ_1 of deflection corresponding to the first dark fringe ($m = 1$) is given by

$$\theta_1 = \sin^{-1}\left(\frac{\lambda}{w}\right) \tag{2}$$

We can now specify the condition that the first three bright fringes originated from the two-slit interference pattern within the angle θ_1 on each of the two sides away from the centre:

$$d \sin(\theta_1) = 3\lambda \tag{3}$$

Using equations (1) and (2), the following equation is obtained:

$$d \frac{\lambda}{w} = 3\lambda \Rightarrow w = \frac{d}{3} = \frac{0.125\,\text{mm}}{3} = 41.7\,\mu\text{m}$$

49. Answers may vary.

1. c

3. c

5. d. Since $r_s = \dfrac{2Gm}{c^2}$, doubling the mass would double the Schwarzschild radius.

7. c. Space-time curvature is greatest when m / r^2 is greatest.

9. d

11. c. $v = \dfrac{0.5c + 0.5c}{1 + \dfrac{(0.5c)(0.5c)}{c^2}} = 0.8c$

13. c

15. We know that $c = \left(\dfrac{1}{\varepsilon_0 \mu_0}\right)^{1/2}$

$$\Rightarrow \mu_0 = \dfrac{1}{\varepsilon_0 c^2} = \dfrac{1}{(8.85 \times 10^{-12})(3.00 \times 10^8)^2} = 1.26 \times 10^{-6} \ \text{N/A}^2$$

17. Total distance $x = 2\pi\left(R_{earth} + 10^6\right) = 2\pi\left(6.37 \times 10^6 + 10^6\right) = 4.63 \times 10^7 \ \text{m}$

(a) $t = \dfrac{x}{v} = \dfrac{x}{0.5c} = \dfrac{4.63 \times 10^7}{(0.5)(3.00 \times 10^8)} = 0.31 \ \text{s}$

(b) $t = \dfrac{x}{v}\sqrt{1 - \dfrac{v^2}{c^2}} = \dfrac{x}{0.5c}\sqrt{1 - \dfrac{(0.5c)^2}{c^2}} = \dfrac{4.63 \times 10^7}{(0.5)(3.00 \times 10^8)}\sqrt{1 - 0.25} = 0.27 \ \text{s}$

19. $x = 4.40 \ \text{ly} = (4.4)(9.47 \times 10^{15}) = 4.17 \times 10^{16} \ \text{m}$

(a) $t = \dfrac{x}{v} = \dfrac{4.17 \times 10^{16}}{(0.75)(3.00 \times 10^8)} = 1.85 \times 10^8 \ \text{s}$

(b) $t = \dfrac{x}{v}\sqrt{1 - \dfrac{v^2}{c^2}} = \dfrac{4.17 \times 10^{16}}{(0.75)(3.00 \times 10^8)}\sqrt{1 - \dfrac{(0.75c)^2}{c^2}} = 1.22 \times 10^8 \ \text{s}$

21. Relativistic kinetic energy is given by

$$K = m_0 c^2 \left[\frac{1}{\sqrt{1 - \dfrac{v^2}{c^2}}} - 1 \right]$$

This can be rearranged to give

$$v = c \left[1 - \left(\frac{m_0 c^2}{m_0 c^2 + K} \right)^2 \right]^{1/2}$$

$$\Rightarrow v = 3.00 \times 10^8 \left[1 - \left(\frac{\left(1.67 \times 10^{-27}\right)\left(3.00 \times 10^8\right)^2}{\left(1.67 \times 10^{-27}\right)\left(3.00 \times 10^8\right)^2 + 75} \right)^2 \right]^{1/2} = 3.0 \times 10^8 \text{ m/s}$$

The speed is not quite equal to the speed of light, but the difference is extremely small.

23. $n \rightarrow p + e^- + \nu_e$

Conservation of total energy with the assumption of zero rest mass of neutrino gives:

$$m_n c^2 = m_p c^2 + \frac{1}{2} m_p v_p^2 + m_e c^2 + \frac{1}{2} m_e v_e^2$$

$$\Rightarrow \frac{1}{2} m_p v_p^2 + \frac{1}{2} m_e v_e^2 = m_n c^2 - m_p c^2 - m_e c^2$$

The left side of this equation represents the total kinetic energy released.

$$\Rightarrow K_{released} = \left(m_n - m_p - m_e \right) c^2$$

$$\Rightarrow K_{released} = \left(1.674927 \times 10^{-27} - 1.672622 \times 10^{-27} - 9.10938 \times 10^{-31} \right) \left(3.00 \times 10^8 \right)^2$$

$$\Rightarrow K_{released} = 1.2547 \times 10^{-13} \text{ J}$$

$$\Rightarrow K_{released} = \frac{1.2547 \times 10^{-13}}{1.6022 \times 10^{-13}} = 0.7831 \text{ MeV}$$

25. For Mei the time interval is zero, that is

$$\Delta t_{Mei} = 0$$

Mei sees two events separated by 300 m in space, that is

$$\left(\Delta x^2 + \Delta y^2 + \Delta z^2 \right)_{Mei} = 300^2$$

Li sees the same events separated by 500 m in space, that is

$$\left(\Delta x^2 + \Delta y^2 + \Delta z^2 \right)_{Li} = 500^2$$

Since the spacetime interval for the two observers must be the same,

$$\left(c\Delta t_{Mei}\right)^2 - \left(\Delta x^2 + \Delta y^2 + \Delta z^2\right)_{Mei} = \left(c\Delta t_{Li}\right)^2 - \left(\Delta x^2 + \Delta y^2 + \Delta z^2\right)_{Li}$$

$$\Rightarrow \left(c\Delta t_{Li}\right)^2 = 0 - 300^2 + 500^2 = 160\ 000$$

$$\Delta t_{Li} = \frac{\sqrt{160\ 000}}{3\times 10^8}$$

$$\Delta t_{Li} = \frac{400}{3\times 10^8}$$

$$\Delta t_{Li} = 1.33\times 10^{-6}\ \text{s}$$

$$\Delta t_{Li} = 1.33\ \mu\text{s}$$

27. Radius of the event horizon can be calculated from

$$R_0 = \frac{2GM}{c^2}$$

Using the given mass of $M = 12M_\odot = (12)\left(1.989\times 10^{30}\right) = 2.387\times 10^{31}$ kg

$$\Rightarrow R_0 = \frac{(2)\left(6.672\times 10^{-11}\right)\left(2.387\times 10^{31}\right)}{\left(3.00\times 10^8\right)^2} = 35391\ \text{m} = 35.4\ \text{km}$$

29. A light year corresponds to the distance travelled by light in one year.

$$\Rightarrow x = ct = \left(3.00\times 10^8\right)\left(1\times 365.25\times 24\times 3600\right) = 9.47\times 10^{15}\ \text{m}$$

31. (a) Since radio signals travel with speed of light, the total time will be

$$t = 2\times\left(25.3\ \text{ly}\right) = 2\times\left(25.3\times 365.25\times 24\times 3600\right) = 1.60\times 10^9\ \text{s} = 50.6\ \text{y}$$

(b) The total round trip distance is given by

$$x_0 = 2c\times\left(25.3\ \text{ly}\right) = 2\left(3.00\times 10^8\right)\left(25.3\times 365.25\times 24\times 3600\right) = 4.79\times 10^{17}\ \text{m}$$

The time taken by the ship in Earth's frame of reference is

$$t = \frac{x_0}{v} = \frac{4.79\times 10^{17}}{\left(0.985\right)\left(3.00\times 10^8\right)} = 1.62\times 10^9\ \text{s (or 51.4 yr)}$$

(c) For an observer on the spaceship, the distance seems contracted, that is

$$x = x_0\sqrt{1 - \frac{v^2}{c^2}}$$

$$\Rightarrow vt = x_0\sqrt{1 - \frac{v^2}{c^2}}$$

$$\Rightarrow t = \frac{x_0}{v}\sqrt{1 - \frac{v^2}{c^2}} = \frac{4.79 \times 10^{17}}{(0.985)(3.00 \times 10^8)}\sqrt{1 - \frac{(0.985)^2 c^2}{c^2}}$$

$$\Rightarrow t = 2.80 \times 10^8 \text{ s (or 8.86 yr)}$$

33. (a) The time for the passengers in the transit line will be dilated according to

$$t_0 = t\sqrt{1 - \frac{v^2}{c^2}}$$

This can be rearranged for the speed of the transit line as

$$v = c\sqrt{1 - \frac{t_0^2}{t^2}}.$$

It is given that the time for the passengers is half the time for observers on

the stations, that is $t_0 = \frac{t}{2}$.

$$\Rightarrow v = (3.00 \times 10^8)\sqrt{1 - \frac{t^2}{(2t)^2}} = 2.60 \times 10^8 \text{ m/s}$$

(b) $t = \frac{x}{v} = \frac{6.0 \times 10^6}{2.60 \times 10^8} = 0.023 \text{ s}$

35. An electron accelerated through a potential difference of V volts gains a kinetic energy of V eV. Therefore, in this case the kinetic energy of the electron is

$$K = 32 \text{ keV} = (32000)(1.602 \times 10^{-19}) = 5.126 \times 10^{-15} \text{ J}$$

(a) $K = \frac{1}{2}mv^2$

$$\Rightarrow v = \sqrt{\frac{2K}{m}} = \sqrt{\frac{(2)(5.126 \times 10^{-15})}{9.109 \times 10^{-31}}} = 1.061 \times 10^8 \text{ m/s}$$

(b) The relativistic kinetic energy is given by

$$K = m_0 c^2 \left(\frac{1}{\sqrt{1 - \dfrac{v^2}{c^2}}} - 1 \right)$$

This can be rearranged to give

$$v = c \sqrt{1 - \left(\frac{m_0 c^2}{m_0 c^2 + K} \right)^2}$$

$$\Rightarrow v = \left(3.00 \times 10^8\right) \sqrt{1 - \left(\frac{\left(9.109 \times 10^{-31}\right)\left(3.00 \times 10^8\right)^2}{\left(9.109 \times 10^{-31}\right)\left(3.00 \times 10^8\right)^2 + 5.126 \times 10^{-15}} \right)^2}$$

$$= 1.014 \times 10^8 \text{ m/s}$$

(c) The relative difference between the two velocities is

$$\varepsilon = \frac{1.061 \times 10^8 - 1.014 \times 10^8}{1.014 \times 10^8} \times 100 = 4.6\%$$

For precision measurements this difference may be significant. However, for CRT electronics it is not significant especially because the electrons travel a very short distance.

37. No, this is not in conflict with special relativity. Neutrinos, having very small interaction probabilities, can pass through dense matter more freely than photons. Therefore, they leave the supernova before the photons. The photons leave with the shockwave and reach Earth a few hours after the neutrinos. While interstellar space is almost empty, the speed of light is not quite that of a vacuum, providing a slight slowing of photon arrival time.

39. (a) In one second, the Sun radiates energy $E = 3.85 \times 10^{26}$ J. We can calculated the mass equivalent of this energy using $E = mc^2$.

$$\Rightarrow m = \frac{E}{c^2} = \frac{3.85 \times 10^{26}}{\left(3.00 \times 10^8\right)^2} = 4.28 \times 10^9 \text{ kg}$$

Hence, the Sun converts a mass of 4.28×10^9 kg per second.

(b) We just calculated the mass being converted by the Sun each second. This can be used to calculate the mass of hydrogen being converted each year.

$$\Rightarrow m_H = \left(4.28 \times 10^9\right)(3600)(24)(365.25) = 1.35 \times 10^{17} \text{ kg/y}$$

Let us assume that the total energy is being produced by the fusion of four hydrogen atoms alone. Since the mass calculated previously is only 0.66% of the mass of hydrogen atoms, the total mass of hydrogen converted each year is

$$m'_H = \frac{1.35 \times 10^{17}}{0.0066} = 2.05 \times 10^{19} \text{ kg/y}$$

(c) The total mass of hydrogen in the Sun available for fusion is

$$m_{H,total} = (2.0 \times 10^{30})(0.1) = 2.0 \times 10^{29} \text{ kg}$$

We calculated above the total amount of hydrogen being burned per year as 2.05×10^{19} kg.

Hence the number of years the hydrogen will last is given by

$$t = \frac{2.0 \times 10^{29}}{2.05 \times 10^{19}} = 9.78 \times 10^9 \text{ y} = 9.78 \text{ billion years}$$

41. Since the spacetime separation must be equal for both observers, that is

$$(c\Delta t)_1^2 - (\Delta x)_1^2 - (\Delta y)_1^2 - (\Delta z)_1^2 = (c\Delta t)_2^2 - (\Delta x)_2^2 - (\Delta y)_2^2 - (\Delta z)_2^2$$

The first observer has a Δx of 3.0 m and a Δt of 1.0×10^{-8} s.

The second observer has Δt of 0 s.

Solving, we find that Δx for the second observer is 0 m.

43. (a) The first observer sees the two events separated by 4.0 m in space and 20.0 ns in time, that is

$$(\Delta x^2 + \Delta y^2 + \Delta z^2)_1 = 4^2 \text{ and } \Delta t_1 = 20 \times 10^{-9} \text{ s}$$

The second observer sees the same events at the same location in space, that is

$$(\Delta x^2 + \Delta y^2 + \Delta z^2)_2 = 0$$

Since the spacetime interval for the two observers must be the same,

$$(c\Delta t_2)^2 - (\Delta x^2 + \Delta y^2 + \Delta z^2)_2 = (c\Delta t_1)^2 - (\Delta x^2 + \Delta y^2 + \Delta z^2)_1$$

$$\Rightarrow (c\Delta t_2)^2 = (3.0 \times 10^8 \times 20 \times 10^{-9})^2 - 4^2 = 20$$

$$\Rightarrow \Delta t_2 = \frac{\sqrt{20}}{3.0 \times 10^8} = 1.49 \times 10^{-8} \text{ s} = 14.9 \text{ ns}$$

(b) Since time is dilated, we can use the formula

$$\Delta t_1 = \frac{\Delta t_2}{\sqrt{1-\dfrac{v^2}{c^2}}}$$

$$\Rightarrow v = c\sqrt{1-\left(\frac{\Delta t_2}{\Delta t_1}\right)^2} = c\sqrt{1-\left(\frac{14.9}{20.0}\right)^2} = \frac{2}{3}c = 2.00\times10^8 \text{ m/s}$$

45. Let v_s and v_m be the speeds of spacecraft and meteorite with respect to the home planet. Then the speed of the meteorite relative to the spaceship is given by

$$v_{rel} = \frac{v_s+v_m}{1+\dfrac{v_s v_m}{c^2}}$$

$$\Rightarrow v_{rel} = \frac{0.75c+0.5c}{1+\dfrac{(0.75c)(0.5c)}{c^2}} = 0.91c$$

For an observer on the spaceship, the meteorite will seem to be travelling at this speed. The time taken by the meteorite to pass the spaceship is then given by

$$t = \frac{L}{v_{rel}} = \frac{250}{(0.91)(3.00\times10^8)} = 9.17\times10^{-7} \text{ s} = 0.917 \text{ } \mu s$$

47. The velocity of the object with respect to C is given as

$$v_{obj} = 0.3c$$

C itself is moving with a velocity of $v_C = 0.4c$ with respect to B, which is moving in the same direction with a velocity of $v_B = 0.5c$ with respect to A.
Therefore the velocity of C with respect to A is given by

$$v_{rel} = \frac{v_B+v_C}{1+\dfrac{v_B v_C}{c^2}} = \frac{0.4c+0.5c}{1+\dfrac{(0.4c)(0.5c)}{c^2}} = 0.75c$$

Now we can calculate the velocity of the object with respect to A.

$$v = \frac{v_{rel}+v_{obj}}{1+\dfrac{v_{rel}v_{obj}}{c^2}} = \frac{0.75c+0.3c}{1+\dfrac{(0.75c)(0.3c)}{c^2}} = 0.86c$$

49. The Schwarzschild radius is given by

$$R_S = \frac{2GM}{c^2}$$

$$\Rightarrow R_{S,min} = \frac{(2)(6.672 \times 10^{-11})(10^{-8})}{(3.00 \times 10^8)^2} = 1.48 \times 10^{-35} \text{ m}$$

Mass density is given by

$$\rho = \frac{M}{\frac{4}{3}\pi R_{S,min}^3} = \frac{10^{-8}}{\frac{4}{3}\pi(1.48 \times 10^{-35})^3} = 7.32 \times 10^{95} \text{ kg/m}^3$$

Similarly, for the maximum mass we have

$$R_{S,max} = \frac{(2)(6.672 \times 10^{-11})(10^{35})}{(3.00 \times 10^8)^2} = 1.48 \times 10^8 \text{ m}$$

$$\rho = \frac{M}{\frac{4}{3}\pi R_{S,min}^3} = \frac{10^{35}}{\frac{4}{3}\pi(1.48 \times 10^8)^3} = 7.32 \times 10^9 \text{ kg/m}^3$$

Note that here we are assuming that the laws of geometry and physics inside a black hole are the same as we currently know them. This may not be true.

51. (a) The relativistic kinetic energy is given by

$$K = m_0 c^2 \left(\frac{1}{\sqrt{1 - \frac{v^2}{c^2}}} - 1 \right)$$

This can be rearranged to give

$$v = c\sqrt{1 - \left(\frac{m_0 c^2}{m_0 c^2 + K} \right)^2}$$

Given: $K = 7.0 \text{ TeV} = 7.0 \times 10^6 \text{ MeV}$

Rest energy of a proton is $m_0 c^2 = 938.257 \text{ MeV}$

$$\Rightarrow v = c\sqrt{1 - \left(\frac{938.257}{938.257 + 7.0 \times 10^6} \right)^2} = 0.999999990992\,c$$

Therefore almost at speed of light.

(b) The time taken by a proton to complete one revolution is

$$t = \frac{x}{v} = \frac{27\,000}{(0.999)(3.00 \times 10^8)} = 9.01 \times 10^{-5} \text{ s}$$

Hence the number of revolutions per second is given by

$$N = \frac{1}{t} = \frac{1}{9.01 \times 10^{-5}} = 11100 \text{ s}^{-1}$$

(c) Relativistic momentum can be calculated from

$$p = \frac{1}{c}\sqrt{E^2 - m_0^2 c^4}$$

Here E is the total energy.

$$\Rightarrow p = \frac{1}{c}\sqrt{(938.257 + 7.0 \times 10^6)^2 - (938.257)^2} = 7.0 \times 10^6 \text{ MeV/c} = 3.7 \times 10^{-15} \text{ kg m/s}$$

53. (a) Each particle is moving with a velocity of $0.8c$. We have to calculate the velocity of one of these particles with respect to the other. For that, we use relative speed formula

$$v_{rel} = \frac{v + u}{1 + \frac{vu}{c^2}}$$

$$\Rightarrow v_{rel} = \frac{0.8c + 0.8c}{1 + \frac{(0.8c)(0.8c)}{c^2}} = 0.976c$$

(b) The kinetic energy of one particle can be calculated as shown below, so the total for the two particles will be 1.33 m_0c^2.

$$K = m_0c^2 \left(\frac{1}{\sqrt{1 - \frac{v^2}{c^2}}} - 1 \right)$$

$$\Rightarrow K = m_0c^2 \left(\frac{1}{\sqrt{1 - \frac{(0.8c)^2}{c^2}}} - 1 \right) = 0.67 m_0c^2$$

(c) $$K = m_0c^2 \left(\frac{1}{\sqrt{1 - \frac{(0.976c)^2}{c^2}}} - 1 \right) = 3.59 m_0c^2$$

55. Conservation of energy gives

$$\frac{m_0 c^2}{\sqrt{1 - \dfrac{V^2}{c^2}}} + \frac{m_0 c^2}{2\sqrt{1 - \dfrac{V^2}{c^2}}} = \frac{m_0' c^2}{\sqrt{1 - \dfrac{u^2}{c^2}}}$$

Here m_0' is the rest mass of the particle produced in the collision and moving with velocity u.

$$\Rightarrow \frac{3 m_0 c^2}{2\sqrt{1 - \dfrac{V^2}{c^2}}} = \frac{m_0' c^2}{\sqrt{1 - \dfrac{u^2}{c^2}}}$$

$$\Rightarrow \frac{m_0}{\sqrt{1 - \dfrac{V^2}{c^2}}} = \frac{2 m_0'}{3\sqrt{1 - \dfrac{u^2}{c^2}}} \qquad (1)$$

This equation can also be written as

$$\Rightarrow m_0' = \frac{3 m_0}{2} \sqrt{\frac{c^2 - u^2}{c^2 - V^2}} \qquad (2)$$

Conservation of linear momentum gives

$$\frac{m_0 V}{\sqrt{1 - \dfrac{V^2}{c^2}}} - \frac{m_0 V}{2\sqrt{1 - \dfrac{V^2}{c^2}}} = \frac{m_0' u}{\sqrt{1 - \dfrac{u^2}{c^2}}}$$

$$\Rightarrow \frac{m_0 V}{2\sqrt{1 - \dfrac{V^2}{c^2}}} = \frac{m_0' u}{\sqrt{1 - \dfrac{u^2}{c^2}}}$$

Substituting the value for $\dfrac{m_0}{\sqrt{1 - \dfrac{V^2}{c^2}}}$ from equation (1) into this equation yields

$$\frac{m_0' V}{3\sqrt{1 - \dfrac{u^2}{c^2}}} = \frac{m_0' u}{\sqrt{1 - \dfrac{u^2}{c^2}}}$$

$$\Rightarrow u = \frac{V}{3}$$

Substituting this into equation (2) gives

$$m_0' = \frac{3 m_0}{2} \sqrt{\frac{c^2 - \left(\dfrac{V}{3}\right)^2}{c^2 - V^2}}$$

Simplifying yields the required rest mass of the particle produced.

$$\Rightarrow m_0' = \frac{m_0}{2}\sqrt{\frac{(3c)^2 - V^2}{c^2 - V^2}}$$

57. Assume an elevator in free fall on Earth. An apparatus on the floor of the elevator emits a photon as the elevator starts falling down. A detector on the elevator's roof measures the frequency of the photon. If the height of the elevator is Δr, the time it will take for the photon to reach the detector in the elevator's frame of reference is given by

$$t = \frac{\Delta r}{c}$$

For an observer on Earth, during this time, the elevator has attained a velocity of

$$v = gt = g\frac{\Delta r}{c}$$

But we know that $g = \frac{GM}{r^2}$.

$$\Rightarrow v = \frac{GM\Delta r}{cr^2}$$

For the observer on Earth, the photon undergoes a shift in frequency, due only to its relative velocity, given by

$$\frac{\Delta f}{f} = \frac{v}{c}$$

This, according to the equivalence principle, becomes

$$\frac{\Delta f}{f} = \frac{GM\Delta r}{c^2 r^2}$$

For small Δf as compared to f and small Δr as compared to r, the above equation can be written as

$$\frac{1}{f}df = \frac{GM}{c^2 r^2}dr$$

Integrating from infinity to R_0 we get

$$\int_{f_0}^{f}\frac{1}{f}df = \int_{\infty}^{R}\frac{GM}{c^2 r^2}dr$$

$$\Rightarrow \ln\left(\frac{f}{f_0}\right) = -\frac{GM}{c^2 R_0}$$

$$\Rightarrow \frac{f}{f_0} = e^{-\frac{GM}{c^2 R_0}}$$

A Taylor series expansion of the right side of the previous equation gives

$$\frac{f}{f_0} = 1 - \frac{GM}{c^2 R_0} + \frac{G^2 M^2}{2c^4 R_0^2} - \cdots$$

In the limit $c^2 R_0 \gg GM$, this can be written as

$$\frac{f}{f_0} = \left(1 - \frac{2GM}{c^2 R_0}\right)^{\!\frac{1}{2}}$$

But we know that frequency is inversely proportional to time.

$$\Rightarrow \frac{\Delta t_\infty}{\Delta t_0} = \left(1 - \frac{2GM}{c^2 R_0}\right)^{\!\frac{1}{2}}$$

59. (a) Tau Ceti f is a habitable exoplanet that is assumed to be orbiting a nearby star Tau Ceti. The planet is about 11.905 light years away from our Sun. It is most likely 2 to 3 times the size of Earth. Not much data are available about its atmosphere, but it may be habitable with a surface temperature that can support complex life.

(b) A reasonable time of travel would be 10 years such that a 30-year-old person would be 40 years old when she reaches the new planet.

(c) The time in the spaceship's frame of reference is given by

$$t = \frac{x}{v}\sqrt{1 - \frac{v^2}{c^2}}$$

Here x is the distance from Earth to the destination.

$$x = 11.905 \text{ ly} = (11.905)(9.46 \times 10^{15}) = 1.13 \times 10^{17} \text{ m}$$

The above equation can be rearranged in terms of velocity.

$$v = \left(\frac{1}{c^2} + \frac{t^2}{x^2}\right)^{\!-1/2}$$

Let us suppose $t = 10$ y in the traveller's frame of reference.

$$\Rightarrow v = \left(\frac{1}{\left(3.00 \times 10^8\right)^2} + \frac{\left(10 \times 365.25 \times 24 \times 3600\right)^2}{\left(1.13 \times 10^{17}\right)^2}\right)^{\!-1/2}$$

$$\Rightarrow v = 2.30 \times 10^8 \text{ m/s}$$

(d) Since the trip takes 10 years, about 200 people should be sufficient for this trip, given that enough people will reach the new planet even after deaths due to diseases, accidents, and natural causes. Moore calculates that about 150 to 180 people are sufficient to sustain a population over many generations.

(e) Since the spaceship is to travel with a speed very close to the speed of light, its mass should be as small as possible. Sputnik 2 had a payload of about 500 kg with a passenger. Even though this underestimates the mass requirements of very long travel times, still we will use this to calculate the payload of the spaceship. Since we have 200 passengers, including crew members, the payload is

$$M = 200 \times 500 = 10^5 \text{ kg}$$

(f) The energy required for this mission would simply be the kinetic energy of the spaceship, that is

$$E = m_0 c^2 \left(\frac{1}{\sqrt{1 - \dfrac{v^2}{c^2}}} - 1 \right)$$

$$\Rightarrow E = \left(10^5\right)\left(3 \times 10^8\right)^2 \left(\frac{1}{\sqrt{1 - \dfrac{\left(2.30 \times 10^8\right)^2}{\left(3 \times 10^8\right)^2}}} - 1 \right) = 5.02 \times 10^{21} \text{ J}$$

Annual energy use in Canada in 2011 was about 1.70×10^{19} J. This implies that the mission will take $\dfrac{5.02 \times 10^{21}}{1.70 \times 10^{19}} \approx 300$ times more energy than the annual energy consumption in Canada.

Note that as for all open problems various reasonable answers are possible.

1. In 1897, J.J. Thomson, postulated the existence of particles smaller than atoms while trying to explain the existence of "cathode rays" that he produced in a glass tube. He postulated that these smaller particles were minuscule pieces of atoms. Later on, Philipp Lenard performed more experiments on these cathode rays and showed that if cathode rays were particles, their mass must be smaller than the masses of atoms.

3. Atoms have concentrated positive charge, a large empty space, and negative charges. Some atoms can spontaneously produce positive, negative, and neutral rays. When subjected to external light, some atoms produce light of different wavelengths.

5. Bohr quantized the angular momentum of electrons bound to atoms.

7. A blackbody spectrum is continuous, whereas a discharge spectrum has peaks at certain wavelengths characteristic of the material.

9. The thermometer works by detecting infrared radiation from the eardrum using a device such as a thermopile.

11. Hydrogen: $q = 1.602 \times 10^{-19}$ C, $m = 1.674 \times 10^{-27}$ kg

$$\Rightarrow \frac{q}{m} = \frac{1.602 \times 10^{-19}}{1.674 \times 10^{-27}} = 9.570 \times 10^{7} \text{ C/kg}$$

Oxygen: $q = 1.282 \times 10^{-18}$ C, $m = 2.657 \times 10^{-26}$ kg

$$\Rightarrow \frac{q}{m} = \frac{1.282 \times 10^{-18}}{2.657 \times 10^{-26}} = 4.823 \times 10^{7} \text{ C/kg}$$

13. (a) The object with higher intensity (black line) is at a higher temperature.

    (b) No.

15. (a) $\tilde{v} = \dfrac{1}{\lambda} = \dfrac{1}{1875 \times 10^{-9}} = 533333.3 \text{ m}^{-1}$

    (b) The equation for the wave number is given by

$$\tilde{v} = \left(1.097 \times 10^{7}\right)\left(\frac{1}{n_2^2} - \frac{1}{n_1^2}\right) \text{ m}^{-1}$$

    This can be rearranged to give

$$\frac{1}{n_2^2} - \frac{1}{n_1^2} = \frac{\tilde{v}}{1.097 \times 10^{7}}$$

$$\frac{1}{n_2^2} - \frac{1}{n_1^2} = \frac{533333.3}{1.097 \times 10^{7}}$$

$$\frac{1}{n_2^2} - \frac{1}{n_1^2} = 0.048617$$

By trial and error, you can determine that for $n_2 = 3$ and $n_1 = 4$,

$$\frac{1}{n_2^2} - \frac{1}{n_1^2} = \frac{1}{9} - \frac{1}{16} = 0.048611$$

which is sufficiently close to the calculated value to conclude that these are the correct values for n_2 and n_1.

(c) The wavelength is 1875 nm which corresponds to light in infrared region of the EM spectrum.

17. (a) The electron experiences a centripetal force equal to the force due to the magnetic field; that is

$$\frac{mv^2}{r} = qvB$$

$$\Rightarrow v = \frac{qrB}{m} = \frac{\left(1.602 \times 10^{-19}\right)\left(0.203\right)\left(1.8 \times 10^{-3}\right)}{9.109 \times 10^{-31}} = 6.4 \times 10^7 \text{ m/s}$$

(b) The electric field must balance the magnetic field; that is

$$qE = qvB$$

$$\Rightarrow E = vB = \left(6.426 \times 10^7\right)\left(1.8 \times 10^{-3}\right) = 1.2 \times 10^5 \text{ V/m}$$

(c) $E = \dfrac{V}{d}$

$$\Rightarrow d = \frac{V}{E} = \frac{150}{1.157 \times 10^5} = 1.3 \times 10^{-3} \text{ m} = 1.3 \text{ mm}$$

19. (a) $E = \dfrac{V}{d} = \dfrac{300}{0.02} = 15000 \text{ V/m}$

(b) If the electron is travelling in a straight line, the force on it due to the magnetic field must be equal to the force due to the electric field. This means that the force due to the magnetic field can be calculated from

$$F = qE = \left(1.602 \times 10^{-19}\right)(15000) = 2.4 \times 10^{-15} \text{ N}$$

(c) The force due to gravity is simply the weight of the electron; that is

$$F = mg = \left(9.109 \times 10^{-31}\right)(9.81) = 8.94 \times 10^{-30} \text{ N}$$

(d) Yes, it is reasonable to neglect gravity since the electric and magnetic forces on the electon are greater than the gravitational force by a factor of $\sim 10^{14}$.

21. The weight with buoyancy accounted for is given by

$$W = \frac{4}{3}\pi r^3 (\rho - \rho_{air}) g$$

The weight without buoyancy accounted for is

$$W' = \frac{4}{3}\pi r^3 \rho g$$

The relative error between the two is

$$\varepsilon = \frac{\frac{4}{3}\pi r^3 \rho g - \frac{4}{3}\pi r^3 (\rho - \rho_{air}) g}{\frac{4}{3}\pi r^3 (\rho - \rho_{air}) g} \times 100$$

$$\Rightarrow \varepsilon = \frac{\rho_{air}}{\rho - \rho_{air}} \times 100$$

(a) $\rho_{oil} = 821 \text{ kg/m}^3$

$\rho_{air} = 1.204 \text{ kg/m}^3$

$$\Rightarrow \varepsilon = \frac{\rho_{air}}{\rho_{oil} - \rho_{air}} \times 100 = \frac{1.204}{821 - 1.204} \times 100 = 0.147\%$$

(b) $\rho_{latex} = 950 \text{ kg/m}^3$

$\rho_{air} = 1.204 \text{ kg/m}^3$

$$\Rightarrow \varepsilon = \frac{\rho_{air}}{\rho_{latex} - \rho_{air}} \times 100 = \frac{1.204}{950 - 1.204} \times 100 = 0.127\%$$

23. (a) $dU = Vdq = \frac{kQr^2}{R^3} \frac{3Qr^2 dr}{R^3} = \frac{3kQ^2 r^4 dr}{R^6}$

(b) $U = \int_0^R \frac{3kQ^2}{R^6} r^4 dr$

$$\Rightarrow U = \frac{3kQ^2}{R^6} \frac{r^5}{5}\Big|_0^R = \frac{3kQ^2}{5R}$$

(c) $U = \frac{3kQ^2}{5R} = \frac{(3)(8.988 \times 10^9)(79 \times 1.602 \times 10^{-19})^2}{(5)(10^{-9})} = 8.6 \times 10^{-16}$ J

25. (a) Given $K = 2.5 \times 10^{-13}$ J

$$\Rightarrow \frac{1}{2} mv^2 = 2.5 \times 10^{-13}$$

$$\Rightarrow mv^2 = 5.0 \times 10^{-13}$$

The stopping distance is given by

$$r_0 = \frac{79 e^2}{\pi \varepsilon_0 mv^2}$$

$$\Rightarrow r_0 = \frac{79 \left(1.602 \times 10^{-19} \right)^2}{\pi \left(8.854 \times 10^{-12} \right) \left(5.0 \times 10^{-13} \right)} = 1.5 \times 10^{-13} \text{ m}$$

(b) If we treat the nucleus as a ball of uniformly distributed positive charge, if the alpha particle has enough energy to enter the nucleus it will pass through the nucleus. This is because inside the nucleus the potential begins to decrease.

27. (a) Kinetic energy gained by electron is

$$K = 1000 \text{ eV} = (1000)\left(1.602 \times 10^{-19} \right) = 1.602 \times 10^{-16} \text{ J}$$

(b) $K = \frac{1}{2} mv^2$

$$\Rightarrow v = \sqrt{\frac{2K}{m}} = \sqrt{\frac{(2)\left(1.602 \times 10^{-16} \right)}{9.109 \times 10^{-31}}} = 1.875 \times 10^7 \text{ m/s}$$

(c) $K = 1000$ eV

29. (a) $E_0 = 2.28 \text{ eV} = (2.28)\left(1.602 \times 10^{-19} \right) = 3.65 \times 10^{-19}$ J

(b) $f_0 = \dfrac{E_0}{h} = \dfrac{3.653 \times 10^{-19}}{6.626 \times 10^{-34}} = 5.51 \times 10^{14}$ Hz

(c) $\lambda_0 = \dfrac{c}{f_0} = \dfrac{3.00 \times 10^8}{5.512 \times 10^{14}} = 5.442 \times 10^{-7}$ m $= 544$ nm

31. (a) $f = \dfrac{c}{\lambda}$

Calculated values are listed in the table on the next page.

(b) $K_{max} = e\Delta V$

Calculated values are listed in the table below.

Wavelength (nm)	Stopping Potential (v)	Frequency (Hz)	K_{max} (J)
260	0.46	1.154E+15	7.369E−20
250	0.66	1.200E+15	1.057E−19
240	0.87	1.250E+15	1.394E−19
230	1.09	1.304E+15	1.746E−19

(c) The plotted curve with the fitted straight line is shown below.

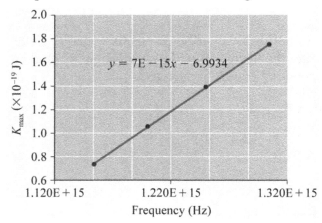

$y = 7E-15x - 6.9934$

(d) The slope of the line is

$$m = \left(7\times10^{-15}\right)\left(10^{-19}\right) = 7\times10^{-34} \text{ J s}$$

This value is fairly close to the Planck's constant of 6.626×10^{-34} J s.

(e) The intercept is the work function in joules, that is

$$E_0 = 6.993\times10^{19} \text{ J} = \frac{6.993\times10^{19}}{1.602\times10^{19}} = 4.37 \text{ eV}$$

33. The speed of an electron in a Bohr orbit is given by

$$v = \frac{nh}{2\pi mr}$$

We know that for $n = 1$ the radius is $r = 0.529\times10^{-10}$ m (also called the Bohr radius).

$$\Rightarrow v = \frac{(1)\left(6.626\times10^{-34}\right)}{(2\pi)\left(9.109\times10^{-31}\right)\left(0.529\times10^{-10}\right)} = 2.19\times10^6 \text{ m/s}$$

35. The period is the time taken by the electron to complete one revolution; that is

$$T = \frac{2\pi r}{v} = \frac{(2\pi)\left(0.529 \times 10^{-10}\right)}{2.188 \times 10^{6}} = 1.52 \times 10^{-16}\ \text{s}$$

37. (a) The wavelength of an incident photon is

$$\lambda = \frac{hc}{E} = \frac{\left(6.626 \times 10^{-34}\right)\left(3.00 \times 10^{8}\right)}{(350000)\left(1.602 \times 10^{-19}\right)} = 3.545 \times 10^{-12}\ \text{m}$$

The wavelength of the scattered photon is

$$\lambda' = \lambda + \frac{h}{m_e c}\left(1 - \cos\theta\right)$$

$$\Rightarrow \lambda' = 3.545 \times 10^{-12} + \frac{6.626 \times 10^{-34}}{\left(9.109 \times 10^{-31}\right)\left(3.00 \times 10^{8}\right)}\left(1 - \cos 40^{0}\right)$$

$$\Rightarrow \lambda' = 4.112 \times 10^{-12}\ \text{m}$$

This wavelength corresponds to an energy of

$$E' = \frac{hc}{\lambda'} = \frac{\left(6.626 \times 10^{-34}\right)\left(3.00 \times 10^{8}\right)}{4.112 \times 10^{-12}} = 4.834 \times 10^{-14}\ \text{J} = \frac{4.834 \times 10^{-14}}{1.602 \times 10^{-16}} = 302\ \text{keV}$$

   (b) The energy lost by an X-ray photon is carried away by the scattered electron.

$$E_e = E - E' = 350 - 301.7 = 48.3\ \text{keV}$$

39. The α-particle will stop at a distance r when the potential energy at this position equals its initial kinetic energy; that is

$$\frac{1}{4\pi\varepsilon_0}\frac{2eQ}{r} = E_\alpha$$

Here Q is the charge on the nucleus.

$$\Rightarrow Q = \frac{4\pi\varepsilon_0 r E_\alpha}{2e} = \frac{(4\pi)\left(8.854 \times 10^{-12}\right)\left(7.0 \times 10^{-15}\right)\left(5.0 \times 1.602 \times 10^{-13}\right)}{(2)\left(1.602 \times 10^{-19}\right)}$$

$$\Rightarrow Q = 1.9 \times 10^{-18}\ \text{C}$$

41.

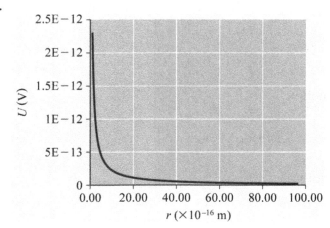

The average separation distance can be calculated by equating potential energy with average thermal energy; that is

$$\frac{1}{4\pi\varepsilon_0}\frac{e^2}{r}=\frac{3}{2}kT$$

$$\Rightarrow r=\frac{1}{6\pi\varepsilon_0}\frac{e^2}{kT}=\frac{1}{(6\pi)(8.854\times10^{-12})}\frac{(1.602\times10^{-19})^2}{(1.381\times10^{-23})(1.5\times10^7)}$$

$$\Rightarrow r=7.4\times10^{-13}\text{ m}$$

Note that we have used average thermal energy for this calculation. There are many particles at the higher energy tail, which would have much lower separation than calculated above.

43. $v_{max}=\dfrac{qBr_{max}}{m_{Ar}}$

r_{max} is the radius of the spectrometer magnet in this case.

$$\Rightarrow v_{max}=\frac{(1.602\times10^{-19})(300\times10^{-3})(0.5)}{(39.948)(1.605\times10^{-27})}=3.7\times10^5\text{ m/s}$$

45. The electric field is given by

$$E=\frac{V}{d}=\frac{200}{0.01}=20000\text{ V/m}$$

When electrons are travelling in a straight line we have

$$eE=evB$$

$$\Rightarrow v=\frac{E}{B}=\frac{20000}{0.8\times10^{-3}}=2.5\times10^7\text{ m/s}$$

When the electric field is removed, the electrons move in a curved path according to

$$\frac{mv^2}{r} = evB$$

$$\Rightarrow r = \frac{mv}{eB} = \frac{\left(9.109 \times 10^{-31}\right)\left(2.5 \times 10^{7}\right)}{\left(1.602 \times 10^{-19}\right)\left(0.8 \times 10^{-3}\right)} = 0.1777 \text{ m} = 17.77 \text{ cm}$$

The maximum length of the plates can be determined by applying the Pythagorean theorem to the triangle shown in the next figure.

$$l_{max} = \sqrt{17.77^2 - 17.27^2} = 4.2 \text{ cm}$$

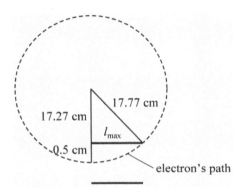

47. We have

$$I = \frac{2hf^3}{c^2} \frac{1}{e^{\frac{hf}{kT}} - 1}$$

This can be written as

$$I = \frac{Af^3}{e^{Bf} - 1}$$

with $A = \frac{2h}{c^2}$ and $B = \frac{h}{kT}$

To find the frequency at which intensity is maximum we will differentiate this with respect to frequency and equate to zero; that is

$$\frac{dI}{df} = \frac{3Af^2}{e^{Bf} - 1} - \frac{ABf^3 e^{Bf}}{\left(e^{Bf} - 1\right)^2} = 0$$

After simplifications, we get

$$Bf = 3\left(1 - e^{-Bf}\right)$$

This equation can be numerically solved to give

$Bf = 2.82$

$\Rightarrow \dfrac{hf}{kT} = 2.82$

The frequency that corresponds to maximum intensity is

$f_{max} = \dfrac{2.82kT}{h}$

49. To probe the electron structure using an electron, its wavelength must be smaller than the radius of the target; that is

$\lambda_e < 10^{-17}$ m

$\Rightarrow \dfrac{h}{p_e} < 10^{-17}$ m

$\Rightarrow p_e > 10^{17} h = \left(10^{17}\right)\left(6.626 \times 10^{-34}\right)$

$\Rightarrow p_e > 6.626 \times 10^{-17}$ kg m/s

The relativistic energy-momentum relation is

$p^2 c^2 = E^2 - m_0^2 c^4$

The condition on momentum can be written as

$p_e^2 c^2 > \left(6.626 \times 10^{-17}\right)^2 c^2$

$\Rightarrow E^2 - m_0^2 c^4 > \left(6.626 \times 10^{-17}\right)^2 c^2$

$\Rightarrow E^2 > \left(6.626 \times 10^{-17}\right)^2 c^2 + m_0^2 c^4$

$\Rightarrow E > \sqrt{\left(6.626 \times 10^{-17}\right)^2 c^2 + m_0^2 c^4}$

$\Rightarrow E > \sqrt{\left(6.626 \times 10^{-17}\right)^2 \left(3.00 \times 10^8\right)^2 + \left(9.109 \times 10^{-31}\right)^2 \left(3.00 \times 10^8\right)^4}$

$\Rightarrow E > 1.988 \times 10^{-8}$ J

Or $E > 124.1$ GeV

51. Let us compare the charge-to-mass ratio of an electron to that of a proton.

$\left(\dfrac{e}{m}\right)_e = \dfrac{1.602 \times 10^{-19}}{9.109 \times 10^{-31}} = 1.758 \times 10^{11}$ C/kg

$\left(\dfrac{e}{m}\right)_p = \dfrac{1.602 \times 10^{-19}}{1.673 \times 10^{-27}} = 9.576 \times 10^7$ C/kg

The charge-to-mass ratio in a crossed-field apparatus is given by

$\dfrac{e}{m} = \dfrac{E}{rB^2}$

Since it would be impractical to change the length of the plates (the radius remains the same), the only variables we can change are the electric and magnetic fields. Dividing the charge-to-mass ratio of the electron by that of proton, we obtain

$$\frac{E_e / rB_e^2}{E_p / rB_p^2} = \frac{1.758 \times 10^{11}}{9.576 \times 10^7} = 1835.84$$

$$\Rightarrow \frac{E_e}{B_e^2} = 1835.84 \frac{E_p}{B_p^2}$$

$$\Rightarrow \frac{B_p^2}{E_p} = 1835.84 \frac{B_e^2}{E_e}$$

Therefore, the ratio of the square of the magnetic field to the electric field must be changed such that this condition is satisfied.

53. Given $\Phi = 1300 \text{ W/m}^2$

Or $\Phi = 1300 \text{ J m}^{-2}\text{s}^{-1}$

This means that every square metre of Earth receives $E = 1300 \text{ J}$ of energy per second.

$$\Rightarrow \frac{Nhc}{\lambda} = 1300$$

Here N is the number of photons.

$$\Rightarrow N = \frac{1300\lambda}{hc} = \frac{(1300)(500 \times 10^{-9})}{(6.626 \times 10^{-34})(3.00 \times 10^8)} = 3.27 \times 10^{21}$$

Hence, about 3.27×10^{21} photons strike the surface of Earth per square metre per second.

55. Hydrogen combustion produces approximately 286 kJ/mol of energy. According to Statistics Canada, in 2007 the average household in Canada consumed about 106×10^9 J of energy. The number of moles of hydrogen needed to produce this much energy is

$$N_{mole} = \frac{106 \times 10^9}{286 \times 10^3} = 3.71 \times 10^5$$

Hence, the number of hydrogen molecules needed per year is

$$N_{hyd} = (3.71 \times 10^5)(6.023 \times 10^{23}) = 2.234 \times 10^{29}$$

Since one water molecule is needed to produce one hydrogen molecule,

$$N_{water} = 2.234 \times 10^{29}$$

This implies that the mass of water needed per year is

$$m_{water} = \frac{(2.234 \times 10^{29})(18)}{6.023 \times 10^{23}} = 6.68 \times 10^6 \text{ g} = 6680 \text{ kg}$$

Chapter 31—INTRODUCTION TO QUANTUM MECHANICS

1. No. A particle at rest would have a definite position and momentum, which violates Heisenberg's uncertainty principle.

3. Yes, since quantum mechanically, all we can do is associate a probability that the electron will be found at some place and time. However, the probability of such events occurring is extremely close to zero.

5. The de Broglie wavelength of a particle at rest does not exist; however, see question 1.

7. (a) No, since the energy of an incident photon is less than the difference between the energy of the ground state and first excited state of hydrogen ($E_1 - E_2 = -13.6\text{ eV} - (-3.4\text{ eV}) = -10.2\text{ eV}$).

   (b) Yes, since the energy of an incident photon is greater than the difference between the energy of the ground state and first excited state of hydrogen.

   (c) Yes, since the energy of an incident photon is greater than the difference between the energy of the ground state and first excited state of hydrogen.

9. $E_n = -\dfrac{13.6}{n^2}\text{ eV}$

   $\Rightarrow \Delta E = 13.6\left(\dfrac{1}{n_1^2} - \dfrac{1}{n_2^2}\right)\text{ eV}$

   For $n = 4$ to $n = 2$: $\Delta E = 2.55\text{ eV}$
   For $n = 6$ to $n = 3$: $\Delta E = 1.13\text{ eV}$
   For $n = 3$ to $n = 1$: $\Delta E = 12.09\text{ eV}$
   For $n = 5$ to $n = 3$: $\Delta E = 0.97\text{ eV}$
   Hence, $\Delta E_d < \Delta E_b < \Delta E_a < \Delta E_c$

11. All of the given quantum numbers for the total angular momentum are possible.

13. Relativistic momentum is given by

$$p = \frac{m_0 v}{\sqrt{1 - \dfrac{v^2}{c^2}}}$$

de Broglie wavelength is given by

$$\lambda = \frac{h}{p}$$

$$\Rightarrow \lambda = \frac{h}{m_0 v}\sqrt{1 - \frac{v^2}{c^2}}$$

(a) $\lambda = \dfrac{6.626 \times 10^{-34}}{\left(9.109 \times 10^{-31}\right)\left(0.9\right)\left(3.00 \times 10^{8}\right)} \sqrt{1 - \dfrac{\left(0.9c\right)^{2}}{c^{2}}} = 1.17 \times 10^{-12}$ m

(b) $\lambda = \dfrac{6.626 \times 10^{-34}}{\left(9.109 \times 10^{-31}\right)\left(0.99\right)\left(3.00 \times 10^{8}\right)} \sqrt{1 - \dfrac{\left(0.99c\right)^{2}}{c^{2}}} = 3.46 \times 10^{-13}$ m

15. Relativistic momentum is given by

$$p = \frac{m_0 v}{\sqrt{1 - \dfrac{v^2}{c^2}}}$$

$$\Rightarrow v = \frac{pc}{\sqrt{p^2 + m_0^2 c^2}}$$

Substituting $p = \dfrac{h}{\lambda}$ in this equation, we get

$$v = \frac{hc}{\sqrt{h^2 + m_0^2 c^2 \lambda^2}}$$

$$\Rightarrow v = \frac{\left(6.626 \times 10^{-34}\right)\left(3.00 \times 10^{8}\right)}{\sqrt{\left(6.626 \times 10^{-34}\right)^{2} + \left(9.109 \times 10^{-31}\right)^{2}\left(3.00 \times 10^{8}\right)^{2}\left(600 \times 10^{-9}\right)^{2}}}$$

$$\Rightarrow v = 1212.35 \text{ m/s}$$

Since this velocity is non-relativistic, we can use classical formula to calculate kinetic energy.

$$K = \frac{1}{2}mv^2$$

$$\Rightarrow K = \frac{1}{2}\left(9.109 \times 10^{-31}\right)\left(1212.35\right)^{2} = 6.694 \times 10^{-25} \text{ J}$$

$$\Rightarrow K = \frac{6.694 \times 10^{-25}}{1.602 \times 10^{-19}} = 4.18 \times 10^{-6} \text{ eV}$$

Therefore, the electron will have to be accelerated through a potential of 4.18 μV.

17. The uncertainty in the x-position of the electron is

$$\Delta x = 1.0 \times 10^{-10} \text{ m}$$

Hence, the uncertainty in the x-component of momentum is

$$\Delta p_x = \frac{\hbar}{2\Delta x} = \frac{1.055 \times 10^{-34}}{(2)\left(1.0 \times 10^{-10}\right)} = 5.3 \times 10^{-25} \text{ kg m/s}$$

The minimum kinetic energy, assuming the electron is constrained to move along the x-axis, is given by

$$K = \frac{(\Delta p_x)^2}{2m} = \frac{(5.27 \times 10^{-25})^2}{(2)(9.109 \times 10^{-31})} = 1.5 \times 10^{-19} \text{ J}$$

19. The uncertainty in the momentum of the object is

$$\Delta p = m\Delta v = (0.001)(0.2) = 0.0002 \text{ kg m/s}$$

Hence, the uncertainty in position is

$$\Delta x = \frac{\hbar}{2\Delta p} = \frac{1.05 \times 10^{-34}}{(2)(0.0002)} = 2.6 \times 10^{-31} \text{ m}$$

21. The uncertainty in time is equal to the mean life, that is

$$\Delta t = 1.44 T_{1/2}$$

(a) $\Delta t = (1.44)(10^{-12}) = 1.44 \times 10^{-12}$ s

The uncertainty in energy is

$$\Delta E = \frac{\hbar}{2\Delta t} = \frac{1.05 \times 10^{-34}}{(2)(1.44 \times 10^{-12})} = 3.64 \times 10^{-23} \text{ J}$$

(b) $\Delta t = (1.44)(10^{-23}) = 1.44 \times 10^{-23}$ s

The uncertainty in energy is

$$\Delta E = \frac{\hbar}{2\Delta t} = \frac{1.05 \times 10^{-34}}{(2)(1.44 \times 10^{-23})} = 3.64 \times 10^{-12} \text{ J}$$

23. $K = \frac{1}{2}mv^2 = \frac{1}{2}(9.109 \times 10^{-31})(10^5)^2 = 4.55 \times 10^{-21}$ J

The maximum allowable uncertainty in energy is

$$\Delta K = \left(\frac{0.002}{100}\right)(4.55 \times 10^{-21}) = 9.1 \times 10^{-26} \text{ J}$$

$$\Rightarrow \Delta t = \frac{\hbar}{2\Delta E} = \frac{1.05 \times 10^{-34}}{(2)(9.1 \times 10^{-26})} = 6 \times 10^{-10} \text{ s}$$

25. Since the square of the wave function integrated over the whole length of the space axis for the one-dimensional case must be a dimensionless quantity, the units of the wave function are $m^{-1/2}$.

27. $E_n = \dfrac{n^2 \hbar^2 \pi^2}{2mL^2}$

$$\Rightarrow E_1 = \frac{(1)^2 \left(1.055 \times 10^{-34}\right)^2 \pi^2}{(2)\left(9.109 \times 10^{-31}\right)\left(10^{-10}\right)^2} = 6.02 \times 10^{-18} \text{ J}$$

$$E_2 = \frac{(2)^2 \left(1.055 \times 10^{-34}\right)^2 \pi^2}{(2)\left(9.109 \times 10^{-31}\right)\left(10^{-10}\right)^2} = 2.41 \times 10^{-17} \text{ J}$$

$$E_3 = \frac{(3)^2 \left(1.055 \times 10^{-34}\right)^2 \pi^2}{(2)\left(9.109 \times 10^{-31}\right)\left(10^{-10}\right)^2} = 5.42 \times 10^{-17} \text{ J}$$

The probability density is given by

$$\psi_n^*(x)\psi_n(x) = \frac{2}{L}\sin^2\left(\frac{n\pi}{L}x\right)$$

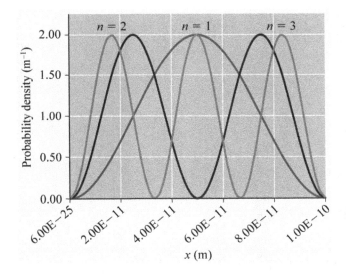

29. $E_n = \dfrac{n^2 \hbar^2 \pi^2}{2mL^2}$

$$\Rightarrow \Delta E = \left(n_2^2 - n_1^2\right)\frac{\hbar^2 \pi^2}{2mL^2} = \left(n_2^2 - n_1^2\right)\frac{\left(1.055 \times 10^{-34}\right)^2 \pi^2}{(2)\left(9.109 \times 10^{-31}\right)\left(50 \times 10^{-9}\right)^2}$$

$$\Rightarrow \Delta E = 2.41 \times 10^{-23} \left(n_2^2 - n_1^2\right)$$

(a) $\Delta E = 2.41 \times 10^{-23} \left(2^2 - 1^2\right) = 7.2 \times 10^{-23} \text{ J}$

(b) $\Delta E = 2.41 \times 10^{-23} \left(3^2 - 1^2\right) = 1.9 \times 10^{-22} \text{ J}$

(c) $\Delta E = 2.41 \times 10^{-23} \left(3^2 - 2^2\right) = 1.2 \times 10^{-22}$ J

(d) $\Delta E = 2.41 \times 10^{-23} \left(1000^2 - 1^2\right) = 2.4 \times 10^{-17}$ J

31. $P = \int_0^{L/2} \psi_n^*(x)\psi_n(x)dx$

$\Rightarrow P = \int_0^{L/2} \frac{2}{L}\sin^2\left(\frac{n\pi}{L}x\right)dx$

Since $n = 1$

$\Rightarrow P = \int_0^{L/2} \frac{2}{L}\sin^2\left(\frac{\pi}{L}x\right)dx$

$\Rightarrow P = \frac{2}{\pi}\int_0^{\pi/2}\sin^2 u\,du$

$\Rightarrow P = \frac{1}{\pi}\left[u - \frac{\sin 2u}{2}\right]_0^{\pi/2} = \frac{1}{\pi}\left[\frac{\pi}{2} - \frac{\sin\pi}{2} - 0 - 0\right]$

$\Rightarrow P = 0.5$

33. $\Delta E = \left(n_2^2 - n_1^2\right)\frac{\hbar^2\pi^2}{2mL^2}$

$\Rightarrow \Delta E = \left(2^2 - 1^2\right)\frac{\left(1.055\times10^{-34}\right)^2\pi^2}{(2)\left(9.109\times10^{-31}\right)\left(0.1\times10^{-9}\right)^2} = 1.81\times10^{-17}$ J

$\lambda = \frac{hc}{\Delta E} = \frac{\left(6.626\times10^{-34}\right)\left(3\times10^8\right)}{1.81\times10^{-17}} = 1\times10^{-8}$ m $= 10$ nm

35. (a) Using a calculator: $\sin(0.1) = 0.099$, $\cos(0.1) = 0.995$, $e^{0.1} = 1.105$

$\sin(0.1) = 0.1 - \frac{0.1^3}{3!} = 0.099$

$\cos(0.1) = 1 - \frac{0.1^2}{2!} = 0.995$

$e^{0.1} = 1 + 0.1 + \frac{0.1^2}{2!} = 1.105$

Using a calculator: $\sin(1) = 0.841$, $\cos(1) = 0.540$, $e^1 = 2.718$

$\sin(1) = 1 - \frac{1^3}{3!} + \frac{1^5}{5!} - \frac{1^5}{7!} = 0.841$

$$\cos(1) = 1 - \frac{1^2}{2!} + \frac{1^2}{4!} - \frac{1^2}{6!} = 0.540$$

$$e^1 = 1 + 1 + \frac{1^2}{2!} + \frac{1^2}{3!} + \frac{1^2}{4!} + \frac{1^2}{5!} + \frac{1^2}{6!} = 2.718$$

Using a calculator: $\sin(10) = -0.544$, $\cos(10) = -0.839$, $e^{10} = 22026.466$

Since $\sin(10) = \sin(10 - 3\pi) \approx \sin(0.5752) = -0.544$, therefore the series expansion gives

$$\sin(10) = (-0.5752) - \frac{(-0.5752)^3}{3!} + \frac{(-0.5752)^5}{5!} = -0.544$$

Since $\cos(10) = -\cos(10 - 3\pi) \approx -\cos(0.5752) = -0.839$, the series expansion gives

$$\cos(10) = -\left[1 - \frac{(-0.5752)^2}{2!} + \frac{(-0.5752)^4}{4!} \right] = -0.839$$

$$e^{10} = 1 + 10 + \frac{10^2}{2!} + \frac{10^3}{3!} + \ldots\ldots\ldots + \frac{10^{37}}{37!} = 22026.466$$

(b) $e^{i\theta} = 1 + i\theta + \frac{(i\theta)^2}{2!} + \frac{(i\theta)^3}{3!} + \frac{(i\theta)^4}{4!} + \frac{(i\theta)^5}{5!} + \ldots\ldots$

$$\Rightarrow e^{i\theta} = 1 + i\theta - \frac{\theta^2}{2!} - \frac{i\theta^3}{3!} + \frac{\theta^4}{4!} + \frac{i\theta^5}{5!} + \ldots\ldots$$

$$\cos\theta + i\sin\theta = \left[1 - \frac{\theta^2}{2!} + \frac{\theta^4}{4!} - \ldots\ldots \right] + i\left[\theta - \frac{\theta^3}{3!} + \frac{\theta^5}{5!} - \ldots\ldots \right]$$

$$\cos\theta + i\sin\theta = 1 + i\theta + - \frac{\theta^2}{2!} - \frac{i\theta^3}{3!} + \frac{\theta^4}{4!} + \frac{i\theta^5}{5!} + \ldots\ldots = e^{i\theta}$$

37. First we normalize the given function

$$A^2 \int_{-2}^{2} \left(4 - x^2\right)^2 dx = 1$$

$$\Rightarrow A^2 \left(16x - \frac{8}{3}x^3 + \frac{1}{5}x^5 \right)\Big|_{-2}^{2} = \frac{512}{15}A^2 = 1$$

$$\Rightarrow A = \sqrt{\frac{15}{512}}$$

The normalized wave function is therefore given by

$$\psi(x) = \begin{cases} \sqrt{\dfrac{15}{512}}\left(4 - x^2\right) & \text{for} \quad -2 \le x \le 2 \\ 0 & \text{everywhere else} \end{cases}$$

39. The non-relativistic energy of a particle is given by

$$E = \frac{p^2}{2m}$$

The time-dependent wave function is give by

$$\psi(x,t) = \psi(x)e^{-i\frac{Et}{\hbar}}$$

$$\Rightarrow \psi(x,t) = \psi(x)e^{-i\frac{p^2 t}{2m\hbar}}$$

Hence, the required wave function is

$$\psi(x) = \begin{cases} \sqrt{\dfrac{15}{512}}(4-x^2)e^{-i\frac{p^2 t}{2m\hbar}} & \text{for} \quad -2 \le x \le 2 \\ 0 & \text{everywhere else} \end{cases}$$

41. Assuming that the particle is in the ground state,

$$P = \int_{0.45L}^{0.55L} \psi_n^*(x)\psi_n(x)dx$$

$$\Rightarrow P = \int_{0.45L}^{0.55L} \frac{2}{L}\sin^2\left(\frac{n\pi}{L}x\right)dx$$

Since $n = 1$

$$\Rightarrow P = \int_{0.45L}^{0.55L} \frac{2}{L}\sin^2\left(\frac{\pi}{L}x\right)dx$$

With the change of variable $\frac{\pi}{L}x = u$ we get

$$\Rightarrow P = \frac{2}{\pi}\int_{0.45\pi}^{0.55\pi} \sin^2 u\, du$$

$$\Rightarrow P = \frac{1}{\pi}\left[u - \frac{\sin 2u}{2}\right]_{0.45\pi}^{0.55\pi} = \frac{1}{\pi}\left[0.1\pi - \frac{\sin 1.1\pi}{2} + \frac{\sin 0.9\pi}{2}\right]$$

$$\Rightarrow P = 0.62$$

43. (a) $E_n = \dfrac{n^2\hbar^2\pi^2}{2mL^2}$

Given: $E_1 = 20$ eV

$$\Rightarrow \frac{\hbar^2\pi^2}{2mL^2} = 20 \text{ eV}$$

$$\Rightarrow E_n = 20n^2 \text{ eV}$$

Hence, the energies of the first three excited states are

For $n = 2$: $E_2 = 20 \times 2^2 = 80$ eV

For $n = 3$: $E_3 = 20 \times 3^2 = 180$ eV

For $n = 4$: $E_4 = 20 \times 4^2 = 320$ eV

(b) Given: $\dfrac{\hbar^2 \pi^2}{2mL^2} = 20 \text{ eV} = (20)(1.602 \times 10^{-19}) = 3.204 \times 10^{-18}$ J

$$\Rightarrow L = \frac{\hbar \pi}{\sqrt{2m(3.204 \times 10^{-18})}} = \frac{(1.055 \times 10^{-34})\pi}{\sqrt{2(9.109 \times 10^{-31})(3.204 \times 10^{-18})}} = 1.4 \times 10^{-10} \text{ m}$$

(c) $\Delta E = \left(n_2^2 - n_1^2\right)\dfrac{\hbar^2 \pi^2}{2mL^2}$

$$\Rightarrow \frac{hc}{\lambda} = \left(n_2^2 - n_1^2\right)\frac{\hbar^2 \pi^2}{2mL^2}$$

$$\Rightarrow \lambda = \frac{8mcL^2}{h\left(n_2^2 - n_1^2\right)} = \frac{8(9.109 \times 10^{-31})(3.00 \times 10^8)(1.37 \times 10^{-10})^2}{(6.626 \times 10^{-34})\left(n_2^2 - n_1^2\right)} = \frac{6.15 \times 10^{-8}}{\left(n_2^2 - n_1^2\right)} \text{ m}$$

The longest wave photon corresponds to the transition of $n = 2$ to $n = 1$.

$$\Rightarrow \lambda = \frac{6.15 \times 10^{-8}}{\left(n_2^2 - n_1^2\right)} = 2.1 \times 10^{-8} \text{ m}$$

45. $P \propto e^{-\frac{2a}{\hbar}\sqrt{2m(V_0 - E)}}$

The exponent is

$$-\frac{2a}{\hbar}\sqrt{2m(V_0 - E)} = -\frac{2(0.1 \times 10^{-9})}{1.055 \times 10^{-34}}\sqrt{2(9.109 \times 10^{-31})(3-2)(1.602 \times 10^{-19})} = -1.025$$

$\Rightarrow P \propto e^{-1.025}$

$\Rightarrow P \approx 0.36$ (since the normalization factor of the wave function is ≈ 1)

47. $P \propto e^{-\frac{2a}{\hbar}\sqrt{2m(V_0 - E)}}$

Since the normalization factor of the wave function is ≈ 1, we can write this as

$P \approx e^{-\frac{2a}{\hbar}\sqrt{2m(V_0 - E)}}$

The exponent is

$$-\frac{2a}{\hbar}\sqrt{2m(V_0 - E)} = -\frac{2(0.5 \times 10^{-9})}{1.055 \times 10^{-34}}\sqrt{2(1.673 \times 10^{-27})(20-10)(1.602 \times 10^{-19})} = -694.3$$

$\Rightarrow P \approx e^{-694.3}$

$\Rightarrow P \approx 3.06 \times 10^{-302}$

Therefore, the number of protons tunnelling per second is given by

$N \approx \left(10^{20}\right)\left(3.06 \times 10^{-302}\right) = 3.06 \times 10^{-282}$ s^{-1}

Hence, the time it will take 1000 protons to tunnel through the barrier is

$t \approx \dfrac{1000}{3.06 \times 10^{-282}} = 3.26 \times 10^{284}$ s $= 1.03 \times 10^{277}$ y

49. $E_n = -\dfrac{1}{n^2}\left(\dfrac{m_e k^2 e^4}{2\hbar^2}\right) = -\dfrac{13.6}{n^2}$ eV

$\Rightarrow E_5 = -\dfrac{13.6}{5^2} = -0.544$ eV

51. The orbital angular momentum quantum numbers are given by

$l = 0, 1, 2, 3, \ldots, n-1$

For $n = 4$, we have

$l = 0, 1, 2, 3$

The magnitude of the orbital angular momentum is given by

$L = \sqrt{l(l+1)}\hbar$

$\Rightarrow L = \hbar, \sqrt{2}\hbar, \sqrt{6}\hbar, \sqrt{12}\hbar$

The z-components of the orbital angular momenta are given by

$L_z = m_l \hbar$ with $m_l = -l, (-l+1), \ldots, (l-1), l$

Thus, for $l = 0$, $L_z = 0$.

Thus, for $l = 1$, $L_z = -\hbar, 0, \hbar$.

Thus, for $l = 2$, $L_z = -2\hbar, -\hbar, 0, \hbar, 2\hbar$.

Thus, for $l = 3$, $L_z = -3\hbar, -2\hbar, -\hbar, 0, \hbar, 2\hbar, 3\hbar$.

53. We know that

$l = 0, 1, 2, \ldots, (n-1)$ and

$m_l = -l, (-l+1), \ldots, (l-1), l$

(a) For $m_l = -3$, the lowest possible values are $l = 3$ and $n = 4$.

(b) For $m_l = 0$, the lowest possible values are $l = 0$ and $n = 1$.

(c) For $m_l = +4$, the lowest possible values are $l = 4$ and $n = 5$.

55. $P_{1s}(r) = \dfrac{4}{a_B^3} r^2 e^{-\frac{2r}{a_B}}$

The most probable distance can be calculated by differentiating the above equation with respect to r and equating it to zero; that is

$\dfrac{dP_{1s}(r)}{dr} = 0$

$\Rightarrow \dfrac{4}{a_B^3}\left[2re^{-\frac{2r}{a_B}} - r^2 \dfrac{2}{a_B} e^{-\frac{2r}{a_B}} \right] = 0$

$\Rightarrow r = a_B$

Hence, the most probable distance is equal to the Bohr radius.

57. (a) $m_\mu = \dfrac{(105.7)(1.602\times10^{-13})}{(3.00\times10^8)^2} = 1.881\times10^{-28}$ kg

$E_n = -\dfrac{1}{n^2}\left(\dfrac{m_\mu k^2 e^4}{2\hbar^2} \right) = -\dfrac{1}{n^2}\left(\dfrac{(1.881\times10^{-28})(8.988\times10^9)^2(1.602\times10^{-19})^4}{2(1.055\times10^{-34})^2} \right)$

$\Rightarrow E_n = -\dfrac{4.5009\times10^{-16}}{n^2}$ J $= -\dfrac{2810}{n^2}$ eV

(b) $\Delta E = 2810\left(\dfrac{1}{n_1^2} - \dfrac{1}{n_2^2} \right)$ eV

$\Delta E = 2810\left(\dfrac{1}{1^2} - \dfrac{1}{2^2} \right) = 2110$ eV

(c) $a_B = \dfrac{\hbar^2}{km_\mu e^2}$

$\Rightarrow a_B = \dfrac{(1.05\times10^{-34})^2}{(8.988\times10^9)(1.881\times10^{-28})(1.602\times10^{-19})^2} = 2.56\times10^{-13}$ m

59. $L = I\omega$

For a solid sphere $I = \dfrac{2}{5}mr^2$.

$\Rightarrow L = \dfrac{2}{5}mr^2\omega$

$$\Rightarrow \omega = \frac{5L}{2mr^2} = \frac{(5)\left(\frac{\sqrt{3}}{2}(1.055\times10^{-34})\right)}{(2)(9.109\times10^{-31})(10^{-18})^2} = 2.51\times10^{32} \text{ rad/s}$$

The linear velocity of any point on the surface is given by

$$v = r\omega = (10^{-18})(2.51\times10^{32}) = 2.51\times10^{14} \text{ m/s}$$

This speed is far greater than the speed of light, which shows that the idea that the electron can be a solid spinning sphere is inconsistent with special relativity.

61. (a) $j = j_1 + j_2, j_1 + j_2 - 1, ..., |j_1 - j_2|$

$$\Rightarrow j = \frac{3}{2} + \frac{3}{2}, \frac{3}{2} + \frac{1}{2}, \frac{3}{2} - \frac{1}{2}, \frac{3}{2} - \frac{3}{2}$$

$$\Rightarrow j = 3, 2, 1, 0$$

(b) $J = \sqrt{j(j+1)}\hbar$

$$\Rightarrow J = \sqrt{12}\hbar, \sqrt{6}\hbar, \sqrt{2}\hbar, 0$$

$$J_z = j\hbar, (j-1)\hbar, ..., -j\hbar$$

For $j = 3$: $J_z = 3\hbar, 2\hbar, \hbar, 0, -\hbar, -2\hbar, -3\hbar$

For $j = 2$: $J_z = 2\hbar, \hbar, 0, -\hbar, -2\hbar$

For $j = 1$: $J_z = \hbar, 0, -\hbar$

For $j = 0$: $J_z = 0$

63. For a given principal quantum number n there are n possible values of the orbital angular momentum quantum number l $(0, 1, 2,..., n-1)$. For each value of l there are $(2l + 1)$ possible of the magnetic quantum number m_l. For each combination of l and m_l there are two possible values of the spin magnetic quantum number (spin is either up or down). Therefore, the maximum number of electrons that can fill a state with a given value of n is

$$N = \sum_{l=0}^{n-1} 2(2l+1) = 2\sum_{l=0}^{n-1}(2l+1)$$

The finite sum can be calculated.

$$\sum_{l=0}^{n-1}(2l+1) = (1 + 3 + 5 + ... + (2n-3) + (2n-1)) = n^2$$

Therefore, the maximum number of electrons allowed in a state with a given n is $2n^2$.

Chapter 32—INTRODUCTION TO SOLID-STATE PHYSICS

1. Decrease

3. c

5. d

7. b

9. Silicon

11. (a) $V = (a)(a)(a) = a^3$

(b) $V = a\hat{i} \times \left(a\cos(30°)\hat{i} + a\sin(30°)\hat{j}\right) \cdot a\sin(30°)\hat{k}$

$\Rightarrow V = a^2 \sin(30°)\hat{k} \cdot a\sin(30°)\hat{k} = \dfrac{a^3}{4}$

13. 1 m^3 of the metal has a mass of $m = 856$ kg

16.8 g of the metal contains 6.023×10^{23} atom.

$\Rightarrow 856$ kg of the metal contains $N = \dfrac{(856)(6.022 \times 10^{23})}{0.0168} = 3.069 \times 10^{28}$ atoms.

$\Rightarrow 3.069 \times 10^{28}$ atoms occupy a volume of 1 m^3.

Hence, one atom occupies a volume of $\dfrac{1}{3.069 \times 10^{28}} = 3.258 \times 10^{-29}$ m^3.

The volume of an ion is given by

$V_{ion} = \dfrac{4}{3}\pi r^3 = \dfrac{4}{3}\pi\left(73.4 \times 10^{-12}\right)^3 = 1.656 \times 10^{-30}$ m^3

Therefore, the volume occupied by the conduction electrons is

$V_e = 3.258 \times 10^{-29} - 1.656 \times 10^{-30} = 3.092 \times 10^{-29}$ m^3

The percentage of the volume occupied by conduction electrons is then given by

$\dfrac{3.092 \times 10^{-29}}{3.258 \times 10^{-29}} \times 100 = 94.9\%$

15. The Fermi energy is given by

$E_f = \dfrac{\hbar^2}{2m}\left(\dfrac{3\pi^2 N}{V}\right)^{\frac{2}{3}}$

$$\Rightarrow \frac{N}{V} = \frac{1}{3\pi^2}\left(\frac{2mE_f}{\hbar^2}\right)^{\frac{3}{2}} = \frac{1}{3\pi^2}\left[\frac{2(9.109\times10^{-31})(9.69\times1.602\times10^{-19})}{(1.054\times10^{-34})^2}\right]^{\frac{3}{2}}$$

$$\Rightarrow \frac{N}{V} = 1.37\times10^{29} \text{ electrons/m}^3$$

Now, the number density of atoms is given by

$$\rho = \frac{(2.44)(6.022\times10^{23})}{33} = 4.45\times10^{22} \text{ atoms/cm}^3 = 4.45\times10^{28} \text{ atoms/m}^3$$

Hence, the number of conduction electrons per atom is

$$N_e = \frac{N/V}{\rho} = \frac{1.37\times10^{29}}{4.45\times10^{28}} = 3 \text{ electrons/atom}$$

17. The effective mass represents inverse of the band curvature.

   (a) Positive as curvature is positive away from the band gap.

   (b) Negative as curvature is negative just below the band gap.

   (c) Positive as curvature is positive just above the band gap.

19. (a) $N = (1.00\times10^{22})(1.00\times10^{-9}) = 1.00\times10^{13}$ conduction electrons

   (b) The total energy transfer rate is given as 1.00 J/s.
   Hence, the average energy transfer rate per electron is

$$Q = \frac{1}{1.00\times10^{13}} = 1.00\times10^{-13} \text{ J/s}$$

21. $\dfrac{N_x}{N_0} \approx e^{-\Delta E/k_B T}$

$$\frac{\Delta E}{k_B T} = \frac{(2.0)(1.602\times10^{-19})}{(1.381\times10^{-23})(500)} = 46.401$$

$$\Rightarrow \frac{N_x}{N_0} \approx e^{-46.401} = 7.05\times10^{-21}$$

23. Since each phosphorus atom contributes one electron, the number of conduction electrons due to phosphorus in silicon is given by

$$N_{Ph} = \frac{6.023\times10^{23}}{30.9758} = 1.9444\times10^{22} \text{ electrons/g}$$

The density of silicon is

$$\rho_{Si} = 2.33 \text{ g/cm}^3 = 2.33\times10^6 \text{ g/m}^3$$

Hence, the number of conduction electrons in 1 g of silicon is

$$N_{Si} = \frac{10^{16}}{2.33 \times 10^6} = 4.29 \times 10^9 \text{ electrons}$$

We need to increase this by a factor of 10^6, that is

$$N' = 4.29 \times 10^{15} \text{ electrons}$$

Hence, the mass of phosphorus can be calculated from

$$(m_{Ph})(1.9444 \times 10^{22}) + 4.29 \times 10^9 = 4.29 \times 10^{15}$$

$$\Rightarrow m_{Ph} = 2.2 \times 10^{-7} \text{ g} = 0.22 \ \mu g$$

25. For forward- and reverse-bias, we have respectively

$$I_+ = I_0 \left(e^{eV_+/kT} - 1 \right) \text{ and } I_- = I_0 \left(e^{eV_-/kT} - 1 \right)$$

$$\Rightarrow \frac{I_+}{I_-} = \frac{e^{eV_+/kT} - 1}{e^{eV_-/kT} - 1}$$

$$\frac{eV}{kT} = \frac{(1.602 \times 10^{-19})(0.3)}{(1.381 \times 10^{-23})(270)} = 12.889$$

$$\Rightarrow \frac{I_+}{I_-} = \left| \frac{e^{12.889} - 1}{e^{-12.889} - 1} \right| = 4.0 \times 10^5$$

27. $P = IV = (0.7)(0.1) = 0.07 \text{ W}$

29. (a) $A = (0.62 \times 10^{-6})(0.62 \times 10^{-6}) = 3.844 \times 10^{-13} \text{ m}^2$

$$d = 0.40 \times 10^{-6} \text{ m}$$

$$C = \frac{\varepsilon_0 \kappa A}{d} = \frac{(8.854 \times 10^{-12})(4.0)(3.844 \times 10^{-13})}{0.40 \times 10^{-6}} = 3.4 \times 10^{-17} \text{ F}$$

(b) The total charge on a plate is given by

$$Q = CV = (3.403 \times 10^{-17})(2.4) = 8.168 \times 10^{-17} \text{ C}$$

And the total number of elementary charges this corresponds to is

$$N = \frac{8.168 \times 10^{-17}}{1.602 \times 10^{-19}} \approx 510$$

31. Given: $I_{tunneling} = Ce^{-2\kappa \Delta z}$, where C is the proportionality constant.

The relative percent change in current is given by

$$\varepsilon = \frac{I - I'}{I} \times 100$$

$$\Rightarrow \varepsilon = \frac{Ce^{-2\kappa\Delta z} - Ce^{-2\kappa\left(\Delta z + 10^{-10}\right)}}{Ce^{-2\kappa\Delta z}} \times 100 = \left(1 - e^{-2\times 10^{-10}\kappa}\right) \times 100$$

$$\kappa = 0.51 \times 10^{10} \sqrt{\phi} = 0.51 \times 10^{10} \sqrt{4.83} = 1.12 \times 10^{10} \text{ m}^{-1}$$

$$\Rightarrow \varepsilon = \left(1 - e^{-\left(2\times 10^{-10}\right)\left(1.12\times 10^{10}\right)}\right) \times 100 = 90\%$$

33. (a) $E = \dfrac{V}{d} = \dfrac{1}{0.001} = 1000$ V/m

    (b) The force experienced by an electron is

    $$F = eE$$

    Hence, the acceleration is given by

    $$a = \frac{eE}{m} = \frac{\left(1.602 \times 10^{-19}\right)\left(1000\right)}{9.109 \times 10^{-31}} = 1.8 \times 10^{14} \text{ m/s}^2$$

    (c) Assuming zero initial velocity, we have

    $$v = 1.8 \times 10^{14} t \text{ m/s}$$

    (d) $K = \dfrac{1}{2}mv^2 = \dfrac{1}{2}\left(9.109 \times 10^{-31}\right)\left(1.758 \times 10^{14} t\right)^2$

    $$\Rightarrow K = 0.014t^2 \text{ J}$$

    (e) $K = 0.014t^2$

    $$\Rightarrow \frac{dK}{dt} = 0.028t \text{ J/s}$$

    (f) $0.028t = 1 \times 10^{-13}$

    $$\Rightarrow t = 3.6 \times 10^{-12} \text{ s}$$

    (g) $x = \dfrac{1}{2}at^2 = \dfrac{1}{2}\left(1.758 \times 10^{14}\right)\left(3.57 \times 10^{-12}\right)^2 = 1.12 \times 10^{-9}$ m $= 1.1$ nm

35. $D(E) = \dfrac{1}{2\pi^2}\left(\dfrac{2m}{\hbar^2}\right)^{3/2}\sqrt{E}$

    $$\Rightarrow D(E) = \frac{1}{2\pi^2}\left(\frac{2\left(9.109 \times 10^{-31}\right)}{\left(1.055 \times 10^{-34}\right)^2}\right)^{3/2}\sqrt{\left(6\right)\left(1.602 \times 10^{-19}\right)} = 1 \times 10^{47} \text{ J/m}^3$$

37. $f(E) = \dfrac{1}{1 + e^{(E - E_f)/kT}}$

   (a) $\dfrac{E - E_f}{kT} = \dfrac{0.3}{(8.617 \times 10^{-5})(0)} \to \infty$

      $\Rightarrow f(E) = 0$

   (b) $\dfrac{E - E_f}{kT} = \dfrac{0.3}{(8.617 \times 10^{-5})(300)} = 11.60$

      $\Rightarrow f(E) = \dfrac{1}{1 + e^{11.60}} = 9.1 \times 10^{-6}$

39. $f(E) = \dfrac{1}{1 + e^{(E - E_f)/kT}}$

    $\Rightarrow E = E_f + kT \ln\left(\dfrac{1}{f} - 1\right)$

   (a) $E = 8.6 + (8.617 \times 10^{-5})(400)\ln\left(\dfrac{1}{0.38} - 1\right) = 8.6 \text{ eV}$

   (b) $E = 8.6 + (8.617 \times 10^{-5})(400)\ln\left(\dfrac{1}{0.84} - 1\right) = 8.5 \text{ eV}$

41. $\rho = \dfrac{m}{ne^2 \tau}$

    $\Rightarrow \tau = \dfrac{m}{ne^2 \rho} = \dfrac{9.109 \times 10^{-31}}{(8.46 \times 10^{28})(1.602 \times 10^{-19})^2 (1.68 \times 10^{-8})} = 2.50 \times 10^{-14} \text{ s}$

43. $C = \dfrac{\varepsilon_0 \kappa A}{d}$

    $\Rightarrow d = \dfrac{\varepsilon_0 \kappa A}{C} = \dfrac{(8.854 \times 10^{-12})(5)(10^{-10})}{1.0 \times 10^{-12}} = 4.4 \times 10^{-9} \text{ m} = 4 \text{ nm}$

45. $s = vt = (0.9 \times 3.0 \times 10^8)(10^{-9}) = 0.27 \text{ m}$

   This distance between the circuit elements on a chip does not seem reasonable. Therefore, the signal takes much less time to propagate than 1 ns.

47. $\tau = RC = (10^{14})(10^{-12}) = 100 \text{ s}$

49. The Fermi wave number depends on the number of electrons that each atom contributes to conduction, which is $\dfrac{N}{N^*}$, for a certain lattice parameter.

51. $k = \dfrac{2\pi}{\lambda} = \dfrac{\pi}{a}$

$\Rightarrow \lambda = 2a$

53. (a) Only those electrons near the Fermi energy have their energies increased. This is because they are the only electrons that have empty states near them in energy.

(b) Number of particles in a given energy interval:

$n(E)\,dE = 2g(E)f(E)dE$

Since at 0 K the energy states below the Fermi level are fully occupied, we have $f(E) = 1$. Hence, the number of electrons within thermal energy of the Fermi energy is given by

$\Rightarrow N' \approx 2g(E_f)kT$

But we know that

$g(E_f) = \dfrac{3N}{4E_f}$

$\Rightarrow N' \approx \dfrac{3NkT}{2E_f}$

$\Rightarrow \dfrac{N'}{N} \approx \dfrac{3kT}{2E_f}$

For copper $E_f = 7$ eV. Therefore, at $T = 300$ K we have

$\dfrac{N'}{N} \approx \dfrac{(3)(8.617 \times 10^{-5})(300)}{(2)(7)} = 0.0055$

(c) No, the electronic contribution to heat capacity in metals is very small since only a fraction of the electrons are within thermal energy of the Fermi energy.

Chapter 33—INTRODUCTION TO NUCLEAR PHYSICS

1.   Nucleon: A particle in an atomic nucleus; specifically, a proton or neutron.
     Nuclide: A species of nuclei with a specific number of neutrons and protons.
     Isotopes: Species of nuclei with the same number of protons and different numbers of neutrons.
     Decay Constant: Proportionality constant that characterizes the relation between the decay rate and the amount of a radioactive substance.
     Half Life: Time it takes for half of the atoms of a sample to decay.
     Decay Diagram: A diagram that specifies the sequence in which some nuclide decays.
     Q-value: Net energy released or absorbed in a nuclear reaction.
     Spontaneous Fission: Spontaneous breaking up of a heavy and unstable nucleus with a release of one or more neutrons.
     Radioactive Dating: A process through which the age of a substance is estimated based on comparison of abundance of some radio-isotope and its decay products.
     Proton-Proton Cycle: Fusion of hydrogen nuclei resulting in production of hydrogen, helium, electrons, neutrinos, and a release of energy.

3. α-decay: In this process the parent nucleus emits an α-particle and transforms into a lighter nucleus.

     β^--decay: In this process a neutron in the nucleus of the parent transforms into a proton with the emission of an electron and an anti-neutrino.

     β^+-decay: In this process a proton in the nucleus of the parent transforms into a neutron with the emission of a positron and a neutrino.

5.   An α-particle has 2 protons and 2 neutrons.
     The decay of $^{238}_{92}$U into $^{206}_{82}$Pb shows that at least 10 protons have been converted into α-particles. The total number of neutrons lost in the decay are therefore given by $238 - 206 - 10 = 22$.
     If 10 protons have converted into 5 α-particles, then 10 neutrons out of 22 must have been used as well. This leaves 12 neutrons.
     We know that neutrons can convert into protons. These 12 neutrons can therefore make a maximum of 3 α-particles.
     Hence, the maximum number of α-particles that can be emitted during this decay is 8.

7. $\dfrac{E_c}{E_g} = \dfrac{kq^2 / r}{Gm^2 / r} = \dfrac{kq^2}{Gm^2}$

$\Rightarrow \dfrac{E_c}{E_g} = \dfrac{\left(8.988 \times 10^9\right)\left(1.602 \times 10^{-19}\right)^2}{\left(6.672 \times 10^{-11}\right)\left(1.673 \times 10^{-27}\right)^2} = 1.235 \times 10^{36}$

9. $N = N_0 e^{-\lambda t}$

Given: $N = 0.25 N_0$

$\Rightarrow 0.25 N_0 = N_0 e^{-\lambda t}$

$\Rightarrow t = -\dfrac{\ln 0.25}{\lambda}$

But $\lambda = \dfrac{\ln 2}{t_{1/2}}$

$\Rightarrow t = -\dfrac{\ln 0.25}{\ln 2} t_{1/2} = 2 t_{1/2}$

11. The binding energy per nucleon for heavy nuclei ($A \geq 200$) is less than the binding energy for nuclei in the range $50 \leq A \leq 120$. Therefore when a heavier nucleus fissions into two lighter nuclei, there is an increase in the total binding energy. This excess binding energy is released in the fission process.

    The iron nucleus has the largest binding energy per nucleon of any other nucleus. Therefore, if two iron nuclei were to fuse into a heavier nucleus (tellurium), the total binding energy of the tellurium nucleus will be less than that of the two iron nuclei. Therefore, there is no release of energy in this process; on the contrary, energy must be absorbed in the reaction for it to occur.

13. If this were the case, all protons would spontaneously decay into neutrons, and so shortly after the big bang there wouldn't be any protons left, and the universe would have been simply a large collection of neutrons.

15. $E = mc^2$

$\Rightarrow E = (0.001)\left(3.00 \times 10^8\right)^2 = 9 \times 10^{13} \text{ J} = \dfrac{9 \times 10^{13}}{1.602 \times 10^{-13}} = 5.6 \times 10^{26} \text{ MeV}$

17. $E = mc^2$

$\Rightarrow E = (0.145)\left(3.00 \times 10^8\right)^2 = 1.31 \times 10^{16} \text{ J} = \dfrac{1.31 \times 10^{16}}{1.602 \times 10^{-19}} = 8.15 \times 10^{34} \text{ eV}$

19. In the beta decay of a neutron, the maximum energy ΔE available to the electron and the neutrino is equal to the difference in the rest mass energies of the neutron and the proton.

$$\Delta E = \left(m_n - m_p\right)c^2 = 1.29 \text{ MeV}$$

Since ΔE is much less than the rest mass energy of a proton, the kinetic energy of the proton can be ignored in this decay.

If all 1.29 MeV of energy goes to the electron, its momentum is given by

$$p_e = \frac{1}{c}\sqrt{\Delta E^2 - m_e^2 c^4} = \sqrt{1.29^2 - 0.511^2} = 1.18 \text{ MeV/c}$$

The maximum momentum of the neutrino corresponds to the situation where both proton and neutron are formed at rest. In this case, the energy of the neutrino is given by (assuming the neutrino is massless)

$$E_v = p_v c = 1.29 - 0.511 = 0.779 \text{ MeV}$$

Hence, the neutrino's maximum momentum is $p_v = 0.779 \text{ MeV/c}$

21. $m_{C12} = 12.000000$ u

$m_{C13} = 13.003354$ u

$\Rightarrow m_C = (12.000000)(0.9889) + (13.003354)(0.0111) = 12.011137$ u

$\Rightarrow m_C = (12.011137)\left(1.6605 \times 10^{-27}\right) = 1.9944 \times 10^{-26}$ kg

23. $B(A,Z) = \left[Zm_p + (A-Z)m_n - M_{nuc}\right]931.5 \text{ MeV}$

$$B_{per}(A,Z) = \frac{B(A,Z)}{A}$$

(a) $B(2,1) = \left[1(1.007276) + (2-1)(1.008665) - (2.014102 - 0.000548)\right]931.5 \text{ MeV}$

$\Rightarrow B(2,1) = 2.2 \text{ MeV}$

$\Rightarrow B_{per}(2,1) = \frac{2.2}{2} = 1.1 \text{ MeV/nucleon}$

(b) $B(4,2) = \left[2(1.007276) + (4-2)(1.008665) - (4.002603 - 2 \times 0.000548)\right]931.5 \text{ MeV}$

$\Rightarrow B(4,2) = 28.3 \text{ MeV}$

$\Rightarrow B_{per}(4,2) = \frac{28.3}{4} = 7.1 \text{ MeV/nucleon}$

(c) $B(6,3) = \left[3(1.007276) + (6-3)(1.008665) - (6.015122 - 3 \times 0.000548)\right]931.5 \text{ MeV}$

$\Rightarrow B(6,3) = 32.0 \text{ MeV}$

$\Rightarrow B_{per}(6,3) = \frac{32.0}{6} = 5.3 \text{ MeV/nucleon}$

(d) $B(56,26) = \big[26(1.007276) + (56-26)(1.008665)$

$$-(55.934939 - 26 \times 0.000548)\big]931.5 \text{ MeV}$$

$$\Rightarrow B(56,26) = 492.2 \text{ MeV}$$

$$\Rightarrow B_{per}(56,26) = \frac{492.2}{56} = 8.8 \text{ MeV/nucleon}$$

(e) $B(208,82) = \big[82(1.007276) + (208-82)(1.008665)$

$$-(207.976627 - 82 \times 0.000548)\big]931.5 \text{ MeV}$$

$$\Rightarrow B(208,82) = 1636.4 \text{ MeV}$$

$$\Rightarrow B_{per}(208,82) = \frac{1636.4}{208} = 7.9 \text{ MeV/nucleon}$$

25. $M_{nuc} = M_{atom} - Zm_e$

(a) $M_{nuc} = 2.014102 - 0.000548 = 2.013554 \text{ u}$

(b) $M_{nuc} = 12.000000 - 6 \times 0.000548 = 11.996712 \text{ u}$

(c) $M_{nuc} = 15.994915 - 8 \times 0.000548 = 15.990531 \text{ u}$

(d) $M_{nuc} = 55.934939 - 26 \times 0.000548 = 55.920691 \text{ u}$

(e) $M_{nuc} = 238.050785 - 92 \times 0.000548 = 238.000369 \text{ u}$

27. $B(A,Z) = 15.8A - 18.3A^{\frac{2}{3}} - 0.72\dfrac{Z^2}{A^{\frac{1}{3}}} - 23.2\dfrac{(A-2Z)^2}{A} + 12.0\dfrac{1}{\sqrt{A}}$

(a) For $_{8}^{4}\text{Be}$

$$\Rightarrow B(8,4) = 15.8(8) - 18.3(8)^{\frac{2}{3}} - 0.72\frac{(4)^2}{(8)^{\frac{1}{3}}}$$

$$-23.2\frac{(8-2\times4)^2}{8} + 12.0\frac{1}{\sqrt{8}} = 1433.4 \text{ MeV}$$

$$\Rightarrow B(8,4) = 89.87 \text{ MeV}$$

(b) For $^{12}_{6}C$

$$\Rightarrow B(12,6) = 15.8(12) - 18.3(12)^{\frac{2}{3}} - 0.72\frac{(6)^2}{(12)^{\frac{1}{3}}}$$

$$-23.2\frac{(12-2\times6)^2}{12} + 12.0\frac{1}{\sqrt{12}} = 1433.4 \text{ MeV}$$

$$\Rightarrow B(12,6) = 85.82 \text{ MeV}$$

(c) For $^{84}_{38}Sr$

$$\Rightarrow B(84,38) = 15.8(84) - 18.3(84)^{\frac{2}{3}} - 0.72\frac{(38)^2}{(84)^{\frac{1}{3}}}$$

$$-23.2\frac{(84-2\times38)^2}{84} + 12.0\frac{1}{\sqrt{84}} = 1433.4 \text{ MeV}$$

$$\Rightarrow B(84,38) = 722.4 \text{ MeV}$$

(d) For $^{208}_{82}Pb$

$$\Rightarrow B(208,82) = 15.8(208) - 18.3(208)^{\frac{2}{3}} - 0.72\frac{(82)^2}{(208)^{\frac{1}{3}}}$$

$$-23.2\frac{(208-2\times82)^2}{208} + 12.0\frac{1}{\sqrt{208}} = 1433.4 \text{ MeV}$$

$$\Rightarrow B(208,82) = 1612 \text{ MeV}$$

(e) For $^{238}_{92}U$

$$\Rightarrow B(238,92) = 15.8(238) - 18.3(238)^{\frac{2}{3}} - 0.72\frac{(92)^2}{(238)^{\frac{1}{3}}}$$

$$-23.2\frac{(238-2\times92)^2}{238} + 12.0\frac{1}{\sqrt{238}} = 1433.4 \text{ MeV}$$

$$\Rightarrow B(238,92) = 1791 \text{ MeV}$$

29. $\lambda = \dfrac{\ln 2}{t_{1/2}} = \dfrac{\ln 2}{(5730)(365.25\times24\times3600)} = 3.833\times10^{-12} \text{ s}^{-1}$

31. $R(0) = 2.2 \times 10^8$ Bq

$$\lambda = \frac{\ln 2}{t_{1/2}} = \frac{\ln 2}{(21)(3600)} = 9.17 \times 10^{-6} \text{ s}^{-1}$$

$$R = R(0)e^{-\lambda t} = 2.2 \times 10^8 e^{-(9.17 \times 10^{-6})(2 \times 24 \times 3600)} = 4.5 \times 10^7 \text{ Bq}$$

33. $R = \lambda_C N$

We know that for carbon-14, $\lambda_C = 3.833 \times 10^{-12} \text{ s}^{-1}$

$$\Rightarrow N = \frac{R}{\lambda} = \frac{10^6}{3.833 \times 10^{-12}} = 2.609 \times 10^{17} \text{ atoms of carbon-14.}$$

$$\Rightarrow m = \frac{2.609 \times 10^{17}}{6.023 \times 10^{23}} \times 14 = 6.06 \times 10^{-6} \text{ g} = 6 \ \mu\text{g}$$

35. $R = \lambda N$

$$\Rightarrow \lambda = \frac{R}{N} = \frac{2.0 \times 10^{10}}{3.0 \times 10^{20}} = 6.67 \times 10^{-11} \text{ s}^{-1}$$

$$t_{1/2} = \frac{\ln 2}{\lambda} = \frac{\ln 2}{6.67 \times 10^{-11}} = 1.0 \times 10^{10} \text{ s} = 330 \text{ y}$$

37. $R = 1000 e^{-(1.925 \times 10^{-5})(30 \times 60)} = 970 \text{ Bq}$

(a) $\lambda = \frac{\ln 2}{t_{1/2}} = \frac{\ln 2}{(10)(3600)} = 1.925 \times 10^{-5} \text{ s}^{-1}$

$R = R(0)e^{-\lambda t}$

(b) $R = 1000 e^{-(1.925 \times 10^{-5})(3 \times 24 \times 3600)} = 6.8 \text{ Bq}$

39. $R(0) = 20.0 \times 10^6$ Bq

$R = 15.0 \times 10^6$ Bq at $t = 240$ s

$R = R(0)e^{-\lambda t}$

$$\Rightarrow \lambda = -\frac{1}{t} \ln\left(\frac{R}{R(0)}\right) = -\frac{1}{240} \ln\left(\frac{15 \times 10^6}{20 \times 10^6}\right) = 1.20 \times 10^{-3} \text{ s}^{-1}$$

$$t_{1/2} = \frac{\ln 2}{\lambda} = \frac{\ln 2}{1.20 \times 10^{-3}} = 578 \text{ s}$$

41. (a)

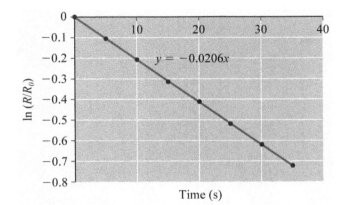

(b)   Slope $= -0.0206$

$\Rightarrow \lambda = 0.0206$ s^{-1}

(c) $t_{1/2} = \dfrac{\ln 2}{\lambda} = \dfrac{\ln 2}{0.0206} = 33.6$ s

43. (a)   Not possible since charge is not conserved.

(b)   Not possible since charge is not conserved.

(c)   Possible

(d)   Not possible since charge is not conserved.

(e)   Not possible since mass number is not conserved.

45. (a) $Q = \left(m_n + m_X - m_Y - m_{He} \right) c^2$

(b) ${}_{0}^{1}n + {}_{8}^{17}O \rightarrow {}_{6}^{14}C + {}_{2}^{4}He$

$Q = \left(m_n + m_O - m_C - m_{He} \right) c^2$

Or $Q = \left(m_n + m_O - m_C - m_{He} \right) 931.5$ MeV

$\Rightarrow Q = \left(1.008665 + 16.999131 - 14.000324 - 4.002603 \right) 931.5$

$\Rightarrow Q = 4.54$ MeV

47. Let us see if Q-value is positive for this reaction.

${}_{6}^{12}C \rightarrow {}_{2}^{4}He + {}_{2}^{4}He + {}_{2}^{4}He$

$\Rightarrow Q = \left(m_C - 3m_{He} \right) 931.5$ MeV

$\Rightarrow Q = \left(12.000000 - 3 \times 4.002603 \right) 931.5$

$\Rightarrow Q = -7.27$ MeV

Since the Q-value is negative, carbon-12 cannot decay through this channel spontaneously.

49. (a) $_{14}^{32}\text{Si} \rightarrow _{12}^{28}\text{Mg} + _{2}^{4}\text{He}$

(b) $_{25}^{56}\text{Mn} \rightarrow _{23}^{52}\text{V} + _{2}^{4}\text{He}$

(c) $_{26}^{60}\text{Fe} \rightarrow _{24}^{56}\text{Cr} + _{2}^{4}\text{He}$

51. (a) $_{25}^{52}\text{Mn} \rightarrow _{24}^{52}\text{Cr} + e^{+} + \nu$

(b) $_{27}^{55}\text{Co} \rightarrow _{26}^{55}\text{Fe} + e^{+} + \nu$

(c) $_{28}^{59}\text{Ni} \rightarrow _{27}^{59}\text{Co} + e^{+} + \nu$

53. (a) $_{1}^{3}\text{H} \rightarrow _{2}^{3}\text{He} + e^{-} + \bar{\nu}$

(b) $Q = \left(m_H - m_e - m_{He} - 2m_e - m_e \right) 931.5 \text{ MeV}$
$\Rightarrow Q = \left(m_H - m_e - m_{He} + 2m_e - m_e \right) 931.5 \text{ MeV}$
$\Rightarrow Q = \left(m_H - m_{He} \right) 931.5 \text{ MeV}$
$\Rightarrow Q = \left(3.016049 - 3.016029 \right) 931.5 \text{ MeV} = 18.6 \text{ keV}$

55. Thorium-232 goes through 4 β-decays and 6 α-decays before converting into stable lead-208 as shown in the decay diagram below.

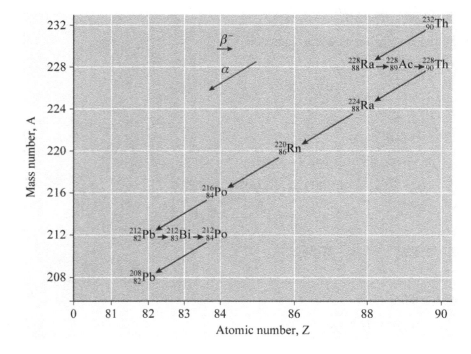

57. (a) $\,_{0}^{1}n + \,_{92}^{235}U \rightarrow \,_{36}^{90}Kr + \,_{56}^{144}Ba + 2\,_{0}^{1}n$

$Q = \left(m_n + m_U - m_{Rb} - m_{Cs} - 2m_n\right)c^2$

or

$Q = \left(m_U - m_{Rb} - m_{Cs} - m_n\right)931.5 \text{ MeV}$

$\Rightarrow Q = \left(235.043924 - 89.919517 - 143.922953 - 1.008665\right)931.5$

$\Rightarrow Q = 179.6 \text{ MeV}$

(b) $\,_{0}^{1}n + \,_{92}^{235}U \rightarrow \,_{37}^{93}Rb + \,_{55}^{141}Cs + 2\,_{0}^{1}n$

$Q = \left(m_n + m_U - m_{Rb} - m_{Cs} - 2m_n\right)c^2$

or

$Q = \left(m_U - m_{Rb} - m_{Cs} - m_n\right)931.5 \text{ MeV}$

$\Rightarrow Q = \left(235.043924 - 92.922042 - 140.920046 - 1.008665\right)931.5$

$\Rightarrow Q = 179.9 \text{ MeV}$

59. $B(A,Z) = \left[Zm_p + (A-Z)m_n - M_{nuc}\right]931.5 \text{ MeV}$

$B_{per}(A,Z) = \dfrac{B(A,Z)}{A}$

(a) $B(2,1) = \left[1(1.007276) + (2-1)(1.008665) - (2.014102 - 0.000548)\right]931.5 \text{ MeV}$

$\Rightarrow B(2,1) = 2.2 \text{ MeV}$

$\Rightarrow B_{per}(2,1) = \dfrac{2.2}{2} = 1.1 \text{ MeV/nucleon}$

$B(6,3) = \left[3(1.007276) + (6-3)(1.008665) - (6.015122 - 3 \times 0.000548)\right]931.5 \text{ MeV}$

$\Rightarrow B(6,3) = 32.0 \text{ MeV}$

$\Rightarrow B_{per}(6,3) = \dfrac{32.0}{6} = 5.3 \text{ MeV/nucleon}$

$B(10,5) = \left[5(1.007276) + (10-5)(1.008665) \right.$

$\left. \qquad\qquad\qquad -(10.012937 - 5 \times 0.000548)\right]931.5 \text{ MeV}$

$\Rightarrow B(10,5) = 64.7 \text{ MeV}$

$\Rightarrow B_{per}(10,5) = \dfrac{64.7}{10} = 6.5 \text{ MeV/nucleon}$

$$B(14,7) = \left[7(1.007276) + (14-7)(1.008665) \right.$$
$$\left. -(14.003074 - 7 \times 0.000548) \right]931.5 \text{ MeV}$$

$$\Rightarrow B(14,7) = 104.7 \text{ MeV}$$

$$\Rightarrow B_{per}(14,7) = \frac{104.7}{14} = 7.5 \text{ MeV/nucleon}$$

$$B(180,73) = \left[73(1.007276) + (180-73)(1.008665) \right.$$
$$\left. -(179.947462 - 73 \times 0.000548) \right]931.5 \text{ MeV}$$

$$\Rightarrow B(180,73) = 1444.6 \text{ MeV}$$

$$\Rightarrow B_{per}(180,73) = \frac{1444.6}{180} = 8.0 \text{ MeV/nucleon}$$

(b) (i) $B(4,2) = \left[2(1.007276) + (4-2)(1.008665) \right.$
$$\left. -(4.002603 - 2 \times 0.000548) \right]931.5 \text{ MeV}$$

$$\Rightarrow B(4,2) = 28.3 \text{ MeV}$$

$$\Rightarrow B_{per}(4,2) = \frac{28.3}{4} = 7.1 \text{ MeV/nucleon}$$

(ii) $B(4,8) = \left[4(1.007276) + (8-4)(1.008665) \right.$
$$\left. -(8.005305 - 4 \times 0.000548) \right]931.5 \text{ MeV}$$

$$\Rightarrow B(4,8) = 56.5 \text{ MeV}$$

$$\Rightarrow B_{per}(4,8) = \frac{56.5}{8} = 7.1 \text{ MeV/nucleon}$$

(iii) $B(12,6) = \left[6(1.007276) + (12-6)(1.008665) \right.$
$$\left. -(12.000000 - 6 \times 0.000548) \right]931.5 \text{ MeV}$$

$$\Rightarrow B(12,6) = 92.1 \text{ MeV}$$

$$\Rightarrow B_{per}(12,6) = \frac{92.1}{12} = 7.7 \text{ MeV/nucleon}$$

(iv) $B(16,8) = \left[8(1.007276) + (16-8)(1.008665) \right.$
$$\left. -(15.994915 - 2 \times 0.000548) \right]931.5 \text{ MeV}$$

$$\Rightarrow B(16,8) = 127.6 \text{ MeV}$$

$$\Rightarrow B_{per}(16,8) = \frac{127.6}{16} = 8.0 \text{ MeV/nucleon}$$

$$\text{(v) } B(182,74) = \left[74(1.007276) + (182-74)(1.008665)\right.$$
$$\left. -(181.948202 - 74 \times 0.000548)\right]931.5 \text{ MeV}$$
$$\Rightarrow B(182,74) = 1459.3 \text{ MeV}$$
$$\Rightarrow B_{per}(182,74) = \frac{1459.3}{182} = 8.0 \text{ MeV/nucleon}$$

61. (a) For $^8_4\text{Be} \rightarrow \,^4_2\text{He} + \,^4_2\text{He}$ we will first calculate the binding energy of beryllium using the liquid drop model.

$$B(A,Z) = 15.8A - 18.3A^{\frac{2}{3}} - 0.72\frac{Z^2}{A^{\frac{1}{3}}} - 23.2\frac{(A-2Z)^2}{A}$$

$$\Rightarrow B(8,4) = 15.8(8) - 18.3(8)^{\frac{2}{3}} - 0.72\frac{(4)^2}{(8)^{\frac{1}{3}}} - 23.2\frac{(8-2\times4)^2}{8} = 47.4 \text{ MeV}$$

$$Q = B(4,2) + B(4,2) - B(8,4)$$
$$\Rightarrow Q = 28.2 + 28.2 - 47.4 = 9.0 \text{ MeV}$$

(b) If Q-value for the reaction $^{12}_6\text{C} \rightarrow \,^4_2\text{He} + \,^4_2\text{He} + \,^4_2\text{He}$ is positive, this decay model will be possible.

We know that the binding energy of carbon-12 is 92.1 MeV. Therefore, the Q-value is
$$Q = B(4,2) + B(4,2) + B(4,2) - B(12,6)$$
$$\Rightarrow Q = 28.2 + 28.2 + 28.2 - 92.1 = -7.5 \text{ MeV}$$
Hence, we conclude that this spontaneous decay mode of carbon-12 is not possible.

63. $$B(A,Z) = a_V A - a_S A^{\frac{2}{3}} - a_C\frac{Z^2}{A^{\frac{1}{3}}} - a_A\frac{(A-2Z)^2}{A}$$

Here we have to calculate the Q-value using the binding energies of parent and daughter nuclei.

$$Q = S_n = B(A,Z) - B(A-1,Z)$$

We will calculate this term by term of the binding energy formula.

1st term: $a_V A - a_V(A-1) = a_V$

2nd term: $a_S A^{\frac{2}{3}} - a_S (A-1)^{\frac{2}{3}} = a_S \left(A^{\frac{2}{3}} - (A-1)^{\frac{2}{3}} \right)$

$$= a_S \left(A^{\frac{2}{3}} - A^{\frac{2}{3}} \left(1 - A^{-\frac{2}{3}} \right)^{\frac{2}{3}} \right)$$

$$= a_S \left(A^{\frac{2}{3}} - A^{\frac{2}{3}} + \frac{2}{3} A^{-\frac{1}{3}} + O\left(A^{-3} \right) \right) \text{ (using a Taylor series expansion)}$$

Now since A is very large, the fourth and higher terms in the bracket can be neglected.

$$\Rightarrow a_S A^{\frac{2}{3}} - a_S (A-1)^{\frac{2}{3}} \approx a_S \frac{2}{3} A^{-\frac{1}{3}}$$

4th term: $a_C \dfrac{Z^2}{A^{\frac{1}{3}}} - a_C \dfrac{Z^2}{(A-1)^{\frac{1}{3}}} \approx 0$ (since all A terms are in the denominator and

the Taylor series expansion will give higher orders in the denominator and

the terms can be neglected since $A \gg 1$.

4th term: $a_A \dfrac{(A-2Z)^2}{A} - a_A \dfrac{(A-1-2Z)^2}{A-1}$

After simplifying, we get

$$a_A \frac{(A-2Z)^2}{A} - a_A \frac{(A-1-2Z)^2}{A-1} = a_A \left[1 - \frac{4Z^2}{A(A-1)} \right]$$

Combining all terms, we get

$$S_n = a_V - a_S \frac{2}{3} A^{-\frac{1}{3}} - a_A \left[1 - \frac{4Z^2}{A(A-1)} \right]$$

Chapter 34—INTRODUCTION TO PARTICLE PHYSICS

1. (a) False; quarks can interact with each other through any of the fundamental forces and can therefore exchange other force mediating particles as well.

   (b) True

   (c) False, since the strong force is mediated by gluons.

   (d) False, since neutrinos are not force carriers.

   (e) False, since gluons are coloured (i.e., have a non-zero colour quantum number), whereas protons and neutrons are colourless (the colour quantum number for composite particles is zero).

3. (a) True

   (b) False, since mesons contain a quark and an antiquark, whereas baryons are made up of three quarks.

   (c) False. Hadrons are classified into baryons and mesons and the lightest meson is the neutral pion with a mass of 134.97 MeV/c^2.

   (d) False, since a pion is a meson and not a baryon.

   (e) True

   (f) True

5. (a) False, since because of momentum conservation half of the kinetic energy of the incident proton goes into the centre of mass motion of the two protons.

   (b) True, if the electron decelerates.

   (c) True, since circular motion is an accelerated motion.

   (d) False; a magnetic field acts in a direction perpendicular to the velocity of a moving charged particle and hence does no work on the charged particle.

   (e) A constant magnetic field can change the velocity of a charged particle; if the charged particle has a component of velocity perpendicular to the direction of the magnetic field, the magnetic field exerts a force on the charged particle, changing its direction of motion.

7. The maximum uncertainty in position is

$$\Delta x = 5 \times 10^{-15} \text{ m}$$

$$\Delta x \Delta p \geq \frac{\hbar}{2}$$

$$\Rightarrow \Delta p \geq \frac{\hbar}{2\Delta x}$$

$$\Rightarrow \sqrt{2m\Delta E} \geq \frac{\hbar}{2\Delta x}$$

$$\Rightarrow \Delta E \geq \frac{1}{2m}\left(\frac{\hbar}{2\Delta x}\right)^2 = \frac{1}{2(9.109 \times 10^{-31})}\left(\frac{1.055 \times 10^{-34}}{2(5 \times 10^{-15})}\right)^2$$

$$\Rightarrow \Delta E \geq 6.10 \times 10^{-11}\text{J} = 381 \text{ MeV}$$

9. electron, neutrino, quark

11. photon, Z^0, gluon

13. Since $mc^2 \sim \dfrac{\hbar c}{2R}$ as shown in the previous question.

$$\Rightarrow R \propto \frac{1}{m}$$

Therefore, if its gauge bosons were massless, the range of the weak force would be infinite.

15. uds: $m = 1110$ GeV/c^2

uus: $m = 1100$ GeV/c^2

dds: $m = 1120$ GeV/c^2

uss: $m = 1300$ GeV/c^2

dss: $m = 1310$ GeV/c^2

sss: $m = 1500$ GeV/c^2

17. The u, d, and s quarks have the following quantum numbers.

	Charge	Baryon Number	Spin
u	$\frac{2}{3}$	$\frac{1}{3}$	$\frac{1}{2}$
d	$-\frac{1}{3}$	$\frac{1}{3}$	$\frac{1}{2}$
s	$-\frac{1}{3}$	$\frac{1}{3}$	$\frac{1}{2}$

This table can be used to find the charge, baryon number, and spin of the baryons.

	Charge	Baryon Number	Spin
uus	+1	+1	1/2 or 3/2
uds	0	+1	1/2 or 3/2
dds	−1	+1	1/2 or 3/2
uss	0	+1	1/2 or 3/2
dss	−1	+1	1/2 or 3/2

19. The quarks have the following quantum numbers:

	Charge	Baryon Number	Spin
u	$\frac{2}{3}$	$\frac{1}{3}$	$\frac{1}{2}$
d	$-\frac{1}{3}$	$\frac{1}{3}$	$\frac{1}{2}$
c	$\frac{2}{3}$	$\frac{1}{3}$	$\frac{1}{2}$
s	$-\frac{1}{3}$	$\frac{1}{3}$	$\frac{1}{2}$
t	$\frac{2}{3}$	$\frac{1}{3}$	$\frac{1}{2}$
b	$-\frac{1}{3}$	$\frac{1}{3}$	$\frac{1}{2}$

Following the same procedure as in previous question we get the following table of quantum numbers and mass ranks.

	Charge	Baryon Number	Spin	Mass ($\times C$ MeV/c^2)	Mass Rank
ttb	+1	+1	$\frac{1}{2}$ or $\frac{3}{2}$	346200	5
tts	+1	+1	$\frac{1}{2}$ or $\frac{3}{2}$	342125	4
tbs	0	+1	$\frac{1}{2}$ or $\frac{3}{2}$	175325	3
tbd	0	+1	$\frac{1}{2}$ or $\frac{3}{2}$	175205	2
tuu	+2	+1	$\frac{1}{2}$ or $\frac{3}{2}$	141005	1

21. $m_{s\bar{u}} = C(125.0 + 2.5) = 127.5C$ MeV/c^2, where C is the proportionality constant.

$m_{s\bar{d}} = C(125.0 + 5.0) = 130.0C$ MeV/c^2

$m_{s\bar{s}} = C(125.0 + 125.0) = 250.0C$ MeV/c^2

$m_{u\bar{s}} = C(2.5 + 125.0) = 127.5C$ MeV/c^2

$m_{d\bar{s}} = C(5.0 + 125.0) = 130.0C$ MeV/c^2

$m_{\bar{c}d} = C(1300.0 + 5.0) = 1305.0C$ MeV/c^2

$\Rightarrow m_{s\bar{u}} = m_{u\bar{s}} < m_{s\bar{d}} = m_{d\bar{s}} < m_{s\bar{s}} < m_{\bar{c}d}$

23. (a) C: $+1 \neq 0 - 1 + 0 \Rightarrow$ conservation of charge violated

(b) C: $0 \neq +1 + 0 \Rightarrow$ conservation of charge violated

(c) conservation of energy violated

(d) L: $-1 \neq -1 + 1 \Rightarrow$ conservation of lepton number violated

(e) B: $+1 \neq 0 + 0 \Rightarrow$ conservation of baryon number violated

(f) B: $+1 + 1 \neq +1 + 0 \Rightarrow$ conservation of baryon number violated

(g) L: $-1 + 0 \neq 0 + 1 \Rightarrow$ conservation of lepton number violated

(h) B: $+1 \neq 0 + 0 \Rightarrow$ conservation of baryon number violated

25. In the centre of mass coordinate system of the electron-positron pair, the total momentum is zero. However, the photon in the final state cannot have zero momentum (otherwise, it will have zero energy i.e., no photons). Therefore, the process

$$e^- + e^+ \rightarrow \gamma$$

is not allowed by momentum and energy conservation.

27. The minimum energy of the neutrino corresponds to zero energies for all other particles.

Conservation of energy then gives

$$E_\nu + m_p c^2 = m_n c^2 + m_e c^2$$

$$\Rightarrow E_\nu = (m_n + m_e - m_p) 931.5 \text{ MeV}$$

$$\Rightarrow E_\nu = (1.008664 + 0.0005486 - 1.0072765) 931.5 = 1.8 \text{ MeV}$$

29. Since the initial kinetic energies of proton and antiproton are zero, momentum conservation gives

$$0 = \vec{p}_{\pi^+} + \vec{p}_{\pi^-}$$

$$\Rightarrow p_{\pi^+} = p_{\pi^-} = p_\pi$$

$$\Rightarrow E_{\pi^+} = E_{\pi^-} = E_\pi$$

Conservation of energy gives

$$2m_p c^2 = 2E_\pi$$

$$\Rightarrow E_\pi = m_p c^2 = 931.5 \text{ MeV}$$

To calculate the momenta, we can use the relation

$$E_\pi^2 = p_\pi^2 c^2 + m_\pi^2 c^4$$

$$\Rightarrow p_\pi = \frac{1}{c}\sqrt{E_\pi^2 - m_\pi^2 c^4}$$

$$\Rightarrow p_\pi = \frac{1}{c}\sqrt{(931.5)^2 - (139.6)^2} = 921.0 \text{ MeV/c}$$

31. Because the proton is the least massive baryon, it cannot decay into another baryon; if it decays into a meson, the law of conservation of baryon number will be violated.

33. Let us assume that the following decay is possible

$$\gamma \rightarrow e^- + e^+$$

Conservation of energy gives

$E_\gamma = 2m_e c^2$, where m_e is the relativistic mass of electron or positron.

Conservation of momentum gives

$p_\gamma = 2p_e \cos\theta$, where p_e is the relativistic momentum of electron or positron and θ is the angle between the direction of motion of photon and the direction of motion of electron or positron.

$$\Rightarrow \frac{E_\gamma}{c} = 2p_e \cos\theta$$

$$\Rightarrow E_\gamma = 2m_e v_e c \cos\theta$$

But $E_\gamma = 2m_e c^2$

$$\Rightarrow 2m_e v_e c \cos\theta = 2m_e c^2$$

$$\Rightarrow v_e = \frac{c}{\cos\theta}$$

$$\Rightarrow v_e \geq c$$

Hence, the reaction is not possible.

35. (a) $\pi^- \rightarrow e^- + \overline{V}_e$

Assuming π^- to be at rest right before the decay, conservation of momentum gives

$$0 = \vec{p}_e + \vec{p}_v$$

$$\Rightarrow p_e = p_v = p$$

Assuming that the electron-neutrino is massless, conservation of energy gives

$$m_\pi c^2 = \sqrt{p^2 c^2 + m_e^2 c^4} + pc$$

$$\Rightarrow p = \frac{1}{c}\frac{m_\pi^2 c^4 - m_e^2 c^4}{2m_\pi c^2} = \frac{1}{c}\frac{(139.6)^2 - (0.511)^2}{2(139.6)} = 69.8 \text{ MeV/c}$$

$$E_v = pc = 69.8 \text{ MeV}$$

$$E_e = \sqrt{p^2 c^2 + m_e^2 c^4} = \sqrt{(69.8)^2 + (0.511)^2} = 69.8 \text{ MeV}$$

(b) $\pi^- \to \mu^- + \bar{\nu}_\mu$

In the rest frame of π^- conservation of momentum gives

$$0 = \vec{p}_\mu + \vec{p}_\nu$$

$$\Rightarrow p_\mu = p_\nu = p$$

Assuming that the muon-neutrino is massless, conservation of energy gives

$$m_\pi c^2 = \sqrt{p^2 c^2 + m_\mu^2 c^4} + pc$$

$$\Rightarrow p = \frac{1}{c} \frac{m_\pi^2 c^4 - m_\mu^2 c^4}{2m_\pi c^2} = \frac{1}{c} \frac{(139.6)^2 - (105.6)^2}{2(139.6)} = 29.9 \text{ MeV/c}$$

$$E_\nu = pc = 29.9 \text{ MeV}$$

$$E_\mu = \sqrt{p^2 c^2 + m_\mu^2 c^4} = \sqrt{(29.9)^2 + (105.6)^2} = 109.7 \text{ MeV}$$

37.

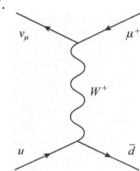

39.

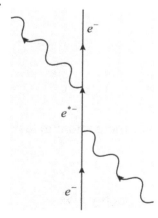

41. Let us assume that the uncertainty in the energy of the exchanged particle is on the order of its rest mass energy, that is

$$\Delta E = m_0 c^2$$

According to Heisenberg's uncertainty relation

$$\Delta E \Delta t \geq \frac{\hbar}{2}$$

Hence, the particle can exist for a time of

$$\Delta t = \frac{\hbar}{2m_0 c^2}$$

and its range is given by

$$R = \frac{\hbar c}{2m_0 c^2}$$

(a) (i) $R = \dfrac{\hbar c}{2m_0 c^2} = \dfrac{\left(1.055\times 10^{-34}\right)\left(3.00\times 10^8\right)}{\left(2\right)\left(135.0\times 1.602\times 10^{-13}\right)} = 7.3\times 10^{-16}$ m

(ii) $R = \dfrac{\hbar c}{2m_0 c^2} = \dfrac{\left(1.055\times 10^{-34}\right)\left(3.00\times 10^8\right)}{\left(2\right)\left(783.0\times 1.602\times 10^{-13}\right)} = 1.3\times 10^{-16}$ m

(iii) $R = \dfrac{\hbar c}{2m_0 c^2} = \dfrac{\left(1.055\times 10^{-34}\right)\left(3.00\times 10^8\right)}{\left(2\right)\left(1020.0\times 1.602\times 10^{-13}\right)} = 9.7\times 10^{-17}$ m

(b) The protons are fermions and the law of conservation of angular momentum does not allow exchange of a fermion between two fermions. Therefore, an electron exchange between two protons is not allowed.

43. The total relativistic energy of a particle is given by

$$K + m_0 c^2 = \frac{m_0 c^2}{\sqrt{1 - \dfrac{v^2}{c^2}}}$$

$$\Rightarrow v = c\sqrt{1 - \left(\frac{m_0 c^2}{K + m_0 c^2}\right)^2}$$

But $v = 2fL_d$

$$\Rightarrow L_d = \frac{c}{2f}\sqrt{1 - \left(\frac{m_0 c^2}{K + m_0 c^2}\right)^2}$$

(a) (i) $L_d = \dfrac{3.0\times 10^8}{\left(2\right)\left(200\times 10^6\right)}\sqrt{1 - \left(\dfrac{931.5}{10 + 931.5}\right)^2} = 0.109$ m $= 10.9$ cm

(ii) $L_d = \dfrac{3.0\times 10^8}{\left(2\right)\left(200\times 10^6\right)}\sqrt{1 - \left(\dfrac{931.5}{100 + 931.5}\right)^2} = 0.322$ m $= 32.2$ cm

(iii) $L_d = \dfrac{3.0\times 10^8}{\left(2\right)\left(200\times 10^6\right)}\sqrt{1 - \left(\dfrac{931.5}{200 + 931.5}\right)^2} = 0.426$ m $= 42.6$ cm

(b) $v = 2fL_d$

Hence, for fixed tube length

$v \propto f$

$\Rightarrow K \propto f$

Therefore, the frequency must be increased to achieve higher kinetic energies.

(c) (i) $L_d = \dfrac{3.0 \times 10^8}{(2)(200 \times 10^6)} \sqrt{1 - \left(\dfrac{0.511}{10 + 0.511}\right)^2} = 0.749 \text{ m} = 74.9 \text{ cm}$

(ii) $L_d = \dfrac{3.0 \times 10^8}{(2)(200 \times 10^6)} \sqrt{1 - \left(\dfrac{0.511}{100 + 0.511}\right)^2} = 0.750 \text{ m} = 75.0 \text{ cm}$

(iii) $L_d = \dfrac{3.0 \times 10^8}{(2)(200 \times 10^6)} \sqrt{1 - \left(\dfrac{0.511}{200 + 0.511}\right)^2} = 0.750 \text{ m} = 75.0 \text{ cm}$

45. (a) $v = c \sqrt{1 - \left(\dfrac{m_0 c^2}{K + m_0 c^2}\right)^2} = \left(3.0 \times 10^8\right) \sqrt{1 - \left(\dfrac{931.5}{500 + 931.5}\right)^2} = 2.28 \times 10^8 \text{ m/s}$

(b) $t = \dfrac{\text{Circumference}}{v} = \dfrac{11.4}{2.28 \times 10^8} = 5.0 \times 10^{-8} \text{ s} = 50 \text{ ns}$

(c) $B = \dfrac{m_0}{\sqrt{1 - \dfrac{v^2}{c^2}}} \left(\dfrac{v}{qR_0}\right)$

$\Rightarrow R_0 = \dfrac{m_0}{\sqrt{1 - \dfrac{v^2}{c^2}}} \left(\dfrac{v}{qB}\right) = \dfrac{931.5 \times 1.602 \times 10^{-13}}{\left(3.0 \times 10^8\right)^2 \sqrt{1 - \dfrac{\left(2.28 \times 10^8\right)^2}{\left(3.0 \times 10^8\right)^2}}} \left(\dfrac{2.28 \times 10^8}{\left(1.602 \times 10^{-19}\right)(2)}\right) = 1.81 \text{ m}$

$\Rightarrow \text{Circumference} = 2\pi R_0 = 11.4 \text{ m}$

47. $$v = c\sqrt{1 - \left(\frac{m_0 c^2}{K + m_0 c^2}\right)^2} = \left(3.0 \times 10^8\right)\sqrt{1 - \left(\frac{931.5}{10^{19} + 931.5}\right)^2}$$

$\Rightarrow v = 2.99....\times 10^8$ m/s (with at least 38 decimal places of 9's)

For the following estimation, we will use a calculator that can handle up to 31 decimal places.

$$B = \frac{m_0}{\sqrt{1 - \frac{v^2}{c^2}}}\left(\frac{v}{qR_0}\right)$$

$$\Rightarrow R_0 = \frac{m_0}{\sqrt{1 - \frac{v^2}{c^2}}}\left(\frac{v}{qB}\right) > \frac{931.5 \times 1.602 \times 10^{-13}}{\left(3.0 \times 10^8\right)^2 \sqrt{1 - \frac{\left(2.99....\times 10^8\right)^2}{\left(3.0 \times 10^8\right)^2}}}\left(\frac{2.99....\times 10^8}{\left(1.602 \times 10^{-19}\right)(10)}\right)$$

$\Rightarrow R_0 > 1.9 \times 10^{31}$ m

$\Rightarrow$ Circumference $= 2\pi R_0 > 10^{32}$ m

Building a synchrotron of this size is simply not possible, as it is larger than the known universe.

$$B = \frac{m_0}{\sqrt{1 - \frac{v^2}{c^2}}}\left(\frac{v}{qR_0}\right)$$

$v = 2.99....\times 10^8$ m/s (with at least 38 decimal places of 9's)

For this estimation, we will use a calculator that can handle up to 31 decimal places.

$$\Rightarrow B > \frac{(931.5)\left(1.602 \times 10^{-13}\right)}{\left(3.0 \times 10^8\right)^2 \sqrt{1 - \frac{\left(2.99....\times 10^8\right)^2}{\left(3.0 \times 10^8\right)^2}}}\left(\frac{\left(2.99....\times 10^8\right)}{\left(1.602 \times 10^{-19}\right)(100000)}\right)$$

$\Rightarrow B > 1.2 \times 10^{11}$ T